Human Performance Modeling in
AVIATION

Human Performance Modeling in AVIATION

Edited by

David C. Foyle
NASA Ames Research Center

Becky L. Hooey
San Jose State University Research Foundation
at NASA Ames Research Center

CRC Press
Taylor & Francis Group
Boca Raton London New York

CRC Press is an imprint of the
Taylor & Francis Group, an **informa** business

CRC Press
Taylor & Francis Group
6000 Broken Sound Parkway NW, Suite 300
Boca Raton, FL 33487-2742

International Standard Book Number-13: 978-0-8058-5964-5 (Hardcover)

Library of Congress Cataloging-in-Publication Data

Human performance modeling in aviation / editors, David C. Foyle and Becky L. Hooey.
    p. ; cm.
    Includes bibliographical references and index.
    ISBN-13: 978-0-8058-5964-5 (alk. paper) 1. Aeronautics--Human factors--Computer simulation. 2. Airplanes--Piloting--Human factors--Computer simulation. 3. Air pilots--Training of--Data processing. 4. Performance technology. I. Foyle, David C. II. Hooey, Becky L.
    [DNLM: 1. Aerospace Medicine--methods. 2. Aviation. 3. Cognition. 4. Psychomotor Performance. WD 700 H918 2008]

    TL553.6.H8646 2008
    629.132'520113--dc22
                                  2007034487

Visit the Taylor & Francis Web site at
http://www.taylorandfrancis.com

and the CRC Press Web site at
http://www.crcpress.com

Contents

PART 1   Goals, Aviation Problems, and Modeling

PART 2   Application of Individual Modeling Tools to the Aviation Problems

PART 3 Implications for Modeling and Aviation

Foreword

These are exciting times in the modeling of human performance in aviation. As this book clearly demonstrates, modeling has advanced to the point that it is able to represent relatively complex aircraft–air traffic control interactions with sufficient realism to assess the performance of alternative systems and procedure designs.

Modeling human performance in aviation actually was initiated in the 1950s— not by human factors professionals (of which there were only a few at the time) or psychologists, but rather by aerodynamicists and control engineers. Duane McRuer was very interested in aircraft handling qualities. He pioneered recasting the dynamics of flight traditionally expressed in partial differential equations into control engineering transfer function terms. But there was one problem. Without a transfer function for the human pilot, there could be no analyses of the complete aircraft control loop. He therefore set about to explore the control engineering representation for what came to be called "manual control"—a model of the dynamical response of the human controller. While there was much human factors research in the aviation world, manual control was the dominant research thrust to human performance modeling in aviation. Beginning in the late 1980s, discrete event simulation, computer-based information processing models a la Newell and Simon, and, eventually, cognitive architectures gradually took over.

There are numerous publications concerning human factors in aviation, including the *International Journal of Aviation Psychology,* Wiener & Nagel (1988), Billings (1997), and the two-volume National Research Council study of human factors in air traffic control (Wickens, Mavor, & McGee, 1997; Wickens, Mavor, Parasuraman, & McGee, 1998). However, this is the first volume devoted entirely to the topic of human performance modeling in aviation. While highlighting modeling is important, the volume is particularly valuable because, similarly to Gluck and Pew (2005), it compared the performance and usefulness of multiple models (five, to be exact) when addressing the same tasks and scenario contexts. As an added benefit, the editors did not stop after articulating a comparison among models; in addition, they sought out and documented the model developers' retrospective reflections on the model-building process and the state of the art in such models. The result is a coherent summary of the capabilities of the five models and an assessment of the state of the art of human performance modeling in general and in the field of aviation in particular.

I agree with the opinions expressed by the editors and authors that the capability for accomplishing this kind of modeling is still in the hands of model development specialists. Before human performance modeling can become a routine part of the system development process, we must simplify model development and provide better capabilities for articulating and visualizing how the models work and for describing why they are worthy of being relied on (i.e., validation). Those are but three of the challenges that lie before what I am sure will be enthusiastic readers of this book.

Richard W. Pew
Principal Scientist, BBN Technologies

REFERENCES

Billings, C. E. (1997). *Aviation automation: The search for a human-centered approach.* Mahwah, NJ: Lawrence Erlbaum Associates.

Gluck, K. A., & Pew, R. W. (Eds.). (2005). *Modeling human behavior with integrated cognitive architectures.* Mahwah, NJ: Lawrence Erlbaum Associates.

Wickens, C. D., Mavor, A. S., & McGee, J. P. (Eds.). (1997). *Flight to the future: Human factors in air traffic control.* Washington, DC: National Academy Press.

Wickens, C. D., Mavor, A. S., Parasuraman, R., & McGee, J. P. (Eds.). (1998). *The future of air traffic control: Human operators and automation.* Washington, DC: National Academy Press.

Wiener, E. L., & Nagel, D. C. (Eds.). (1988). *Human factors in aviation.* New York: Academic Press.

Preface

David C. Foyle and Becky L. Hooey

This volume chronicles the research activities of the 6-year National Aeronautics and Space Administration Human Performance Modeling (NASA HPM) project conducted as an element of the NASA Aviation Safety and Security Program (AvSSP). The overall goals of the HPM project were: (1) to develop and advance the state of the art of cognitive modeling of human operators for use on aviation problems; (2) during the course of that development, to investigate and inform specific solutions to actual aviation safety problems; and (3) to propose and explore methods for using and integrating human performance modeling into the design process in aviation. Each of five teams from academia and industry applied a different cognitive model to two aviation problems. In so doing, they put the HPM project in a position to shed light on the unique approaches and associated consequences that each modeling tool brings to bear on two qualitatively different aviation problems.

The NASA HPM project addressed two different classes of aviation problems: (1) pilot navigation errors during airport taxi, and (2) approach and landing performance with synthetic vision systems (SVSs). The first problem addresses pilot navigation errors during airport taxi and involves current-day technology and resultant pilot errors. The second problem is aimed at an emerging future technology (SVS) and addresses issues of conceptual design and procedures associated with its use. For both problem sets, NASA supplied a description of the problem, tasks, information requirements, and human-in-the-loop (HITL) simulation data to the modeling teams to produce working cognitive models of the two problems. These problems, a short history of human performance modeling, and an overview of the NASA HPM project are presented in the first three chapters.

In the second section of this book, the five modeling teams describe how they applied their model to these two aviation problems. They detail their results in terms of the aviation problems they addressed, the modeling challenges they faced, and the solutions they developed to address these complex, real-world aviation problem areas. The five models included four mature cognitive architectures (ACT-R, IMPRINT/ACT-R, Air MIDAS, and D-OMAR), as well as a new component model of situation awareness (A-SA). Using these models, the teams were able to develop explanatory mechanisms for observed behavior, suggest operational or procedural changes, conduct "what if" simulations of proposed design changes, and propose new implementations and metrics for the modeled systems. In addition, each of the modeling teams addressed the issue of model validation and the approach that it took toward that goal.

The third and final section of this book brings us to the heart of the matter. In this section, we will "eavesdrop" on a virtual roundtable among the modeling team members as they each address a set of challenging questions regarding modeling, the use of human performance models in complex domains such as aviation, and the difficulties they faced and the solutions they developed throughout the course of the project. In another chapter of this final section, we explore comparisons of the five modeling tools' structures, approaches, and outputs. In the final chapter, we attempt to pull it all together with a discussion of what we learned from this project from the perspectives of the aviation and the human performance modeling communities, as well as what the findings mean for both research fields.

The focus of the NASA HPM project was on computational frameworks that facilitate the use of modeling and simulation for the predictive analysis of pilot behaviors in actual aviation environments. To this end, the project developed and demonstrated cognitive models of human performance that can aid aviation system designers in understanding the pilot's performance in complex aviation scenarios. Such models will also enable the development of technologies and procedures that will support pilots' tasks, be easier to use, and be less susceptible to error. In addition, the NASA HPM project was able to demonstrate the value of coupling HITL simulation with human performance modeling to provide deep understanding of the aviation problem addressed, assess the efficacy of procedure and operation concepts, determine the potential for underlying error-precursor conditions, and conduct "what if" system redesigns in an efficient manner.

This synergistic combination of HITL and HPM as a method proved a more powerful tool than either of the two approaches alone. It is our hope that the results and advances of the NASA HPM project and the specific HPM models developed can be directly applied or will enable the development of tools and techniques that lead to better aviation systems and improved aviation safety.

Acknowledgments

The National Aeronautics and Space Administration Human Performance Modeling (NASA HPM) project was a large, multiyear effort that took the work of many people associated with many different organizations to bring about a successful completion. The editors are indebted to all who were directly and indirectly involved in the project. As will be seen in the forthcoming chapters, the project was organized such that the NASA Ames Research Center was responsible for the general and financial management of the project, for defining its technical direction, and for providing aviation-related information and human-in-the-loop (HITL) data to the investigators who conducted the actual modeling efforts.

The NASA HPM project was an element of NASA's Aviation Safety and Security Program (AvSSP) led by Mike Lewis and George Finelli. Herb Schlickenmaier, the current Aviation Safety Program Director, supported the completion of this volume. Other activities not directly related to the taxi error and synthetic vision system (SVS) usage modeling efforts were supported under the NASA AvSSP HPM element and are not reported in this volume. The editors wish to thank Tina Beard, Sandy Hart, and Jim Johnston at the NASA Ames Research Center for their programmatic help and support; Terry Allard, Patty Jones, and Rose Ashford of the NASA Ames Research Center Human Systems Integration Division; and Kevin Jordan, of San Jose State University (SJSU), and Marlene Hernan, of QSS Group, Inc., for their managerial support and general encouragement of the HPM project and book.

The editors wish to acknowledge the research efforts of Lynne Martin and Allen Goodman of San Jose State University for the analysis, preparation, and packaging of materials and data that were provided to the modeling teams for the first problem addressed in the HPM project—that of modeling taxi operations errors. For the second aviation topic—the use of synthetic vision systems for approach and landing—we are grateful to John Keller, Ken Leiden, Jon French, and Ron Small of Micro Analysis & Design, who conducted the cognitive task analyses, and to Allen Goodman and John Wilson of San Jose State University, who developed the scenarios and conducted the SVS HITL experiment. Their efforts were crucial to the success of the SVS modeling effort, and we are very grateful for their contributions. The editors also acknowledge and thank Lissa Webbon of QSS Group, Inc. for her help with the final preparation of this volume.

As managing center, the NASA Ames Research Center was responsible for the disbursement of financial support via contracts and grants to on-site personnel and outside agencies. Grants and cooperative agreements included: Rice University (NCC 2-1219, NCC 2-1321), San Jose State University (NCC 2-1155, NCC 2-1378, NAG2-1563), and University of Illinois (NAG2-1535, NAG2-1609). BBN Technologies and Micro Analysis & Design were funded via a subcontract to Battelle Memorial Institute (NAS2-99091, NNA05AC07C, Irv Statler, COTR). Simulator support was conducted under contract to Monterey Technologies, Inc. (NAS2-02081) and analysis and publication support to QSS Group, Inc. (NNA04AA18B).

The efforts conducted at NASA Ames Research Center were, of course, only part of the HPM project. The editors are grateful to the five modeling teams for their scientific and technical contributions throughout the 6-year project. The participating modeling team members conducted top-notch and cutting-edge work that we believe has moved the field forward. We are especially grateful for their overall willingness and eagerness to address the technical challenges posed by the HPM project, as well as those more mundane and onerous issues that one must endure (e.g., contractual reporting requirements) while doing so. We cannot thank them enough for their efforts. Specifically, we thank all of the research investigators of those five modeling teams: Mike Byrne, Alex Kirlik, and Michael Fleetwood from Rice University and the University of Illinois (ACT-R); Christian Lebiere, Rick Archer, Brad Best, and Dan Schunk of Micro Analysis & Design (IMPRINT/ACT-R); Kevin Corker, Koji Muraoka, Savvy Verma, Amit Jadhav, and Brian Gore of San Jose State University (Air MIDAS); Steve Deutsch and Dick Pew of BBN Technologies (D-OMAR); and Chris Wickens, Jason McCarley, Amy Alexander, Lisa Thomas, Michael Ambinder, and Sam Zheng of the University of Illinois (A-SA). Clearly, the efforts of these individuals formed the backbone of the HPM project.

A special note of thanks goes to Ken Leiden of Alion Science and Technology Micro Analysis & Design Operations. Ken's work proved to be the cornerstone of the HPM project. In the early phases, he led reviews on the role of human error in aviation and of current human performance models. As mentioned, for the SVS phase of the project, Ken's team developed a cognitive task analysis of landing and approach that served as a key resource in developing the models. Additionally, Ken made direct contributions to this volume: In chapter 4, with representatives of the modeling teams, he provides an overview of the five modeling tools; in chapter 10 (with Brad Best), he describes the various modeling tools and their results along a variety of dimensions. We are extremely indebted to Ken and his team members at Alion Science and Technology Micro Analysis & Design Operations for providing these key components of the HPM project.

Finally, we collectively thank the entire group of researchers, support personnel, managers, and others for allowing us to take the NASA HPM project in the directions that we saw fit, for conducting such technically excellent work, for doing so in a professional and amiable manner and in the spirit of jointly advancing the state of the art, and for supporting the documentation of the HPM project in the form of this volume.

David C. Foyle
Becky L. Hooey

The Editors

David C. Foyle is a Senior Research Psychologist and member since 1988 of the Human Systems Integration Division at NASA Ames Research Center, directing activities in the Human-Centered Systems Laboratory. He received a Ph.D. degree in cognitive/mathematical psychology from Indiana University in 1981. Dr. Foyle has authored more than 80 papers on attentional and design issues with head-up display (HUD) superimposed symbology, pilot operations and flight deck technology during airport taxi surface operations, and related topics. He led the Taxiway Navigation and Situation Awareness (T-NASA) system research team that developed a flight deck display suite consisting of HUD "scene-linked symbology" in which conformal and perspective symbology are projected virtually into the world, and a perspective moving map display depicting route and other traffic information. Currently, he is investigating advanced display concepts and operations for time-based taxi clearances for civil transport pilots. Dr. Foyle recently led the 6-year NASA Human Performance Modeling project under NASA's Aviation Safety and Security Program (AvSSP), the goal of which was to develop models of human performance and error in aviation contexts to enable the development and evaluation of error mitigation technologies and procedures. This book chronicles that effort.

Becky L. Hooey is a Senior Research Associate for San Jose State University Research Foundation at NASA Ames Research Center. She has served as a principal investigator in NASA's Human-Centered Systems Laboratory since 1997 and has over 12 years of experience applying human factors engineering research and principles to the design and evaluation of complex systems including advanced traveler information systems and aircraft cockpit displays. Ms. Hooey received her Masters of Science degree in Psychology (Human Factors) from the University of Calgary in 1995 and is a Ph.D. candidate in the Mechanical and Industrial Engineering Department at the University of Toronto. Ms. Hooey has authored over 50 publications on the topics of aviation safety, advanced display concepts, and human-centered design and evaluation methods. Her design and evaluation work have been honored by national and international awards including the SAE Wright Brothers Memorial Medal.

The Contributors

Amy L. Alexander[1]
Aptima Inc.
Woburn, Massachusetts

Michael Ambinder[1]
Department of Psychology
University of Illinois at
  Urbana–Champaign
Champaign, Illinois

Rick Archer
Alion Science and Technology
  Corporation
Boulder, Colorado

Brad Best[2]
Adaptive Cognitive Systems
Boulder, Colorado

Michael D. Byrne
Department of Psychology
Rice University
Houston, Texas

Kevin M. Corker
Industrial & Systems Engineering
  Department
San Jose State University
San Jose, California

Stephen E. Deutsch
BBN Technologies
Cambridge, Massachusetts

Michael D. Fleetwood
Department of Psychology
Rice University
Houston, Texas

David C. Foyle
Human Systems Integration Division
NASA Ames Research Center
Moffett Field, California

Brian F. Gore[3]
Human Systems Integration Division
San Jose State University Research
  Foundation at NASA Ames Research
  Center
Moffett Field, California

Becky L. Hooey
Human Systems Integration Division
San Jose State University Research
  Foundation at NASA Ames Research
  Center
Moffett Field, California

Amit Jadhav
Industrial & Systems Engineering
  Department
San Jose State University
San Jose, California

Alex Kirlik
Human Factors Division and Beckman
  Institute
University of Illinois at
  Urbana–Champaign
Urbana, Illinois

Christian Lebiere[2]
Human–Computer Interaction Institute
Carnegie Mellon University
Pittsburgh, Pennsylvania

Kenneth Leiden
Alion Science and Technology
  Corporation
Boulder, Colorado

Jason S. McCarley
Human Factors Division, Institute of
  Aviation
University of Illinois at
  Urbana–Champaign
Savoy, Illinois

Koji Muraoka[3]
Air Safety Technology Center
Institute of Space Technology and
  Aeronautics
Japan Aerospace Exploration Agency
Tokyo, Japan

Richard W. Pew
BBN Technologies
Cambridge, Massachusetts

Dan Schunk[2]
Ariba Inc.
Portland, Oregon

Lisa C. Thomas[1]
Boeing Phantom Works
Seattle, Washington

Savita Verma[3]
Aviation Systems Division
San Jose State University Research
Foundation at NASA Ames Research
  Center
Moffett Field, California

Christopher D. Wickens[1]
Alion Science and Technology
  Corporation
Boulder, Colorado

Sam Zheng[1]
Siemens Corporation
Princeton, New Jersey

The work reported was conducted at:
[1]Human Factors Division, Institute of Aviation, University of Illinois at Urbana–Champaign, Savoy,
  Illinois
[2]Micro Analysis & Design, Boulder, Colorado
[3]Industrial & Systems Engineering Department, San Jose State University, San Jose, California

Part 1

Goals, Aviation Problems,
and Modeling

1 The NASA Human Performance Modeling Project
Goals, Approach, and Overview

David C. Foyle and Becky L. Hooey

CONTENTS

THE NASA HUMAN PERFORMANCE MODELING PROJECT

As part of the Aviation Safety and Security Program (AvSSP), the National Aeronautics and Space Administration Human Performance Modeling (NASA HPM) project was aimed at advancing the state of the art of human performance modeling and developing applicable tools for the study of human performance in aviation. This project followed the approach of applying multiple cognitive modeling tools to a common set of problems. As will be seen in the chapters to follow, the five modeling architectures attempted to predict human error and behavior, given changes in system design, procedures, and operational requirements. The NASA HPM project focused on modeling the performance of highly skilled and trained operators (commercial airline pilots) in complex aviation tasks. Leveraging existing NASA data and simulation facilities, we were able to offer rich data sets of highly skilled operators performing complex operational aviation tasks to the modeling teams for use in model development and validation.

NASA HPM PROJECT GOALS

The primary goal of the project was to develop and extend human modeling capabilities while gaining knowledge regarding aviation operations and supporting emerging capabilities and technologies that increase aviation safety. Since previous modeling efforts (e.g., the U.S. Air Force's Agent-Based Modeling and Behavioral Representation [AMBR] project; Gluck & Pew, 2005) and other NASA programs were investigating the role of the air traffic controller (ATC), the NASA HPM project concentrated on modeling pilot performance with regard to safety issues under current-day operations and when using emerging future technologies. Two task-problem domains representing different types of aviation safety problems and spanning NASA's charter were chosen for study and application of the modeling efforts. The two aviation domain problems addressed by the modeling teams of the HPM project and their characterizations are:

- Airport surface (taxi) operations—problem time frame: current-day operations; problem class: errors (taxi navigation errors); and
- Synthetic vision system (SVS) operations—problem time frame: future operations; problem class: conceptual design, concept of operations development. Note: SVS is a new display technology for a visual virtual representation of the airport environment from a digital database via computer-generated imagery.

These two problems are described in a following section and in detail in chapter 3. By having the NASA HPM project modeling teams address these two aviation problems in a sequential manner, we were able to investigate various aspects associated with improving the state of the art of HPM tools in addressing actual complex-environment aviation problems. Since the two problems target two problem classes that have been historically treated differently—those of human error for the airport taxi problem and performance and usage for the SVS problem—this challenged the individual modeling architectures by requiring them to address both problem classes. Finally, because the NASA HPM project was a large U.S. government-funded project, we were able to apply five unique operator models to these two problems. This allowed for comparisons among the models regarding capabilities, strengths, weaknesses, data requirements, and output, just to name a few comparison dimensions.

It should be noted here that the NASA HPM project was never structured as a "bake off": There was never an explicit or implicit agenda to determine a "winner." For a given specific problem or domain (e.g., workload prediction of ATC aircraft handoffs) and with a particular use in mind (e.g., comparing various concepts of operations), a particular model may prove more useful than others. In fact, given the state of the art of HPMs, we believe that proclaiming a general cognitive architecture or modeling tool the winner is misguided—premature at best. However, as will be seen in chapter 10, different modeling architectures have different strengths, weaknesses, and assumptions, and they approach the same problem differently.

NASA HPM PROJECT APPROACH

The NASA HPM project was organized along the development timeline shown in Figure 1.1. In this project, five predictive models of human performance simultaneously addressed two well-specified problems in aviation safety.

As shown in Figure 1.1, initial NASA HPM project review efforts consisted of (1) determining the context of human error in aviation, and (2) reviewing the state of the art of appropriate modeling tools. In order to understand the role of human error in commercial aviation (Leiden, Keller, & French, 2001), existing literature and techniques were reviewed regarding the conditions surrounding human error in aviation. In general, the concept of error chains and existing taxonomies of human error were reviewed, with discussions of initial conditions and event sequences and with respect to various existing error taxonomies. With this aviation safety context in mind, the NASA HPM project reviewed past efforts in human performance modeling

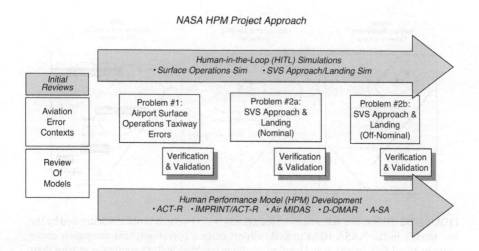

NASA HPM Project Approach

FIGURE 1.1 NASA HPM project approach.

(Leiden, Laughery, et al., 2001). It made clear that no single modeling architecture or framework had the scope to address the full range of interacting and competing factors driving human actions in dynamic, complex environments.

As a consequence, the decision was made to develop and expand multiple modeling efforts to extend the current state of the art within a number of HPM tools. Five modeling frameworks were selected in response to a proposal call for computational approaches for the investigation and prediction of operator behaviors associated with incidents or accidents in aviation. This was, in essence, a request for analytic techniques that employed cognitive modeling and simulation. The five modeling frameworks from industry and academia were selected with selection criteria including model theory, scope, maturity, and validation.

NASA HPM PROJECT AVIATION PROBLEMS AND SCOPE

The approach used in the NASA HPM project involves applying different cognitive modeling tools to the analysis of well-specified aviation operational problems for which empirical data of pilot performance in the task are available (see Figure 1.2).

In the first phase of the project (represented by the left panel of Figure 1.2), the five modeling tools were used to develop a deeper understanding of the causes of taxi errors and to explore potential mitigating solutions. To support this effort, the modelers were provided with a task analysis of rollout and taxi operations, airport layout information, and a rich data set that included a series of land-and-taxi-to-gate scenarios taken from a high-fidelity, full-mission, human-in-the-loop (HITL) simulation study. This HITL simulation provided an extensive data set of pilot performance data, including taxi speed, navigation errors, intracockpit communications, pilot-ATC communications, workload, and situation awareness, as well as a robust set of taxi navigation error data (Hooey, Foyle, & Andre, 2000).

In the second phase of the project (see Figure 1.2, center and right panels), the five human performance modeling tools were extended to the more complex problem

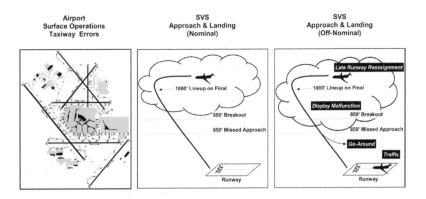

FIGURE 1.2 Illustrative schematics of the aviation safety-related problems addressed by the five models in the NASA HPM project: airport surface operations, taxi navigation errors (left); nominal SVS approach and landing condition (middle); and off-nominal scenarios (four examples shown in white text on black bars) for SVS approach and landing (right).

of modeling pilot behaviors during approach and landing operations with an SVS display. A part-task HITL simulation study was conducted to investigate the effect of proposed new SVS displays on pilot performance, visual attention, and crew roles and procedures during low-visibility instrument approaches (Goodman, Hooey, Foyle, & Wilson, 2003). The five modeling teams were provided with a cognitive task analysis of the approach and landing phase of flight, as well as human performance data that included eye movements, communications, and control panel responses from this NASA part-task simulation. The five HPMs were applied to nominal (routine and uneventful) approach and landing scenarios (Figure 1.2, middle panel) with and without the SVS displays, as well as to "off-nominal" conditions. These included such non-normal events as late runway reassignment; SVS display malfunction; and aborted landings, or "go-arounds," because of cloud cover and runway traffic (shown notionally in the right panel of Figure 1.2).

HUMAN PERFORMANCE MODELING IN AVIATION: EFFORTS AND FINDINGS

NASA's Human Performance Modeling project effort applied five human performance modeling architectures to two classes of specific aviation problems. As a result of this process, we are in a unique position to characterize some of the ways in which specific models interact with the problem, and to note similarities and differences in how model representations affect the characterization of the modeling problem. As will be seen in the chapters to follow, the NASA HPM project efforts resulted in both design solutions and procedural recommendations to enhance the safety of aviation systems. Significant advancements to the state of human performance modeling were achieved by broadening the scope of the five models to include the aviation domain and through the augmentation and expansion of specific modeling capabilities.

This volume is organized into three separate sections. First, a short history and discussion of the state of the art of human performance modeling is provided, and the specific aviation problems are introduced. Next, the efforts and results from the five individual modeling teams are presented. Finally, a cross-model comparison, roundtable discussion, and a discussion of findings and implications for both human performance modeling and aviation safety comprise the final section.

PART 1. GOALS, AVIATION PROBLEMS, AND MODELING

"Goals, Aviation Problems, and Modeling" provides a brief introduction to the history of HPMs, state-of-the-art assessments, and the aviation problems modeled in NASA's HPM project.

WHY MODEL?

In chapter 2, "Using Human Performance Modeling in Aviation," we provide a short history of human performance modeling, review current thinking about the nature of human error and performance, and suggest an approach coupling HITL testing (incorporating off-nominal scenarios) with human performance modeling. In addition, the value of such an approach to providing understanding of error causes, discovering latent error conditions, and determining and assessing system redesigns is discussed.

AVIATION PROBLEMS: AIRPORT SURFACE (TAXI) NAVIGATION AND APPROACH AND LANDING WITH SVS

Chapter 3, "Human-in-the-Loop Simulations in Support of Human Performance Model Development and Validation," is dedicated to a detailed description of the two aviation problems that the modeling teams addressed in the HPM project. First, we describe in detail the full-mission surface operations HITL simulation study (Hooey, Foyle, & Andre, 2000), in which pilots were required to follow taxi clearances at O'Hare International Airport in Chicago, Illinois. Under current-day procedures and equipment and under moderately low-visibility conditions, pilots made navigation errors on nearly one fourth of the trials. The taxi environment, procedures, tasks, to-be-followed taxi route clearances, and the specific errors made were supplied to the teams to model in this first phase of the HPM project.

In the second phase of the project, the HPM modeling teams addressed the aviation problem of using proposed advanced (SVS) displays for approach and landing. We present a cognitive task analysis (Keller & Leiden, 2002a, 2002b) describing the pilot's cognitive tasks during approach and landing with such a system. Finally, a detailed description of the cockpit layout and information and nominal and off-nominal scenarios for a part-task HITL simulation of approach and landing at the airport at Santa Barbara, California, are presented, along with a description of the supplied data.

PART 2. APPLICATION OF INDIVIDUAL MODELING TOOLS TO THE AVIATION PROBLEMS

"Application of Individual Modeling Tools to the Aviation Problems" offers a description of each of the five modeling tools applied to the two aviation problems, specific modifications and enhancements needed for the model to address these problems, the model's ability to describe or predict the human performance data, validation efforts, and implications for the model and the aviation problem.

INTRODUCTION TO THE MODELS

The second section of this book begins with chapter 4, "Overview of Human Performance Modeling Tools," by Kenneth Leiden, Michael Byrne, Kevin Corker, Stephen Deutsch, Christian Lebiere, and Christopher Wickens. This chapter provides an introduction to the five human performance modeling tools used in the NASA HPM project:

- Adaptive Control of Thought-Rational (ACT-R);
- Improved Performance Research Integration Tool/ACT-R hybrid (IMPRINT/ACT-R);
- Air Man–machine Integration Design and Analysis System (Air MIDAS);

- Distributed Operator Model Architecture (D-OMAR); and
- Attention-Situation Awareness (A-SA).

Leiden and coauthors briefly describe the history and evolution of each modeling tool, provide a general introduction of each models' architecture, and discuss the types of human performance problems that each modeling tool has typically addressed. This chapter provides a description of the models' starting point for the HPM project.

ACT-R (ADAPTIVE CONTROL OF THOUGHT-RATIONAL)

In chapter 5, "An ACT-R Approach to Closing the Loop on Computational Cognitive Modeling: Describing the Dynamics of Interactive Decision Making and Attention Allocation," Michael Byrne, Alex Kirlik, and Michael Fleetwood describe their findings from applying ACT-R to the NASA HPM aviation problems. Beginning with an historical perspective on cognitive modeling, one of the themes of their chapter is the need for incorporating the environmental task characteristics. In fact, Byrne et al. state that one of their many challenges was how to bring "a cognitive architecture applied and validated primarily in laboratory tasks to bear on realistically complex, operational scenarios." They proceeded through the use of three central principles that guided their modeling approach: (1) the need for a dynamic, closed-loop model of pilot cognition in interaction with the cockpit, aircraft, and environment; (2) the presumption that pilots are knowledgeable and adapted operators; and (3) the importance of allocation of visual attention in yielding important design and training-related insights.

For the taxi navigation problem, the three environmental task systems (aircraft control, visual information available, and airport runway–taxiway environment) were computationally modeled and integrated with the ACT-R pilot model. This approach allowed the authors to apply their model to an investigation of the decision heuristics and strategies that drive pilot performance during taxi navigation. For the SVS approach and landing problem, this emphasis on the multiple environmental components led them to "close the loop" formed between their ACT-R model and the environment by connecting a real-time aircraft simulation (X-Plane) to their pilot model. Their approach produced high-level predictions of gaze time that fit well with HITL simulation data. Additionally, the model proved sensitive to the local properties of the SVS display, demonstrating that the type and format of presented flight symbology are strong determinants of SVS usage.

IMPRINT/ACT-R (IMPROVED PERFORMANCE RESEARCH INTEGRATION TOOL/ADAPTIVE CONTROL OF THOUGHT-RATIONAL) HYBRID

In chapter 6, "Modeling Pilot Performance With an Integrated Task Network and Cognitive Architecture Approach," Christian Lebiere, Rick Archer, Brad Best, and Dan Schunk describe their hybrid modeling approach. They detail the unique integration of the low-level cognitive architecture, ACT-R, with the task network simulation tool, IMPRINT, to provide a viable approach for modeling complex domains. This approach allowed Lebiere and colleagues to characterize more simply the key

environmental aspects and decomposed mission tasks within IMPRINT, while applying ACT-R to the cognitive tasks. For the taxi navigation problem, the IMPRINT/
ACT-R model was applied to clearance errors involving memory, producing both
errors of omission and commission. For the SVS approach and landing problem,
the IMPRINT/ACT-R model permitted sensitivity analyses of mission success rates
to global parameters regarding latency of procedural, visual, motor, and auditory
actions, as well as stochastic manipulations of decision-making times.

These analyses provided important inferences regarding effective design objectives for both information display and procedures. Among other findings, the model's
outcomes suggest that pilot performance is very sensitive to the speed of visual shifts
between widely separated information sources. Similarly, pilot performance proved
highly sensitive to the overhead of communications; increases in the number and/
or duration of communications acts rapidly deteriorated performance. Noteworthy
in these modeling analyses was the apparent "performance tipping point" in which
near-perfect mission success rates would suddenly plummet with only the slightest increase in parameter latency. Multiple model verification and validation techniques were
employed, including an analysis of ACT-R's high-level trace output and correlation of
model predictions and HITL data on the dimension of visual allocation of attention.

AIR MIDAS (AIR MAN-MACHINE INTEGRATION DESIGN AND ANALYSIS SYSTEM)

Next, in chapter 7, "Air MIDAS: a Closed-Loop Model Framework," Kevin Corker,
Koji Muraoka, Savita Verma, Amit Jadhav, and Brian Gore describe their efforts to
apply the Air MIDAS modeling architecture to the NASA HPM aviation problems.
They first discuss the modeling and simulation framework of Air MIDAS, describing
how information flows through structures that include an "Updateable World Representation" and perceptual, decision-making, scheduling, and queue mechanisms.
Applying Air MIDAS to the taxi navigation problem, they modeled a representation
of the pilot's interaction with ATC and the airport environment, also modeling the
landing and taxi procedures. Their results show that pilots' taxi clearance errors
were produced by a combination of high workload while receiving the taxi clearance and memory decay of that clearance while taxiing on the airport surface. For
the SVS approach and landing problem, the Air MIDAS visual sampling model was
calibrated and verified with an extensive alternate empirical data set. Air MIDAS
was then used to predict performance on three approach and landing scenarios from
the NASA SVS HITL data set. Additionally, the Air MIDAS team conducted a third
simulation in which Air MIDAS was connected in a closed-loop manner to an external aircraft flight simulation program (PC Plane).

Output from these model simulations permitted detailed inspection of the executed task sequences for both the pilot flying (PF) and the pilot not flying (PNF).
Analyses of these sequences indicate differences in task completion ordering, timing,
and success between scenario conditions, thus suggesting possible vulnerabilities in
crew coordination and timing resulting from specific situational demands. In another
finding, the authors note that the Air MIDAS model output suggests that the addition of an SVS display may delay flight control action due to display redundancy and
the need for additional cross-checks, but seemingly not enough to have an adverse
affect.

D-OMAR (DISTRIBUTED OPERATOR MODEL ARCHITECTURE)

Stephen Deutsch and Richard Pew, in chapter 8, "D-OMAR: an Architecture for Modeling Multitask Behaviors," describe their efforts to model the taxi navigation and SVS approach and landing problems within the D-OMAR simulation framework. Deutsch and Pew explain how D-OMAR does not provide a single, constrained cognitive architecture; rather, it is a flexible modeling environment comprising of a discrete event simulator and a suite of representation languages that can be used to determine a cognitive architecture and construct models of human performance. The cognitive architecture evolved for this application focuses on multitask behavior, the role of vision, and working memory. In the taxi navigation problem, they use D-OMAR to model and understand conditions leading to two types of taxi errors observed in the HITL data: error driven by expectation based on partial knowledge and error driven by habit.

For the SVS approach and landing problem, D-OMAR models of the captain, first officer, and ATC working in concert demonstrated considerable robustness by executing successful landings across different scenarios. The model's prediction that the availability of the SVS display would reduce time devoted to other displays (the navigation display) was matched in the HITL data. Deutsch and Pew further explored this finding by conducting a "what if" modeling exercise in which they developed and modeled an "enhanced SVS" display as the single primary flight display, to produce predicted data that were more representative of the observed HITL instrument-scanning strategies. In modeling one of the off-nominal conditions—the "SVS misalignment" condition—these authors were able to demonstrate that, with D-OMAR, they were able to modify the modeled pilot's decision-making behaviors to match the ambivalence observed in the HITL pilots' go-around behavior.

A-SA (ATTENTION-SITUATION AWARENESS)

The final chapter in the second section is chapter 9, "Attention-Situation Awareness (A-SA) Model of Pilot Error," by Christopher Wickens, Jason McCarley, Amy Alexander, Lisa Thomas, Michael Ambinder, and Sam Zheng. In this chapter, the authors describe the modification of their algorithmic SEEV (salience, effort, expectancy, value) model of attentional allocation in dynamic environments to the prediction of errors of situation awareness (SA). Two versions of the A-SA model are developed and applied to the data: a static equation/analytic model version predicting average dwell time on visual regions and a dynamic version that also predicts the dynamic visual scan path (transition frequencies).

For the taxi navigation problem, Wickens and coauthors produce sensitivity analyses showing that the model predicts loss of SA with time and distraction, recovery of SA after relevant events, and improved SA with augmented navigation displays. Turning to SVS approach and landing, the authors apply both static and dynamic versions of the A-SA model to a more extensive SVS approach and landing experiment that they conducted. One noteworthy finding is that individual pilots who conform to the optimal scanning model were reliably better on flight path tracking and somewhat better on traffic detection, as one would expect from the latter stages of SA.

PART 3. IMPLICATIONS FOR MODELING AND AVIATION

"Implications for Modeling and Aviation" provides an analysis of cross-model comparisons, aviation implications and modeling lessons learned, and a roundtable question and answer discussion of issues related to human performance modeling.

CROSS-MODEL COMPARISONS

Starting off Part 3, which analyzes the models, findings, and implications of the models and summarizes the results of the HPM project, is chapter 10, "A Cross-Model Comparison," by Ken Leiden and Brad Best. These authors provide a review and analysis of the five NASA HPM project models regarding the following capabilities: error prediction, external environment, crew interactions, scheduling and multitasking, memory, visual attention, workload, situation awareness, learning, resultant and emergent behavior, and verification and validation. This is presented to enable the reader to understand the assumptions and philosophy underlying each modeling architecture, to gain insight into their predictive and explanatory capabilities, and to make more informed choices in choosing a modeling strategy for future projects.

HUMAN PERFORMANCE MODELS IN AVIATION: A ROUNDTABLE DISCUSSION

In chapter 11, "Human Performance Modeling: a Virtual Roundtable Discussion," we pose a series of questions to each of the five modeling teams regarding their individual models, challenges, and reflections on the NASA HPM project. In this chapter, we have the unique ability to join a virtual roundtable discussion with the five modeling teams, gaining some perspective of the take-home lessons and insights that they gained after completing the NASA HPM project. Some of the specific issues discussed include:

- Model architecture and structures: To what extent do the specific architectures and structures in the various models impact the user's choice of a modeling tool, the ability to describe or predict the data, and the validation of results?
- Role of the external environment: How is the external environment captured in the model and how does the model interact with that environment?
- Model predictive ability: To what extent are the specific modeling tools accurately modeling or predicting behavior, producing emergent behavior, acting as prediction versus simulation in nature, and allowing for the extrapolation to other untested display or procedural conditions?
- Usefulness and implications of the modeling results: To what extent are the models able to discover latent errors or rare, unsafe conditions?
- Model validation: How does one proceed in validating modeling results of complex tasks such as pilot performance or error?

Additionally, the future directions and challenges for human performance modeling in aviation are discussed by each modeling team.

IMPLICATIONS AND LESSONS LEARNED FROM THE NASA HPM PROJECT

Finally, in chapter 12, "Advancing the State of the Art of Human Performance Models to Improve Aviation Safety," we summarize the findings of the NASA HPM project. In this final chapter we discuss some of the major findings and implications of the project and attempt to put these into perspective. We characterize the findings of the project in terms of demonstrated contributions for the aviation community and the significant advances in HPMs that were required to realize those contributions. Subsequently, insights gleaned from this modeling effort with regard to model selection, development, interpretation, and validation are offered.

Within the scope and domain of aviation safety, we discuss how HPMs and HITL simulations can be integrated synergistically in the system design and evaluation process. Drawing from examples across the NASA HPM project, we discuss the advantages and disadvantages of HITL simulations and HPMs across the system design and evaluation life cycle. In the process of addressing the two aviation domain problems, each modeling architecture underwent significant augmentations. Advances in the state of the art of HPMs that are highlighted in this chapter include new techniques and capabilities to model human-environment interactions, visual attention, situation awareness, and human error.

As a result of the application of multiple models to two aviation safety topics, the NASA HPM project was in a unique position to characterize specific considerations associated with the use of HPMs in aviation. The four considerations were:

- Model selection: Selecting a model for a particular aviation problem is a nontrivial task that requires knowledge of the underlying model assumptions and philosophies.
- Model development: Appropriate domain knowledge that includes not only normative task sequences and timing data but also insights into operator strategies is required by model development.
- Model interpretation: For models to be of true value to the aviation community, modelers must take steps to ensure that end-users understand the underlying assumptions and constraints, rather than blindly accepting the model output.
- Model validation: Determining methods of model validation is often a challenge in aviation research, given limited data sets, especially for new concepts and technologies. Validation techniques used in the NASA HPM program to meet different validation requirements are discussed. The techniques include sensitivity analyses, comparing high-level outcomes and task traces to HITL data, and quantitative results validation.

ACKNOWLEDGMENTS

The authors thank Alex Kirlik and Dick Pew for their thoughtful, insightful comments and suggestions on an earlier version of this chapter.

REFERENCES

Gluck, K. A., & Pew, R. W. (2005). *Modeling human behavior with integrated cognitive architectures: Comparison, evaluation, and validation*. Mahwah, NJ: Lawrence Erlbaum Associates.

Goodman, A., Hooey, B. L., Foyle, D. C., & Wilson, J. R. (2003). Characterizing visual performance during approach and landing with and without a synthetic vision display: A part-task study. In D. C. Foyle, A. Goodman, & B. L. Hooey (Eds.), *Proceedings of the 2003 Conference on Human Performance Modeling of Approach and Landing with Augmented Displays* (NASA/CP-2003-212267) (pp. 71–89). Moffett Field, CA: NASA Ames Research Center.

Hooey, B. L., Foyle, D. C., & Andre, A. D. (2002). A human-centered methodology for the design, evaluation, and integration of cockpit displays. *Proceedings of the NATO RTO SCI and SET Symposium on Enhanced and Synthetic Vision Systems*. NATO.

Keller, J. W., & Leiden, K. (2002a). *Information to support the human performance modeling of a B757 flight crew during approach and landing: RNAV* (Tech. Rep.). Boulder, CO: Micro Analysis & Design.

Keller, J. W., & Leiden, K. (2002b). *Information to support the human performance modeling of a B757 flight crew during approach and landing: SVS addendum* (Tech. Rep.). Boulder, CO: Micro Analysis & Design.

Leiden, K., Keller, J. W., & French, J. W. (2001). *Context of human error in commercial aviation (Tech. Rep.)*. Boulder, CO: Micro Analysis & Design.

Leiden, K., Laughery, K. R., Keller, J. W., French, J. W., Warwick, W., & Wood, S. D. (2001). *A review of human performance models for the prediction of human error (Tech. Rep.)*. Boulder, CO: Micro Analysis & Design.

2 Using Human Performance Modeling in Aviation

David C. Foyle and Becky L. Hooey

CONTENTS

INTRODUCTION

The aviation community is continuously striving to increase the efficiency and safety of the entire aviation system, which includes human operators (e.g., commercial airline pilots, general aviation pilots, and air traffic controllers [ATCs]), technology (i.e., hardware, software, automation, displays, and controls), and the interactions among human operators and technology. Addressing these complicated system interactions requires research toward the design and development of new system concepts; investigation of procedural, training, and communication issues; and the design and evaluation of avionics, equipment, and interfaces.

This chapter focuses on the uses of human performance models (HPMs) to address aviation-related issues. First, a brief historical perspective is presented describing the ways that HPMs have changed as increasingly complex behaviors, tasks, and environments have been addressed. Second, recent state-of-the-art assessments of the human performance modeling field are reviewed. In the third section, the main domain of interest is introduced with a discussion of aviation safety, pilot error and performance, and system latent error. Finally, an approach of coupling human-in-the-loop (HITL) simulation and human performance modeling is proposed to gain insight into the underlying causes of pilot performance and error and, more importantly, to determine procedural or system designs that mitigate these problems, ultimately leading to increased aviation safety.

As will be seen in later chapters, these actual aviation problems (as opposed to laboratory test problems developed for model evaluation or theoretical development) represent a class of real-world problems that are of sufficient complexity that the simple application of these existing models was not possible without significant extensions to the HPM state of the art. The following section briefly reviews the history of human performance modeling regarding the application of HPMs to actual, complex-environment problems.

HUMAN PERFORMANCE MODELING: FROM THE LABORATORY TO THE WORLD

In the years immediately after World War II and into the 1970s, researchers in the field of mathematical and cognitive psychology primarily used formal mathematical techniques such as Markov models or other mathematical representations (e.g., Green & Swets, 1966/1988; Restle & Greeno, 1970) to model and understand human performance. In the area of cognition, the applications addressed were typically psychology laboratory studies, including discrimination learning and concept identification, to name a few, although some real-world subtasks (i.e., detection) were also addressed. In other areas, such as control theory, however, these techniques were quite developed (for example, the Optimal Control Model, OCM; see Baron, 2005; Baron, Kruser, & Huey, 1990, for an historical perspective).

With the explosion of computer technology in the 1960s and 1970s and the increasing popularity of so-called "box models" of information processing in which cognition was represented essentially as a computer flow chart with each element being a cognitive/perceptual process, Monte Carlo computer simulations became popular as a technique to model cognitive processes. By coding perceptual/cognitive processes as mechanisms with specifically defined characteristics (e.g., encoding, decay, retrieval) and associated probabilities, the entire stream of behavior (from stimulus to perception to cognition to the final response output) could now be modeled as separate "processes" with computer modules for each process. Feigenbaum's (1959) Elementary Perceiver And Memorizer (EPAM) verbal learning model is likely the earliest example of such a computer simulation model. A more recent example of such a model is the Search of Associative Memory (SAM) developed by Shiffrin and his colleagues (e.g., Raaijmakers & Shiffrin, 1981), which models long-term memory recall and recognition laboratory results.

In the aviation domain, a computer simulation model of note is the Procedure-Oriented Crew (PROCRU) model, a derivative of OCM developed by Baron and colleagues (1990). Within the control-theoretic framework, PROCRU modeled pilot crew activities and performance for a typical instrument landing system (ILS) approach and landing. Baron (2005) cites two main reasons as to why such control-theoretic approaches became less popular in the mid-1980s. First, such continuous-time representational approaches were viewed as computationally inefficient for very slowly changing systems and long operating times. Second, new methods were emerging to approach issues in supervisory control: the emergence of new cognitive modeling architectures and the exploding field of artificial intelligence.

Since the early 1980s, these integrated cognitive modeling architectures have been developed and successfully used to model, understand, and effect design changes in subsystems of complex environments (some noteworthy reviews will be described in a later section). One particularly well known early example was Project Ernestine, in which a proposed keypad layout for telephone operators evaluated using the Goals, Operators, Methods and Selection rules (GOMS) model was shown actually to slow down operator input (Gray, John, & Atwood, 1993). This counterintuitive prediction of design inefficiency with the proposed keypad layout supported a similar outcome observed in field tests, giving weight to the veracity of the counterintuitive outcome seen in the field trials. Without diminishing the value of this landmark application of a human performance modeling tool to a real-world task (a solution that reportedly saved the telephone company, NYNEX, millions of dollars), this application was still relatively simple, representing only one task (albeit the one of interest) in the entire set of duties of a telephone operator.

Beginning in the 1990s, the importance of the role of the human operator in the development of new concepts of air traffic operations in the United States and in Europe has prompted the modeling and simulation community to tackle increasingly complex real-world problems and situations using integrated cognitive modeling architectures. Most notably, these have been in the area of air traffic control (ATC) and air traffic management (ATM) in the United States (e.g., Corker & Pisanich, 1998; Ippolito & Pritchett, 2000) and by joint efforts of the Federal Aviation Administration (FAA) in the United States and EUROCONTROL in Europe (see Blom et al., 2001, for an overview). With a similar goal of extending human performance models to complex, real-world problems, the U.S. Air Force Research Laboratory (AFRL) recently completed the Agent-based Modeling and Behavioral Representation (AMBR) project (Gluck & Pew, 2005). The specific problem that served as the study domain for the AMBR project was that of an existing part-task ATC sector control simulation. Although the AMBR project used an experimental ATC task and did not use actual ATC operators, this ATC part-task simulation approach did allow for the collection of a large HITL data set for use in addressing model development and validation issues.

HUMAN PERFORMANCE MODELING:
STATE-OF-THE-ART REVIEW AND CHALLENGES

There are many books on the domain of cognitive modeling as a field, including survey overviews (e.g., Polk & Seifert's *Cognitive Modeling*, 2002) and specific modeling architectures (e.g., Anderson & Lebiere's *Atomic Components of Thought*, 1998), as well as on more narrowly defined aspects of modeling (e.g., Miyake & Shah's *Models of Working Memory*, 1999). A comprehensive review of the state of the art of human performance modeling is beyond the scope of this book, but to help the reader navigate through the existing state-of-the-art reviews, some of the more comprehensive review efforts are listed and briefly discussed (also see Gray's *Integrated Models of Cognitive Systems*, 2007, for a recent comprehensive review of cognitive modeling techniques). To guide the reader, differences and similarities between those existing reviews and the efforts documented in this volume are noted.

In the late 1980s, the National Research Council (NRC) commissioned a series of panels to assess current trends and to shape the future direction of human performance modeling. The first of these panels, Pilot Performance Models for Computer-Aided Engineering (Elkind, Card, Hochberg, & Messick, 1989), was requested by James Hartzell at the NASA Ames Research Center to provide guidance for the Army-NASA Aircrew/Aircraft Integration (A³I) program. This program was developing a prototype of a human factors computer-aided engineering workstation for designing helicopter cockpits. (In fact, the Air MIDAS model described in this volume is a descendent of that prototype.)

Three problems were identified: the difficulty of integrating component-level models (e.g., a visual contrast model) into a larger cognitive architecture; the presence of "gaps" between these component-level models (i.e., all needed functions have not been modeled, so the exhaustive set of component models from which to draw does not exist); and the failure to incorporate perception. Elkind et al. (1989) concluded that attempting to develop such engineering workstations incorporating human performance models would "put pressure on researchers to extend models in directions that are likely to lead to interesting theoretical (e.g., overcoming integration problems) and ultimately practical developments" (1989, p. 300). Presumably, this recommendation was made within the historical context of the emergence of integrated cognitive architectures such as Newell and Simon's Soar in the early 1980s (see Laird & Rosenbloom, 1996, for a history of Soar), which would later influence other architectures such as Anderson's ACT-R (see Anderson & Lebiere, 1998).

About the same time, another NRC-commissioned panel produced the review *Quantitative Modeling of Human Performance in Complex, Dynamic Systems* (edited by Baron et al., 1990). The authors stated that at the time of publication, human performance models tended to be designed or selected for specific situations and used to simulate a single–function such as search models, signal detection models, game theory models, and tracking models. The authors noted:

> Most existing human performance models have been developed only for relatively simple situations. Many of the real-world person–machine systems of interest today are highly complex, involving multiple operators, multiple tasks, and variable environmental or equipment contexts. Preferred methods for developing human performance models for these systems have not been identified. (p. 72)

A decade later, the NRC commissioned another review of the state of the art of human modeling. With a focus on military organizational units, the resultant review, *Modeling Human and Organizational Behavior: Application to Military Simulations* (Pew & Mavor, 1998), provided a comprehensive resource covering issues relating to the modeling of the human and organizational behavior, especially with regard to attention, learning, memory, decision-making, perception, situation awareness, and planning. However, complex system-level issues such as system design, multi-operator interactions, and procedural and operational issues were mostly not addressed in that review.

Pew and Mavor (1998) set the stage for later efforts in that they provided a description of many modeling architectures, including earlier versions of some of those that were used in this NASA modeling effort (i.e., MIDAS, ACT-R, D-OMAR). However, since 1998, when Pew and Mavor's review was published, there have been significant advances in computing technology and in the state of the art of the modeling tools. This volume offers to update the modeling and simulation community on some significant advancements that have been made in these modeling tools, and also to highlight two new modeling tools that have since been developed. Additionally, this volume goes beyond a description of each tool and also provides concrete examples of how the models were implemented, the results that were obtained, and the validation techniques that were employed.

Pew and Mavor concluded that the state-of-the-art modeling tools were capable only of situation-specific models and that "at [the time of publication] we are a long way from having a general-purpose cognitive model" (1998, p. 320). Toward this goal of developing such general-purpose cognitive models, the NASA HPM project sought to extend and apply existing cognitive architectures to aviation safety. In the chapters to follow, we will demonstrate advances in the development of cognitive models as evidenced by the application of each model to two different types of aviation problems (an operational error analysis and an equipment display design effort).

In 2003, as a Human Systems Integration Information Analysis Center (HSIAC) State Of the Art Report (SOAR), Ritter et al. (2003) updated the state of the art in modeling, again with a focus largely on modeling synthetic forces and organizational behavior for military simulation applications. As well as reviewing existing models usable in the domain of synthetic force simulation (i.e., human agents in simulations), they make the case for needed improvements in the field of modeling. Specifically, Ritter et al. make the statement that "one of the most important aspects of human performance, which has often been overlooked in models of behavior and problem solving, is errors" (p. 15). The present volume attempts to follow this suggestion and to highlight the importance of, and relative lack of attention that has been paid to, the modeling of human error and error recovery, both of which are addressed in the chapters to follow. In the NASA HPM project, each modeling team modeled taxi navigation errors on the airport surface in an attempt to predict error vulnerabilities and identify causal factors and mitigating solutions.

Ritter et al. also challenged the modeling community by stating that "tying cognitive models to synthetic environments in psychologically plausible ways should be easier" and suggested the need for model augmentations in this area (2003, p. 62). This was a challenge that our modeling effort needed to address because of the complex task domain within which we were working. The model-specific chapters in this book will directly address this challenge by outlining the strategies that each model team employed to tie the models to a simulated environment and will share their lessons learned from this process. This is a prime example of a significant advancement in the knowledge of human performance modeling that was realized because of the challenge of modeling complex, real-world tasks and environments.

AVIATION SAFETY: THE ROLE OF THE HUMAN OPERATOR

One challenge that the NASA HPM project aimed to address was in direct support of the fundamental goals of the NASA Aviation Safety and Security Program (AvSSP): to develop knowledge and technologies that improve aviation safety from the perspective of the flight deck. As part of its mission, NASA conducts research on current-day operations with existing technologies as well as investigating the impact of proposed future operational changes. The investigation of future operational changes includes developing conceptual designs of emerging and proposed procedural, operational, and technological solutions, and assessing the impacts of such changes on aviation safety and efficiency. As will be described in a later section and seen in subsequent chapters, the NASA HPM project addresses each of these, applying HPM tools to (1) a current-day safety problem of airport taxi navigation errors, and (2) an emerging-technology problem of using synthetic vision systems (SVS) for approach and landing. In the following section, issues associated with aviation safety will be reviewed.

HUMAN ERROR AND ACCIDENTS IN AVIATION

More than two thirds of all aircraft accidents have been attributed to pilot error (FAA, 1990). However, one must be careful not to mistake such a statistical attribution as representing the "smoking gun" pointing to the source or root cause of the majority of aviation accidents. The attribution of two thirds of the accidents to pilot error is a superficial classification that confuses correlation with causation; it does not help us to gain a deep understanding of the hierarchy of causal factors that led to the accidents and does not represent any obvious indicator of potential solutions.

In recent years, the nature as well as the very definition of "human error" has been the source of many discussions in the human factors and engineering communities. Topics have included the causes, role, and implications of human error in complex systems—specifically, that of pilot error in aviation accidents and incidents. Hollnagel (1993) and Dekker (2003) have pointed out the fallacy in confusing error classification with the understanding of root causes of error. They note that in today's classification systems, the term "error" is used in three distinctly different ways. Specifically, it has been used as the *cause* of failure (e.g., an event was due to "human error"); as the *failure itself* (e.g., the decision was an "error"); and as a *process*, when referring to failure to follow standards or operating procedures.

Similarly, when discussing the fallacy of referring to human error as the *cause* of failure, Woods and Cook (2004) note that implicit in the misuse of the term "error" in this sense is that "error is some basic category or type of human behavior that precedes and generates a failure" (p. 98) and that the idea of error-as-cause trivializes expert human performance by categorizing it simply as either acts that are errors or acts that are nonerrors. Woods and Cook argue that this dichotomization of performance into errors and nonerrors "evaporates in the face of any serious look at human performance" (p. 99). In fact, this was pointed out 40 years earlier by B. F. Skinner (1966), who stated that "the term error does not describe behavior, it passes judgment on it" (p. 219). We agree with this view of the relationship between human error and human performance—that human error is not a special type of behavior, but rather is a result of the range of normal human performance interacting with the human's cognitive and sensory capabilities (what a person *can* do) and response proclivities (what a person *does* do) (see Watson, 1973) as shaped by the environment, tasks, and procedures (see Hollnagel, 1993). In fact, one might consider that analyzing accidents that have been attributed to human error and analyzing human performance in the field or in simulation allow one to work the human–system problem from two different, complementary directions.

LATENT ERRORS

Recent discussions of safety, accidents, and system design have found it useful to distinguish between the concepts of *active errors* and *latent errors* (Rasmussen & Pedersen, 1984; Reason, 1990). Active errors involve the actions of the operators of the complex system, such as the control room operators, pilots, and air traffic controllers, and their impact on the system is immediate. Active errors involve the direct control interface or human–computer interface and would include flipping the wrong switch or ignoring a warning tone. Latent errors, in contrast, are produced by actions that are removed in space and time from those of the actual operator directly interfacing with the system. As general examples of latent errors, Reason lists poor design, faulty maintenance, incorrect installation, and bad management decisions. Thus, such producers of latent errors might include designers, maintainers, and high-level decision makers of the complex system. Reason uses a "resident pathogen metaphor" of latent error, in which the problem (e.g., design flaw) may lie dormant until the right set of interacting conditions emerge allowing the error to be expressed (p. 197). Reason points out that it may be more efficient to concentrate on these latent errors, and that "to discover and neutralise these latent failures will have a greater beneficial effect upon system safety than will localised efforts to minimise active errors" (p. 173).

Reason's suggestion that working to discover and eliminate latent errors might be an effective safety strategy finds support in a recent analysis conducted by Dismukes, Berman, and Loukopoulos (2007). They analyzed the 19 airline accidents that occurred in the United States between 1991 and 2000 that were attributed primarily to crew error. Their analysis indicated that, in two thirds of these accidents, equipment failures or design flaws started the chain of events leading to the accident or inhibited recovery by making diagnosis of the problem difficult. Thus, two thirds of these recent airline accidents might have been mitigated by the elimination of the inherent latent errors.

A "NEW LOOK" AT ERROR

Hobbs (2004) makes the point that there are many contributing factors supporting human error in accidents, but notes that one must look beyond the temporal proximity of the error to the accident. He warns:

> A conclusion of "human error" is never an adequate explanation for an accident because although the action of a person may be the immediate precursor of an accident, a mix of task, equipment, and environmental factors may also be involved. (p. 340)

In fact, we believe that this temporal proximity of the pilot to the accident has led many to treat the pilot as part of the error chain of events to be reckoned with, rather than taking the fundamentally different view of the pilot as the last line of defense in the safety chain. Such a view is not merely philosophical, but also impacts training, operations, and the selection and design of accident-mitigating systems. Researchers have argued for a "new look" at human error (Reason, 1990; Woods, Johannesen, Cook, & Sarter, 1994), where the determination of human error is not the stopping point as has traditionally been the case, but instead should be the starting point for learning about the many interactions of equipment design, operational procedures, and organizational pressures contributing to error.

COUPLING HUMAN-IN-THE-LOOP TESTING
AND HUMAN PERFORMANCE MODELING

Recognizing that human error and human performance result from an underlying performance continuum allows for an organizing principle for study: Through the investigation of human performance in the domain of interest, one can identify and study factors that act as performance stressors, environmental impacts of behavior, task demands, and procedural implications on behavior.

Identifying when equipment and procedures do not fully support the operational needs of pilots is critical to reducing error and improving flight safety (Leiden, Keller, & French, 2001). This becomes especially relevant in the development of new flight deck technologies that have traditionally followed a design process more focused on component functionality and technical performance than pilot usage and operability. A recent white paper prepared for the FAA by Parasuraman, Hansman, and Busso-lari (2002) addresses the problems and issues associated with the technical transfer of systems from the laboratory to the field within the context of aviation systems for surface operations safety. They argue for early human factors input, not only into the display interface as is more typical, but also into the very system functional requirements. They suggest that failure to do so may lead to

> a mismatch between the functionality as specified by the designer, the operating environment (i.e., procedures), and the user's requirements for the system or his or her mental model of system functionality ... [resulting in] inefficient system performance, errors, and possible adverse performance including accidents. (Parasuraman et al., 2002, p. 7)

To help counter this bias and to better understand the potential for human error associated with the deployment of new and complex systems (thus creating more error-tolerant designs), advanced tools and methods are needed for predicting pilot performance in real-world operational environments.

The human-centered design and evaluation process method shown in Figure 2.1 has been employed successfully to identify system and display designs and associated procedures that, if not discovered, might have led to the inclusion of latent design flaws. This method (a modification of Foyle et al., 1996, and Hooey, Foyle, & Andre, 2002) integrates task analyses, technology and operational assumptions, and

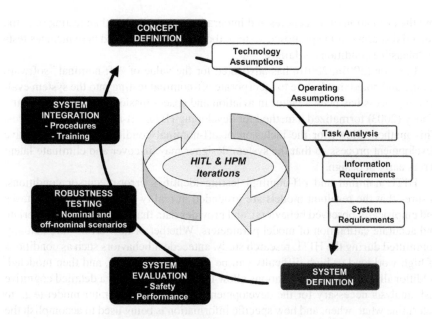

FIGURE 2.1 Human-centered design and evaluation process incorporating iterative human-in-the-loop (HITL) testing and human performance model (HPM) loops. Nominal and off-nominal scenarios enable discovery and mitigation testing of latent system design flaws. (Modification of Foyle, et al., 1996; and Hooey, Foyle, & Andre, 2002.)

information requirements analysis into system requirements that then are instantiated as the defined system. The major feature of the process is that there is an iterative evaluation and validation loop that, through the use of nominal and off-nominal scenarios, allows for performance evaluation and definition and integration of procedures. These off-nominal scenario evaluations are the tools by which the latent design flaws can be discovered and mitigated early during the design phase by identifying error precursors such as high-demand attention displays or high workload tasks and conditions. Although this human-centered design and evaluation process was originally developed in support of HITL testing, HPMs can be applied at every phase of design represented to eliminate latent error and improve system design.

As will be seen in the chapters to follow, coupling HITL and HPM proves to be a powerful approach, especially when both incorporate nominal and off-nominal scenarios. In contrast to a nominal scenario, in which operations follow normal, standard, or formalized procedures, an off-nominal scenario is one in which the unexpected occurs (Leveson, 2001a, 2001b). The unexpected may range from minor deviations that occur frequently in an operational setting up to the catastrophic failure of the system or subsystems. The inclusion of off-nominal scenarios allows for a full exercise of the system, a determination of where and why a system fails, the exploration of interactions with other user agents, a deeper understanding of usage, and the evaluation of procedures and integration issues. Rather than testing only the potential success of the proposed system, the approach advocated here includes tests of "plausible conditions of failure."

Leveson (2001a, 2001b) has advocated for the value of "off-nominal" software testing and noted that failure to incorporate off-nominal testing into the system evaluation process has been a factor in aviation and space-mission accidents. Foyle and Hooey (2003) formalized a method for developing HITL off-nominal test scenarios. This method allows for the inclusion of off-nominal conditions into the iterative development process so that the system designer may discover and eliminate latent errors early in design.

HITL nominal and off-nominal testing identifies error-precursor conditions, ensures that the resultant models are grounded in reality (i.e., capture all the tasks and empirically observed behaviors), and provides data that allow for the appropriate and accurate calibration of model parameters. Whether or not "errors" are actually committed during the HITL research study, antecedent behaviors such as conditions of high workload or high difficulty can be identified, analyzed, and then modeled. Additionally, such HITL evaluations allow for the production of a detailed cognitive task analysis necessary for the development of the human operator model (e.g., to determine what, when, and how specific information is being used to accomplish the goals of the operator; although, see Kieras & Meyer, 2000, and chapter 12 of this volume for some of the challenges in doing so).

As noted in the literature on aviation safety, serious piloting errors and the resultant accidents are rare events (for a review, see Leiden et al., 2001). The low probability of occurrence makes the study of serious pilot errors difficult to investigate in the field and in the laboratory (although antecedent precursor conditions can be identified). These errors characteristically result from a complex interaction among unusual circumstances, subtle "latent" flaws in system design and procedures, and limitations and biases in human performance. This can lead to the fielding of equipment that puts flight safety at risk, particularly when operated in a manner or under circumstances that may not have been envisioned or tested. When combined with HITL testing incorporating nominal and off-nominal scenarios (Foyle & Hooey, 2003), human performance modeling provides a complementary technique to develop systems and procedures tailored to the pilots' tasks, capabilities, and limitations. Because of its fast-time nature, human performance modeling is a powerful technique to assess the impact of latent errors in which a system contains a design flaw that may induce pilot error only under some low-probability confluence of precursors, conditions, and events.

Coupled together, HITL testing and human performance modeling can be a powerful tool set to identify potential latent error conditions or latent design flaws prior to implementation. Human performance models can then be applied to "look inside" the human's cognitive processes (i.e., investigate the internal modeled processes), determine and assess the impact of the antecedent ("error precursor") behaviors on system safety, and, in some instances, extrapolate to proposed display or procedural conditions not tested.

Human performance modeling using fast-time simulation offers a powerful technique to examine human interactions with existing and proposed aviation systems across a wide range of possible operating conditions. It provides a flexible and economical way to manipulate aspects of the task-environment, the equipment and procedures, and the human for simulation analyses. In particular, human performance modeling can suggest the nature of likely pilot errors, as well as highlight precursor conditions to error such as high levels of memory demand, mounting time pressure and workload, attentional tunneling or distraction, and deteriorating situation awareness. Fast-time modeling permits the generation of very large sample sizes from which low-rate-of-occurrence events are more likely to be revealed. An additional advantage associated with the use of human performance modeling includes the ability to propose and evaluate display and procedural changes. This is especially useful in that these evaluations of proposed changes can be done early in the design cycle, without the need to fabricate expensive prototype hardware. Finally, the careful characterization and formal thinking of the assumptions and processes involved in the problem can, by themselves, lead the modeler to gain new insights into system development and usage.

Thus, HITL testing with off-nominal scenarios and HPM techniques are two powerful tools that when coupled together provide a much more powerful approach toward the development of error-tolerant, human-centered system designs. This approach can lead to a detailed understanding of operator performance, provide insight into the root causes of human error, and determine conditions of latent error in which system design conditions, if left unchecked, may lead to errors. Such an approach during the design or redesign phases of a system will produce systems that are safer, more efficiently used by the operator, more robust to error and inadvertent misuse, and more likely to bridge the gap when moving from a laboratory prototype to a fielded, operational system. With these goals in mind, the NASA HPM project used this approach of coupling HITL and HPM techniques as the five modeling teams addressed the two aviation tasks discussed in detail in chapter 3.

ACKNOWLEDGMENTS

The authors thank Alex Kirlik and Dick Pew for their thoughtful and insightful comments and suggestions on an earlier version of this chapter.

REFERENCES

Anderson, J. R., & Lebiere. C. (1998). *Atomic components of thought*. Mahwah, NJ: Lawrence Erlbaum Associates.

Baron, S. (2005). Control systems R&D at BBN. *IEEE Annals of the History of Computing, 27,* 52–64.

Baron, S., Kruser, D. S., & Huey, B. M. (Eds.). (1990). *Quantitative modeling of human performance in complex, dynamic systems*. Washington, DC: National Academy Press.

Blom, H. A. P., Bakker, G. J., Blanker, P. J. G., Daams, J., Everdij, M. H. C., & Klompstra, M. B. (2001). *Accident risk assessment for advanced air traffic management* (NLR-TP-2001-642). Amsterdam: National Aerospace Laboratory (NLR).

Corker, K. M., & Pisanich, G. (1998). A cognitive system model for en route traffic management. In D. Harris (Ed.), *Engineering psychology and cognitive ergonomics* (pp. 121–129). London: Ashgate.

Dekker, S. W. A. (2003). Illusions of explanation: A critical essay on error classification. *International Journal of Aviation Psychology, 13,* 95–106.

Dismukes, R. K., Berman, B. A., & Loukopoulos, L. D. (2007). *The limits of expertise: Rethinking pilot error and the causes of airline accidents.* Aldershot, England: Ashgate.

Elkind, J. I., Card, S. K., Hochberg, J., & Messick, B. (Eds.). (1989). *Human performance models for computer-aided engineering.* Washington, DC: National Academy Press.

FAA (Federal Aviation Administration). (1990). *The national plan for aviation human factors.* Washington, DC: Author.

Feigenbaum, E. A. (1959). *An information processing theory of verbal learning* (RAND Report P-1817). Santa Monica, CA: RAND Corporation.

Foyle, D. C., Andre, A. D., McCann, R. S., Wenzel, E., Begault, D., & Battiste, V. (1996). Taxiway Navigation and Situation Awareness (T-NASA) System: Problem, design philosophy and description of an integrated display suite for low-visibility airport surface operations. *SAE Transactions: Journal of Aerospace, 105,* 1411–1418.

Foyle, D. C., & Hooey, B. L. (2003). Improving evaluation and system design through the use of off-nominal testing: A methodology for scenario development. *Proceedings of the Twelfth International Symposium on Aviation Psychology* (pp. 397–402). Dayton, OH: Wright State University.

Gluck, K. A., & Pew, R. W. (2005). *Modeling human behavior with integrated cognitive architectures: Comparison, evaluation, and validation.* Mahwah, NJ: Lawrence Erlbaum Associates.

Gray, W. D. (2007). *Integrated models of cognitive systems.* New York: Oxford University Press.

Gray, W. D., John, B. E., & Atwood, M. E. (1993). Project Ernestine: Validating a GOMS analysis for predicting and explaining real-world performance. *Human–Computer Interaction, 8*(3), 237–309.

Green, D. M., & Swets, J. A. (1966/1988). *Signal detection theory and psychophysics.* Los Altos, CA: Peninsula Publishing.

Hobbs, A. (2004). Human factors: The last frontier of aviation safety? *The International Journal of Aviation Psychology, 14,* 335–341.

Hollnagel, E. (1993). *Human reliability analysis: Context and control.* London: Academic Press.

Hooey, B. L., Foyle, D. C. & Andre, A. D. (2002). A human-centered methodology for the design, evaluation, and integration of cockpit displays. *Proceedings of the NATO RTO SCI and SET Symposium on Enhanced and Synthetic Vision Systems.* NATO.

Ippolito, C. A., & Pritchett, A. R. (2000) Software architecture for a reconfigurable flight simulator. *Proceedings of the AIAA Modeling and Simulation Technologies Conference* (AIAA-2000-4501).

Kieras, D. E., & Meyer, D. E. (2000). The role of cognitive task analysis in the application of predictive models of human performance. In J. M. C. Schraagen, S. E. Chipman, & V. L. Shalin (Eds.), *Cognitive task analysis* (pp. 237–260). Mahwah, NJ: Lawrence Erlbaum Associates.

Laird, J. E., & Rosenbloom, P. (1996). The evolution of the Soar cognitive architecture. In D. M. Steier & T. M. Mitchell (Eds.), *Mind matters: A tribute to Allen Newell* (pp. 1–50). Mahwah, NJ: Lawrence Erlbaum Associates.

Leiden, K., Keller, J. W., & French, J. W. (2001). *Context of human error in commercial aviation (Tech. Rep.)*. Boulder, CO: Micro Analysis & Design.

Leveson, N. (2001a). The role of software in recent aerospace accidents. *Proceedings of the 19th International System Safety Conference, System Safety Society.* Unionville, VA.

Leveson, N. (2001b). Systemic factors in software-related spacecraft accidents. *AIAA Space 2001 Conference and Exposition* (Paper AIAA 2001-4763). Reston, VA: AIAA.

Miyake, A., & Shah, P. (1999). *Models of working memory.* Cambridge, UK: Cambridge University Press.

Parasuraman, R., Hansman, J., & Bussolari, S. (2002). *Framework for evaluation of human–system issues with ASDE-X and related surface safety systems* (White Paper for AAR-100). Washington, DC: Federal Aviation Administration.

Pew, R. W., & Mavor, A. S. (Eds.). (1998). *Modeling human and organizational behavior: Application to military simulations.* Washington DC: National Academy Press.

Polk, T. A., & Seifert, C. M. (2002). *Cognitive modeling.* Cambridge, MA: MIT Press.

Raaijmakers, J. G. W., & Shiffrin, R. M. (1981). Search of associative memory. *Psychological Review, 88,* 93–134.

Rasmussen, J., & Pedersen, O. M. (1984). Human factors in probabilistic risk analysis and risk management. In *Operational safety of nuclear power plants* (Vol. 1). Vienna: International Atomic Energy Agency.

Reason, J. (1990). *Human error.* Cambridge, UK: Cambridge University Press.

Restle, F., & Greeno, J. (1970). *Introduction to mathematical psychology.* Reading, MA: Addison–Wesley.

Ritter, F. E., Shadbolt, N. R., Elliman, D., Young, R. M., Gobet, F., & Baxter, G. D. (Eds.). (2003). *Techniques for modeling human performance in synthetic environments: A supplementary review.* Wright-Patterson Air Force Base, OH: Human Systems Information Analysis Center.

Skinner, B. F. (1966). An operant analysis of problem solving. Reprinted in A. C. Cantania and S. Harnad (Eds.) (1988), *The selection of behavior: The operant behaviorism of B. F. Skinner* (pp. 218–236). Cambridge, UK: Cambridge University Press.

Watson, C. S. (1973). Psychophysics. In B. B. Wolman (Ed.), *Handbook of general psychology* (pp. 275–306). Englewood Cliffs, NJ: Prentice Hall.

Woods, D. D., & Cook, R. I. (2004). Mistaking error. In B. J. Youngberg & M. J. Hatlie (Eds.), *The patient safety handbook* (pp. 95–108). Sudbury, MA: Jones & Bartlett.

Woods, D. D., Johannesen, L. J., Cook, R. I., & Sarter, N. B. (1994). *Behind human error: Cognitive systems, computers and hindsight.* Columbus, OH: CSERIAC.

3 Aviation Safety Studies
Taxi Navigation Errors and Synthetic Vision System Operations

Becky L. Hooey and David C. Foyle

CONTENTS

INTRODUCTION

The National Aeronautics and Space Administration Human Performance Modeling (NASA HPM) Project focused on two specific problems that are representative of general classes of problems in the aviation industry and that have significant implications for aviation safety. First, five different modeling teams were asked to address a current-day aviation problem within the realm of airport surface operations safety by identifying causal factors of navigation errors during taxi. This aviation problem contains several elements that are pervasive throughout every phase of flight, including the need to understand causal factors of errors, predict error vulnerabilities, and identify potential error mitigations. Second, the same modeling teams were then asked to model pilot performance during the approach and landing phase of flight for both a baseline configuration representing today's glass cockpit and a future cockpit that incorporates a synthetic vision system (SVS). The research issues inherent in this problem are common to the design, development, and integration of any advanced cockpit display technology and include system design and procedural issues.

Effective human performance models (HPMs) require extensive understanding of the task and the domain environment in order to produce valid and meaningful results. To enable model development, information was provided to the modeling teams, including task analyses and objective and subjective ratings data from human-in-the-loop (HITL) simulations that were conducted at NASA Ames Research Center. The HITL data were used in two different ways in this project. In some cases the data were used by the modeling teams to populate and develop their models, while in other cases, the data were used to validate the model output.

The information and data that were made available for use in the model development and validation process are presented next. First, a detailed description and analysis of the tasks required to navigate the airport surface (including aircraft control, navigation, communication, and monitoring), the results of a full-mission HITL simulation, and an analysis of navigation errors will be presented. Second, a description of SVS technology, a task analysis of approach and landing, and the results of an HITL simulation comparing pilots' performance with and without SVS technology in clear visibility and low visibility will be provided.

STUDY 1: AIRPORT SURFACE (TAXI) OPERATIONS

INTRODUCTION

On the airport surface, pilot navigation errors (i.e., failure to follow the taxi clearance issued by air traffic control [ATC]) are often simply considered an inconvenience that increases time to taxi and fuel burn, and adversely affects airline on-time performance. Although these navigation errors certainly slow airport operations, they may also be indicative of an underlying safety threat. In general, pilots who commit navigation errors either do not know where they are on the airport surface or have misunderstood where they are supposed to be. Either of these erroneous knowledge states may lead a pilot inadvertently to taxi onto, or across, an active runway, creating a runway incursion safety hazard. Runway incursions are defined as any occurrence on an airport runway involving an aircraft, vehicle, person, or object on the ground that creates a collision hazard or results in a loss of required separation with an aircraft taking off, landing, or intending to land (FAA, 2002a). From the years 1998 to 2001, there were 1,460 runway incursions at U.S. airports. Pilot deviations accounted for 58% of the runway incursions reported, operational errors (ATC) accounted for 23%, and vehicle/pedestrian errors accounted for 19% (FAA, 2002a).

In contrast to en-route and approach phases of flight, the aviation research community has devoted only minimal resources to understanding the factors that contribute to pilot error during surface operations. In fact, airport surface operations have been referred to as "the forgotten phase of flight." However, in recent years the rate of runway incursions has risen to a level of sufficient concern that the FAA has developed a runway safety blueprint (FAA, 2002b) and conducted regional and national workshops to identify solutions to the increasing runway incursion problem. In the FAA blueprint, numerous suggestions were raised, including procedural and operational changes, improvements to pavement markings and signage, and in-cockpit navigation technologies. The FAA Safe Flight 21 program was established with the goal of providing pilots with cockpit-based tools to reliably increase the pilots' awareness of aircraft position on the airport surface (FAA Safe Flight 21, 2002). However, in order to devise, prioritize, and predict the success of potential solutions, it is imperative that we first understand the nature of the runway incursion problem and the factors that contribute to navigation errors on the airport surface.

Past work at NASA Ames Research Center has focused on understanding and characterizing the task of pilots during airport surface (taxi) operations, including the nature of the in-cockpit duties, intracrew communications, and pilot–ATC interactions. The work has included in-flight observation of commercial operations (Andre, 1995), two medium-fidelity simulations (McCann, Andre, Begault, Foyle & Wenzel, 1997; McCann, Foyle, Andre, & Battiste, 1996), two full-mission simulations (Hooey, Foyle, & Andre, 2000; McCann et al., 1998) and a flight test at Atlanta Hartsfield International Airport (Andre, Hooey, Foyle, & McCann, 1998). These HITL evaluations addressed a variety of environmental conditions (airport complexity, and visibility) and technical interventions (data link, electronic moving maps, and head-up displays) and have produced a wealth of data with which to draw inferences about taxi efficiency (taxi speed and time stopped) and safety (navigation errors, workload, and situation awareness).

Previously, in support of a NASA effort on enhancing terminal area efficiency, an analysis of HITL data had been conducted in order to identify the factors that contribute to pilot navigation errors and to evaluate the efficacy of potential mitigating solutions that directly address the contributing factors. Data regarding taxi performance, navigation errors, and crew roles and procedures revealed that navigation errors are not simply random errors and are not due to pilot inattention, but rather that a number of contributing factors support or allow these errors (Hooey & Foyle, 2006). One goal of the taxi modeling efforts was to identify and better understand these contributing factors.

UNDERSTANDING AIRPORT SURFACE (TAXI) OPERATIONS

The first step in this modeling effort was to conduct a thorough task analysis of the taxi task as conducted by two-crew commercial airline pilots; the captain served as the pilot flying (PF) and the first officer (FO) served as the pilot not flying (PNF).

The task analysis (conducted by Goodman & Martin, 2001, for the NASA HPM project) was generated through interviews with commercial airline pilots and observation of pilots in actual commercial flights and full-mission HITL simulations. The task analysis (presented in appendix A) included task time and attentional resource information for a nominal set of procedures and was not intended to reflect those of any particular airline. A brief synopsis follows.

For the purposes of this chapter, the task analysis begins just after landing as the crew embarks on the landing rollout procedure and taxies to their gate. (The modelers actually received a task analysis that started in the approach sequence with the pilots receiving a clearance to land and a preferred runway exit from the approach controller, and a gate assignment from their company.) Throughout the landing rollout and taxi procedure, the captain adjusts aircraft speed using throttle and brakes and steers to negotiate turns and maintain centerline position using a tiller to control nose wheel direction. As the captain completes the landing rollout procedure, he or she slows the aircraft to a safe runway exit speed while monitoring the environment for an appropriate runway exit. The captain's main priority is to exit the runway as quickly and safely as possible and to ensure the aircraft is completely clear of the runway before stopping (if necessary) to wait for a taxi clearance.

During the landing rollout procedure, the FO calls out aircraft speed and the distance to the appropriate runway exit. When approaching the runway exit, the FO notifies the tower controller that the aircraft is clearing the runway, switches the radio frequencies to ground control, and then notifies the ground controller that the aircraft has cleared the runway. The ground controller responds with a taxi clearance in the form of a list of taxiways and hold instructions. Generally, the FO writes down the clearance (or commits it to memory), acknowledges the clearance by a full read-back, and communicates the clearance to the captain. This process is often hindered by congested radio frequencies in that communications may be "stepped on" by another pilot sharing the same radio frequency, and ATC commands are often difficult to hear and understand because of the fast rate of speech and poor radio clarity (Andre, 1995).

After the taxi route is communicated, pilots formulate a taxi plan of their intended route by integrating the ATC-issued taxi clearance into their knowledge of the airport layout using either a mental map developed by previous experience or a standard paper, north-up chart. This includes a plan for the taxiways, holds, and runway crossings required to navigate to their destination. While navigating the cleared taxi route, both pilots work to identify their assigned taxiways, determine the direction of required turns (not provided in most clearances), identify runways for which they are required to hold, and monitor traffic around them to avoid incidents. Pilots make a series of navigation decisions (when and where to turn) based on their understanding of where they are on the airport surface and the distance to, and direction of, their next turn in the clearance.

Local guidance information (i.e., signs and markings that indicate where they are on the airport surface) is consulted to enable the task of maneuvering the aircraft along the cleared route (Lasswell & Wickens, 1995; McCann et al., 1998). Global awareness is also generated and maintained; this includes awareness of the environment and the airport layout including runways, hold locations, traffic,

and concourses (Lasswell & Wickens, 1995; McCann et al., 1998). Global awareness is important for navigation, at large, busy airports since ATC does not typically include directional information in the taxi clearances in order to minimize
radio frequency congestion and save time. Therefore, to follow a taxi clearance of
"Alpha, Bravo, Charlie," a pilot would have to know which way to turn onto Bravo,
in order to reach Charlie.

The task of navigating on the airport surface is made difficult by the complexity of the airport layout, which often consists of a tangled network of taxiways and
runways. Since navigation signs cannot be placed directly overhead (as with our road
networks), they are necessarily placed in awkward positions such as on grass and
cement islands to the side of the taxiways and at a distance before the intersection
(Andre, 1995), making it difficult to identify taxiways, especially at multiway intersections. These problems with navigating on the airport surface are exacerbated in
low-visibility and night conditions, when the degraded visibility of signs and markings contributes to lower taxi speeds, increased navigation errors, and increased workload (McCann et al., 1998).

Currently, there are no cockpit technologies in wide use to assist pilots in
navigating the airport surface, except for compass heading indicators. Pilots use a
paper chart (often referred to as a Jeppesen chart) that depicts the airport layout and
denotes runways, taxiways, and concourses. Pilots report that the paper taxi charts
can be confusing, cluttered, and difficult to read, promoting excessive head-down
time (Andre, 1995). Further, pilots must translate information on the chart to the
environment, which often requires mental rotation from the north-up chart to their
actual heading.

In addition to maneuvering the aircraft, communicating with ATC, navigating,
and monitoring the environment, the pilots must also complete tasks such as cockpit
reconfiguration (e.g., resetting flaps), radio communications to determine their often
changing gate assignment, and preparations for upcoming flights. The culmination
of all of the required tasks, the difficult operating environment, and the lack of aiding technologies result in a very high-workload, time-pressured phase of flight.

AIRPORT SURFACE (TAXI) OPERATIONS HITL SIMULATION

A high-fidelity HITL simulation study was conducted at NASA Ames Research Center
to assess pilot taxi performance under current-day conditions at a complex airport.
The study, conducted in low-visibility conditions, compared pilots' taxi performance
during current-day taxi operations and future operations that included data-link and
advanced navigation displays such as electronic moving maps and head-up displays.
Performance data including navigation errors, taxi velocity, runway exit efficiency,
taxi clearance processing time, communication rates, workload, and situation awareness were collected for both current-day and future-technology conditions, under both
nominal and off-nominal scenarios (see Hooey, Foyle, & Andre, 2000). However,
the modelers focused primarily on navigation errors that occurred during current-
day operations and as such, only data from the nominal current-day operations trials
will be reported here. The reader is directed to Hooey, Foyle, and Andre (2000) and
Hooey, Foyle, Andre, and Parke (2000) for other details and results.

Participants

Eighteen crews, each consisting of one captain and one FO from the same airline and who flew the same type aircraft, participated in this high-fidelity simulation. Pilots represented six different commercial airlines. All pilots were current on modern glass-equipped aircraft with a mean of 2,645 flying hours logged.

Flight Deck Simulator

The simulation was conducted in NASA Ames Research Center's high-fidelity Advanced Concepts Flight Simulator (ACFS; see Figure 3.1), which is a full-motion, generic glass cockpit simulator with a high-fidelity representation of Chicago O'Hare Airport (KORD) replicating the airport layout, signage, painted markings, lights, concourses, and structures. The flight deck, a configurable generic glass cockpit, contains advanced flight systems, including touch-sensitive electronic checklists and programmable flight displays. A Flight Safety International VITAL VIIIi image generator, providing a 180° field of view with full cross-cockpit viewing capability, generated the out-the-window (OTW) view. An experimental ATC facility allowed for a highly realistic representation of current-day surface operations by integrating confederate (experimenter) local and ground controllers. Confederate "pseudopilots" provided background party-line communications that were synchronized to the movement of the computer-controlled airport traffic visible to the pilots from the cockpit.

FIGURE 3.1 Advanced Concept Flight Simulator (ACFS) at NASA Ames Research Center.

Procedure

Each crew completed three nominal trials under baseline, current-day operating conditions. (Also see Hooey, Foyle, & Andre, 2000, for off-nominal and future-technology scenario descriptions.) Each trial began approximately 12 nautical miles (nmi) out on a level approach into Chicago O'Hare Airport (see airport layout diagram in Figure 3.2). Pilots performed an autoland and taxied to the gate in low visibility (runway visual range [RVR] 1,000 ft conditions). Prior to each scenario, the runway and gate destination were provided. While airborne, pilots received a clearance to land and a preferred runway exit from ATC. Pilots were encouraged to take the preferred exit when it was safe, but it was emphasized that pilots could refuse the exit and select another if they

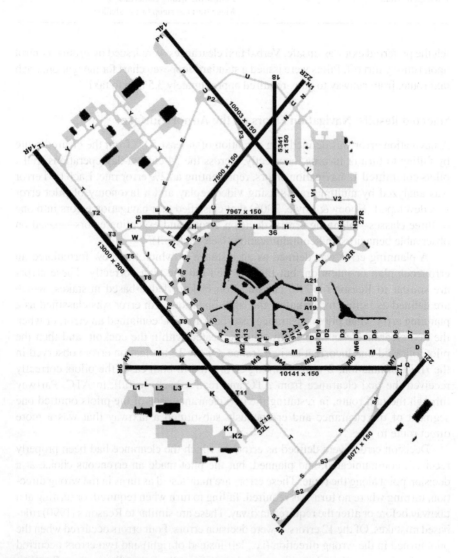

FIGURE 3.2 Airport diagram of Chicago O'Hare (KORD) Airport.

TABLE 3.1
Taxi Navigation Errors

Type of Error	Number of Errors (12 total)	Examples
Planning errors	2	Commit error in clearance read-back Omit or substitute one segment of the taxi clearance
Decision errors	6	Turn in the wrong direction Fail to turn Turn when not required
Execution errors	4	Follow the wrong centerline Make turn too steep or too shallow

felt the preferred exit was unsafe. Verbal taxi clearances were issued by ground control upon runway turnoff. Pilots were issued a standard Jeppesen chart for navigation. Each taxi route, from runway to gate, required approximately 3.5 min to taxi.

Selected Results: Navigation Errors on the Airport Surface

A navigation error was defined as a deviation of at least 50 ft from the cleared route by failing to turn or turning incorrectly. Across the 54 current-day operations trials, pilots committed 12 navigation errors, representing a 22% error rate. Each taxi error was analyzed by multiple raters using video replay and a taxonomy of pilot error was developed (Hooey & Foyle, 2006) that classified taxi navigation errors into one of three classes—planning errors, decision errors, and execution errors—based on observable behaviors and communications (see Table 3.1).

A planning error was defined as an instance in which the crew formulated an erroneous plan or intention, but then carried out the plan correctly. These errors are similar to Reason's (1990) classification of knowledge-based mistakes, which are defined as failing to formulate the right intentions. An error was classified as a planning error when pilots' clearance read-back to ATC contained an error, or when the taxi clearance was incorrectly communicated within the cockpit, and then the pilot followed that incorrect taxi clearance. Of the 12 navigation errors observed in the HITL simulation, 2 were planning errors and in both cases the pilots correctly received the taxi clearance from ATC and read it back, in full, to ATC. Partway through the taxi route, in restating the taxi clearance, one of the pilots omitted one segment of the clearance and erroneously substituted a taxiway that was a more direct route to gate.

Decision errors were defined as errors in which the clearance had been properly received, communicated, and planned, but the pilot made an erroneous choice at a decision point along the route. These errors are manifested as turns in the wrong direction, turning where no turn was required, failing to turn when required, or turning at a taxiway before or after the required taxiway. These are similar to Reason's (1990) rule-based mistakes. Of the 12 errors, 6 were decision errors. Four errors occurred when the pilot turned in the wrong direction (i.e., left instead of right) and two errors occurred when the pilot continued taxiing straight, missing the turn all together. Post-hoc analyses revealed that three of the six errors occurred at the first decision choice point after clearing the runway, when the FO was occupied with cockpit tasks and ATC communication. In the remaining three errors, crew communications demonstrated a lack of global awareness or an inaccurate mental map of the airport layout (i.e., thinking the concourses were to the left, when they were actually to the right).

Execution errors were those in which the clearance was correctly communicated and pilots identified the correct intersection and direction of the turn; however, they erred in carrying out the maneuver. These are akin to Reason's (1990) classification of "slips" in which the right intention is carried out incorrectly. Of the 12 errors observed in this HITL study, 4 were execution errors. In these cases, the pilot followed the wrong taxi line at a multiway intersection, or took a hard right turn instead of a soft right turn. These errors were most frequently attributed to confusing environmental cues such as lighting, signage or centerline markings, and complicated taxiway geometry with many closely spaced taxiways and multiway intersections.

Subsequent analyses were conducted to further explore the effect of taxiway complexity on taxi navigation errors (Goodman, 2001). Across all 18 crews, each completing three baseline trials, there was a total of 582 intersection crossings, each representing a unique navigation decision point. Goodman categorized these intersection crossings across two main geometric factors: complexity (two-way, three-way, four-way, or five-way intersections) and directional fit of next route segment (towards gate, away from gate).

First, the error rate was calculated as a function of intersection complexity. The error rate increased as the intersection complexity increased, with a 1.6% error rate in the 192 instances of two-way intersections, a 2.1% error rate in the 336 instances of three-way intersections, and a 5.6% error rate in the 36 instances of four-way intersections. There were no errors observed in the relatively rare (18) occurrences of a five-way intersection. Second, the error rate was calculated as a function of directional fit—that is, whether the clearance required a turn toward or away from the gate. Of the 534 instances in which the turn direction was toward the gate, the error rate was 1.3%, whereas for the 48 instances in which the turn direction was away from the gate, the error rate was 10.4%.

SUMMARY

This full-mission simulation allowed for the observation of navigation errors on the airport surface and produced a unique data set from which to draw inferences. These data suggest that factors such as pilot workload caused by intracockpit tasks and ATC-communication requirements, airport layout, route complexity, and the extent to which a taxi clearance conforms to pilots' expectations may contribute to taxi navigation errors. Although post-hoc analyses were conducted to understand the causal factors (see Hooey & Foyle, 2006), it is expected that HPMs will enable a richer analysis allowing for a more complete understanding of the cognitive factors that played a role in these errors. To enable these modeling efforts, the following data were provided to each of the five modeling teams:

- Nominal task analysis with task times (see appendix A);
- Chicago O'Hare Airport (KORD) charts and signage maps;

- Visual scene depictions (video tape of the taxi environment as viewed from the cockpit during taxi);
- Description of navigation errors based on an error taxonomy and geometry analyses;
- Communication data (frequency and nature of dialogue) with ATC and within the cockpit;
- Workload and situation awareness ratings as determined by pilot subjective ratings and objective "over-the-shoulder" ratings conducted by a retired airline captain;
- Pilot demographic data; and
- Aircraft performance data, including mean taxi speed, speed at runway turnoff, and time spent stopped on the airport surface.

HUMAN PERFORMANCE MODELING CHALLENGES TO IMPROVE AIRPORT SURFACE OPERATIONS

Using the data from the taxi navigation simulation, the modeling teams were challenged to demonstrate how HPMs can be used to predict situations in which pilots might be vulnerable to error, and to help distinguish among potential mitigating strategies to determine which may be most successful. It is important to note that the modelers were not tasked to predict the taxi error data set, the observed error rate, or proportion of errors for each error category. Rather, they were asked to use the data set and other available information to help build solid models of pilot behavior that would be the foundation of a set of tools with which taxi navigation error causes and mitigations could be explored. As will be seen in the coming chapters, for most of the modeling teams, this translated into developing models that could replicate one or more of the error classes, and then proceeding to test causal theories that might explain these errors. Specifically, the modelers were challenged to address important safety-related questions such as:

- Can HPMs replicate the types of errors made by the pilots in the HITL simulation?
- Can HPMs offer deeper insights and understanding about the nature and cause of errors?
- Can HPMs offer solutions such as technologies, procedures, or operational changes that might mitigate the errors in the operational setting?
- Ultimately, what HPM advancements are required to predict the occurrence or likelihood of errors?

STUDY 2: APPROACH AND LANDING WITH SYNTHETIC VISION SYSTEMS

INTRODUCTION

A synthetic vision system (SVS; see example in Figure 3.3) is an integrated flight deck display concept that provides pilots with a continuous view of terrain, combined with integrated guidance symbology, in an effort to increase situation awareness

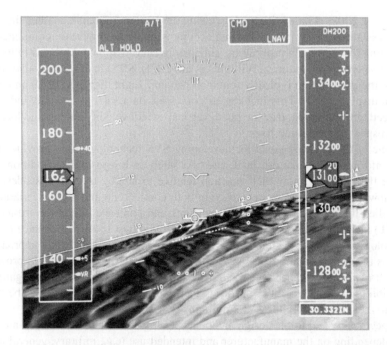

FIGURE 3.3 Example of a synthetic vision system (SVS): Enhanced view of terrain with conventional primary flight display symbology overlaid. (Courtesy of NASA Langley Research Center.)

(SA) and decrease workload during low-visibility operations, or instrument meteorological conditions (IMC; Bartolone, Glaab, Hughes, & Parrish, 2005). Williams et al. (2001) describe three components of SVS technology: an enhanced intuitive view of the flight environment, hazard and obstacle detection and display, and precision navigation guidance.

The enhanced intuitive view is derived from terrain database images and provides a picture of the environment that replicates what the pilot would see out the window in clear visibility. The hazard and obstacle detection component of an SVS depicts terrain and hazardous obstacles, and provides warning and avoidance alerting. Williams et al. (2001) described several display concepts under consideration, including those that depict terrain (land, vegetation), ground obstacles (aircraft, towers, vehicles, construction, wildlife), air obstacles (traffic, birds), atmospheric phenomena (weather, turbulence, wind shear, icing, wake vortices), restricted airspace, and noise-sensitive areas. The third SVS component, precision navigation guidance (described by Williams et al., 2001) proposes the use of virtual renditions of navigation cues that allow pilots to view their location accurately and rapidly correlate their position to terrain and other prominent features. For example, SVS designs may offer three-dimensional information regarding the forward flight path in the form of a pathway, tunnel, or "highway in the sky" (see Alexander, Wickens, & Hardy, 2003; Beringer, 2000; Fadden, Ververs, & Wickens, 1998;

Williams, 2002). The addition of precision navigation guidance enables the pilot to access and monitor path-following accuracy and is an important component in achieving low approach minimums, supporting curved approaches, and following noise abatement procedures (Williams et al., 2001). SVS displays may also incorporate the information typically found on primary flight displays (PFD) and navigation displays (ND). This includes aircraft state data such as altitude, indicated airspeed, ground speed, true airspeed, vertical speed, velocity vector, and location with respect to navigation fixes.

One of the primary goals for developing SVS technology is to provide sufficient information to take off, land, and taxi safely in low-visibility conditions and enable the pilots to know their position relative to nearby terrain, obstacles, and the runway. The continued occurrence in both commercial and general aviation of controlled-flight-into-terrain (CFIT) accidents and controlled-flight-toward-terrain (CFTT) incidents (Flight Safety Foundation, 1998) attest to the importance of this technology. Also, SVS technology is expected to enable more flexible approaches such as the area navigation (RNAV) approach, particularly since the SVS provides for in-trail and lateral spacing to be transferred from ATC to the aircrew regardless of visibility and any terrain or obstacles such as runway traffic would also be "visible" to the crew through the SVS (Williams et al., 2001).

Specifics related to the design and concept of operations (CONOPS) of an SVS vary depending on the manufacturer and intended use (e.g., military, general aviation, or commercial aviation) and are continually being refined throughout the development and certification process. Though considerable research has been devoted to implementation and design aspects of synthetic vision, such as display size, field of view, optimal scene texturing, and appropriate symbology (e.g., Alexander et al., 2005; Comstock, et al., 2001; Norman, 2002), little is known about the impact that these systems might have on general flight deck operations.

Motivating the present part-task study and corresponding modeling efforts is the recognition that the deployment of new flight deck technologies alters the manner in which pilots carry out tasks, often in ways that were not fully anticipated by designers (Billings, 1996; Woods & Dekker, 2000). These unanticipated usage patterns can lead to adverse operational consequences. Clearly, the introduction of an SVS display changes the information available to pilots on the flight deck by providing a new and rich source of visual–spatial information. For this reason, an appropriate research issue concerns how pilots adjust their visual performance to incorporate this new information. That is, what changes, if any, do pilots make in where they look, how long they look, and the sequence in which they look at available informational sources? What is the consequence of these changes?

Gaining a better understanding of how pilots alter their visual performance when flying with an SVS display should help inform the design of effective operational procedures and highlight potential sources of error. Research questions remain regarding what information should be presented, in what format, on which display, and how the information and display technology should be used and incorporated into current-day standard operating procedures. This provides an excellent opportunity for human performance modeling to impact this process.

UNDERSTANDING THE APPROACH AND LANDING TASK

Before embarking on the modeling process, a thorough task analysis was completed to identify the tasks involved in an RNAV approach and landing with and without SVS technology (see Keller & Leiden, 2002; Keller, Leiden, & Small, 2003). RNAV is a method of navigation that uses the global positioning system (GPS), instead of ground-based equipment (e.g., very high frequency omnidirectional range [VOR]), and allows aircraft to follow any desired flight path such as geographic waypoints expressed by latitude and longitude instead of requiring aircraft to fly over ground-based navigation aids (Meyer & Bradley, 2001).

The RNAV approach was chosen as it represents a trend in future flight operations toward aircraft-based rather than ground-based precision guidance. Meyer and Bradley (2001) highlighted a number of advantages of RNAV over conventional forms of navigation, including: user-preferred routes that account for pressure altitude and wind, more direct routes for shorter flight distances, dual or parallel routes to accommodate greater traffic flow, bypass routes to overfly high-density terminal areas, alternative or contingency routes, and improved locations for holding patterns.

The RNAV approach and landing task analysis conducted by Keller and Leiden (2002) is presented in appendix A, which lists the tasks, the operators involved (ATC, PF, PNF), and whether it was a discrete or continuous task. The task analysis, for a nominal airport and air carrier, begins as the aircraft transitions into the approach phase (approximately 15 nmi from the runway threshold flying level at 2,500 ft above ground level [AGL], at 200 kts). It assumes that the initial approach flow and approach checklist have been completed, the approach is programmed into the flight management system (FMS), both the autopilot and autothrottle are engaged, and the flight mode annunciators (FMAs) indicate that lateral navigation (LNAV) and vertical navigation (VNAV) are engaged. A high-level summary of the RNAV approach and landing task from Keller and Leiden is presented next.

The crew receives an approach clearance from approach control, which is usually accompanied by a speed instruction (i.e., slow to 160 kts). The crew verifies that LNAV and VNAV are engaged, sets the speed on the mode control panel (MCP), and adjusts the flaps. Throughout the approach, the crew makes a series of speed and flap adjustments in order to maintain the necessary descent rate and to slow the aircraft, and lowers the landing gear as they proceed to descend toward the final approach fix (FAF) altitude, and begin to align with the designated runway.

The crew monitors the systems, the progression of the flight plan, and the attitude and flight path of the aircraft, and completes the landing checklist. The tasks are shared (usually according to predefined procedure) between the PF and the PNF; the primary tasks of the PF involve all aspects of the aircraft attitude, position, and function and the PNF performs necessary communications with ATC; responds to requests from the PF for speed, flap, and gear settings; and cross-checks the actions of the PF. Both pilots monitor the flight control displays and the environment.

Typically, before reaching the FAF, the crew sets the decision altitude (DA). The DA is the altitude at which the aircraft must be configured and within established tolerances for speed, descent rate, altitude, and engine thrust and the pilots must have the runway in sight in order to initiate a landing. By the time the aircraft reaches the FAF, the aircraft should be descending at an appropriate rate and speed and be aligned with the runway centerline in the proper landing configuration.

At about 5 miles from the runway threshold, the crew makes their final approach preparations, changes their communication radios to the tower frequency, and requests a landing clearance. By the DA, the crew makes their final landing determination. If the approach is unstable (e.g., the aircraft's speed does not fall within predetermined parameters), if the pilots cannot see the runway, or if there is an obstruction on the runway, they will initiate a missed approach; otherwise, the landing continues and the PF prepares to take control and assume manual flying.

Synthetic Vision System HITL Simulation

A medium-fidelity HITL study was conducted to generate empirical data and qualitative information regarding pilot performance during approach and landing, with and without the use of an SVS display, for the development of human performance models (see Goodman, Hooey, Foyle, & Wilson, 2003). The simulation data characterized pilot performance during the approach and landing phase of flight using conventional and augmented displays under both IMC and visual meteorological conditions (VMC). The test plan, rather than emphasizing statistical power, focused on a limited number (three) of subject pilots operating across a variety of conditions from which performance estimates could be derived. This approach was intended to challenge the modeling teams to develop meaningful models of human performance despite limited data—as is often the case in the aviation domain, particularly when researching future systems or concepts that have not yet been developed or implemented. The focus was on providing the modeling teams with a range of pilot strategies and behaviors with which they could build relevant and valid models, rather than a complete data set that the models were required to predict.

Participants

Three current airline pilots (two captains and one FO) with 13,000 mean logged flight hours participated in the simulation study. All three pilots fulfilled the role of captain during the simulation trials.

Flight Deck Simulator

A PC-based part-task simulator (see Figure 3.4) was constructed using the PC Plane software package developed at NASA (Palmer, Abbott, & Williams, 1997), which approximated the instruments, controls, and flight characteristics of a Boeing-757. This was linked with an X-IG© 3-D image generation system employing a visual database of Santa Barbara Municipal Airport (SBA) and its surrounding terrain. Santa Barbara was chosen because of the challenging terrain environment surrounding the airport, especially for a "go-around" aborted landing. The visual OTW scene was presented on a large front-projection screen measuring 8 ft horizontally (H) and 6 ft vertically (V) located approximately 8 ft in front of the pilot and providing a

FIGURE 3.4 Synthetic vision system (SVS) simulation flight deck layout.

54.6° H and 34.9° V field of view of the forward external world (after accounting for the obscuration caused by the front panel of the simulated flight deck).

The simulated flight deck avionics displays included the conventional glass cockpit PFD and ND as well as a new-concept SVS. The conventional PFD (see Figure 3.5, top) provided information specifying airspeed, attitude, current and targeted altitude, vertical speed, engine pressure ratios (EPRs), distance to next waypoint, and flight mode annunciation. The conventional ND (see Figure 3.5, bottom) provided information specifying the aircraft's current heading, latitude/longitude of current position, previous and next fix, distance to next fix, range arcs, and descent crossing arcs.

The SVS was implemented on a large-format, head-down display and presented computer-generated three-dimensional color imagery of terrain and cultural features overlaid with flight-director symbology, including a flight path predictor (see Figure 3.6). The field of view of the presented imagery was set at 31° H and 23° V, providing a somewhat "wide-angle" perspective relative to unity. This fixed field of view was chosen as a good compromise setting, falling between the wide-angle 60° horizontal field of view, which research (Comstock et al., 2001) had shown was preferred by pilots during early approach phases, and the unity field of view desired at landing.

The participant made control inputs using a joystick with throttle lever. Touch-screen software buttons were used to make inputs into a B757-type MCP

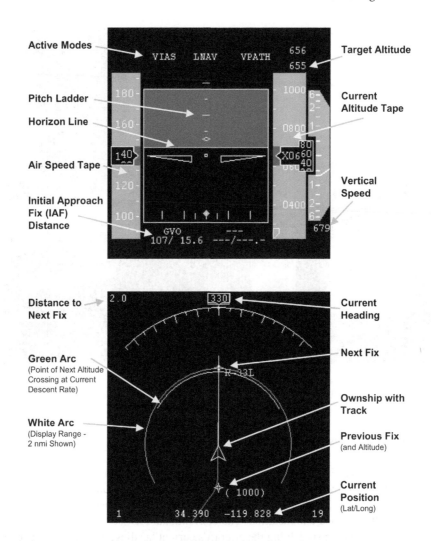

FIGURE 3.5 Primary flight display (PFD; top) measured 5.25 in. (H) × 5.25 in. (V) and was located 34 in. from pilot eye-point (3.1° H, 3.1° V). Navigation display (ND; bottom) measured 5.25 in. (H) × 5.25 in. (V) and was located 34 in. from pilot eye-point (3.1° H, 3.1° V).

(see Figure 3.7). The MCP was used by the confederate FO to activate autoflight control functions in response to the captain's commands.

Also, a somewhat unconventional touch screen was implemented in the simulator, situated below the MCP, in order to control activation of landing gear, flap setting, speed brake, and throttle setting (see Figure 3.8).

General Scenario Description

A flight-qualified member of the NASA HPM project research team acted as FO, which allowed for a close approximation of realistic crew procedures and allocation

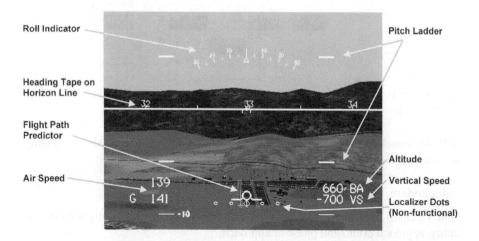

Roll Indicator

Pitch Ladder

Heading Tape on Horizon Line

Flight Path Predictor

Air Speed

Altitude

Vertical Speed

Localizer Dots (Non-functional)

FIGURE 3.6 SVS display with symbology overlay shown on approach to runway 33L at Santa Barbara Municipal Airport (SBA). The SVS display measuring 10 in. (H) × 7.5 in. (V) was located 34 in. from pilot eye-point (16.7° H, 12.6° V).

of duties. These duties included carrying out all MCP and control inputs specified by the captain, making appropriate speed and altitude callouts, and handling ATC communications. A second research team member assumed the role of ATC and provided approach and landing clearances for each trial and, once per testing block, a late reassignment of runway.

In all scenarios, pilots performed an RNAV approach to runway 33L at SBA under calm winds. Since the RNAV approach is not currently implemented at SBA, an approach plate was constructed for the simulation based on other published RNAV plates and briefed to pilots (see Figure 3.9). The FMS was preprogrammed by the study team to reflect the approach so that no pilot interactions with the FMS were required during the simulation trials.

All trials began 36 nmi inbound from the northwest at 10,000 ft and 250 kts awaiting ATC clearance for approach. Pilots were required to fly the cleared approach fully coupled to the autopilot, using the LNAV and VNAV automated flight modes down to the 650 ft DA, at which point they took full manual control. Depending on scenario circumstances, pilots either continued the landing (trials terminating

FIGURE 3.7 Mode control panel (MCP) as presented in SVS simulation measured 13.5 in. (H) and 3.0 in. (V) and was located 39 in. from pilot eye-point (9.2° H, 2.0° V).

FIGURE 3.8 Controls panel as presented in SVS simulation measured 10.5 in. (H) and 1 in. (V) (7.2° H, 0.7° V)

at 50 ft) or declared a missed approach and executed a go-around (trials terminating when ascent reached 3,000 ft). Each trial lasted approximately 12 min.

Independent Variables

Four variables of interest were investigated in the study: display configuration, visibility, approach event, and phase of approach.

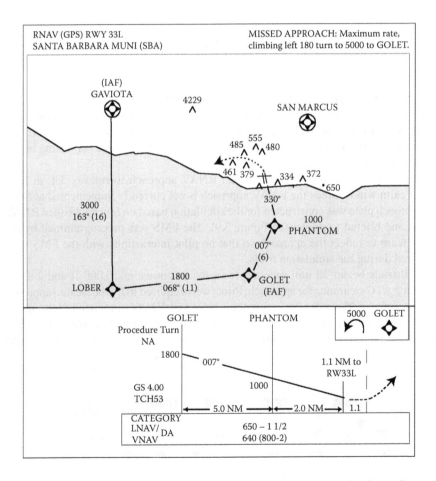

FIGURE 3.9 (Hypothetical) area navigation (RNAV) approach plate for Santa Barbara Municipal Airport (SBA) created for use in the human-in-the-loop (HITL) simulation.

Display configuration. The simulator allowed rapid reconfiguration of the flight deck display suite between trials. Two display configurations were tested. The baseline configuration represented current-day operations with the PFD, ND, MCP, and controls panel. The SVS configuration included all displays presented in the baseline configuration with the addition of the SVS display.

Visibility. Trials were conducted in either VMC, in which the entire trial was conducted in clear day, unlimited visibility, or IMC, with zero visibility due to dense cloud ceiling down to 800 ft, at which point, during nominal trials, the aircraft broke out into unlimited visibility.

Approach event. Trials consisted of either nominal landings, during which the aircraft was cleared first for approach and then for landing without incident, or one of the following off-nominal events: late runway reassignment, missed approach, or terrain mismatch. In the late runway reassignment event trials, the aircraft was cleared first for approach and then landing, but at 1,000 ft, ATC requested that the pilot conduct a side-step maneuver to the adjacent parallel runway (33R) due to traffic remaining on runway 33L. (Note: This event was tested in the study during IMC conditions with the SVS display, although current ATC operations would allow such a maneuver only in VMC.) In the missed approach event trials, the aircraft was cleared first for approach and then landing. In the VMC condition, the confederate FO called out traffic on runway at 600 ft; in the IMC condition, dense cloud cover extended to the ground and there was no breakout as anticipated at 800 ft. Both conditions required the captain to declare a missed approach and execute a go-around. In the terrain mismatch event trials, the aircraft was cleared first for approach and then landing, but instruments (including SVS display) were laterally offset by 250 ft from the OTW scene. Previously during training, pilots were instructed to declare a missed approach and execute a go-around if instruments were determined to be unreliable.

Phase of approach. Each trial was divided into four phases (see Table 3.2), which differed not only in duration but also in external circumstances and required pilot activities. From start to initial approach fix (IAF), pilots were focused on obtaining approach clearance from ATC and setting up that approach within the autoflight system through directed actions with the MCP. From IAF to final approach fix (FAF), pilots closely monitored the progress of the approach and configured the aircraft (i.e., set landing gear, adjust flaps, and trim). During FAF to DA, pilots flew in IMC conditions with "breakout" from the cloud ceiling into full visibility (except for missed approach trials, in which pilots did not break out of IMC conditions). In this phase,

TABLE 3.2

Approach Phase

Phase	Start and End Points	Altitudes	Duration
Start–IAF	Start of trial–initial approach fix	Crossing at 10,000 ft	(1.0 min)
IAF–FAF	Initial approach fix–final approach fix	10,000–1,800 ft	(7.5 min)
FAF–DA	Final approach fix–decision altitude	1,800–650 ft	(2.5 min)
DA–end	Decision altitude–scenario end	650–50 ft	(1.0 min)

Notes: DA = decision altitude; FAF = final approach fix; IAF = initial approach fix.

TABLE 3.3
Test Conditions

Approach Event	Baseline Displays VMC	Baseline Displays IMC	Baseline + SVS IMC
Nominal approach (nominal landing)	Scenario 1	Scenario 4	Scenario 7
Late reassignment (sidestep and land)	Scenario 2		Scenario 8
Missed approach (go around)	Scenario 3	Scenario 5	Scenario 9
Terrain mismatch (go around)		Scenario 6	Scenario 10

Notes: IMC = instrument meteorological conditions; SVS = synthetic vision system; VMC = visual meteorological conditions.

pilots were attempting to visually acquire the runway and confirm proper alignment. By DA, if the runway had been visually acquired and alignment confirmed, pilots transitioned to manual control while maintaining proper descent rate and runway alignment. Otherwise, pilots initiated a missed approach.

Test Conditions

The four approach events listed previously were flown in three display/visibility configurations: baseline VMC (current-day display configuration with clear visibility), baseline IMC (current-day display configuration with dense cloud ceiling to 800 ft), and SVS IMC (SVS display configuration with dense cloud ceiling to 800 ft). This yielded 10 viable test conditions, or scenarios (as shown in Table 3.3), which were each flown once by the three subject pilots. Data collected from all trials were segmented into the four progressive phases of flight for analysis.

Procedure

Simulation displays, controls, and procedures were reviewed with participating pilots during an orientation briefing. Thereafter, a training flight was flown in baseline VMC conditions to familiarize participants with aircraft handling, crew coordination issues, and the SBA approach. On initiation of each trial, participants instructed the confederate FO to arm the autopilots, set the altitude, and engage LNAV and VNAV. The participant's task, then, was to monitor and supervise the programmed FMS descent and approach, commanding such actions as flaps, speed brakes, landing gear, and altitude settings. At DA (650 ft), participants took full manual control (stick and throttle) of the aircraft and either attempted the landing (i.e., descent to 50 ft) or declared a missed approach and executed a go-around.

The six scenarios conducted with the baseline display configuration were randomly divided into two testing blocks of three scenarios each, while the four SVS scenarios were randomly ordered into a third testing block. During data collection, pilots first flew a block of baseline trials, followed by the block of SVS trials, and concluded with the remaining baseline block. Prior to the start of each trial, pilots were told only whether they would be in VMC conditions or IMC conditions (with ceiling reported to 800 ft) and whether the SVS display would be available.

Selected Results: Eye-Movement Data

A complete data set was provided to the modelers that included data for all nominal and off-nominal scenarios for visual fixation frequency and duration, visual fixation transition matrices, aircraft performance and state data, pilot control inputs, and measures of workload and situation awareness (Goodman, Foyle, & Hooey, 2002). Of particular interest in this study were the eye-movement data and what they revealed about changes in visual performance associated with the availability of an SVS display. A summary of measures (from Goodman et al., 2003) characterizing the distribution, frequency, and duration of the pilots' visual attention during the nominal landing trials (a subset of the scenarios tested) is presented here to illustrate the types of analyses that the study data supported and that were available to the modeling teams.

This summary reflects data collected from Scenarios 1, 4, and 7, which differ only in the display/visibility conditions in which participating pilots flew (i.e., baseline VMC vs. baseline IMC vs. SVS IMC). All measures are specified in terms of the six visual regions of interest: OTW, SVS, PFD, ND, MCP, and controls. As these six regions constituted the principal sources of flight information within the simulation, fixations recorded outside these regions were ignored. For this summary, the distribution of visual attention was measured as the percentage dwell time in one region relative to the total dwell time in all six display regions. Reported dwell counts indicate the number of eye-gaze "visits" to a particular display region as defined by one or more continuous fixations within a region until visual disengagement. Lastly, reported dwell durations reflect the mean length of time per visit to a display region. From this summary, regularities in scanning behavior with and without the SVS display are noted, as are localized differences in individual usage strategies.

The results for each of the measures are presented graphically for each pilot by display condition and phase of approach in Figures 3.10a, 3.10b, 3.11a, and 3.11b. The figures are arranged into horizontal panels corresponding to phase of approach and consist of three separate graphs, one for each participating pilot. Figures 3.10a and 3.10b present the distribution of visual attention of pilots during the nominal landing trials. Here the horizontal axis of each graph shows the display region of interest while the vertical axis represents the percentage of overall dwell time spent in that region. Figures 3.11a and 3.11b present the durations and dwell counts. The horizontal axis for these graphs, again, shows the display region of interest while the vertical axis represents the mean duration in seconds per dwell. Additionally, the number shown next to each data point marker indicates the dwell count or number of visits made to the display region during that phase of flight.

Based on the observed data, Goodman et al. (2003) offered a brief discussion highlighting consistencies and differences in visual performance between approach phases, display conditions, and individual pilots. This discussion is presented next, organized in terms of display region.

Out-the-window usage. A review of OTW percent dwell times (see Figures 3.10a and 3.10b) shows that regardless of display configuration, all three pilots devoted an increasing percentage of their visual attention OTW as the aircraft progressed

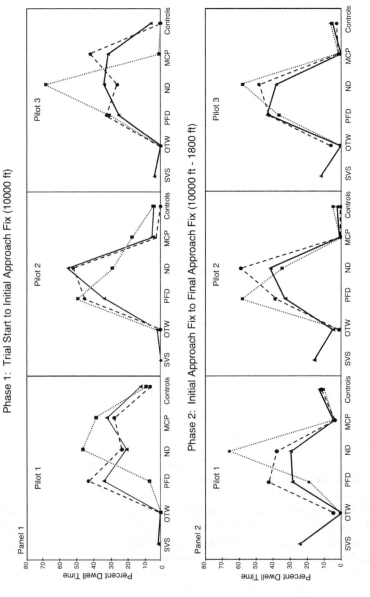

FIGURE 3.10A Pilot dwell time percentages across six areas of interest during approach Phases 1 and 2 of nominal landing trials. Three display/visibility configurations are compared: baseline VMC, baseline IMC, and SVS IMC.

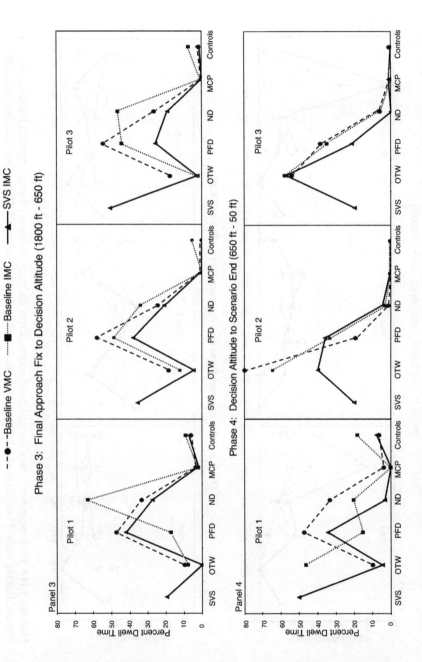

FIGURE 3.10B Pilot dwell time percentages across six areas of interest during approach Phases 3 and 4 of nominal landing trials. Three display/visibility configurations are compared: baseline VMC, baseline IMC, and SVS IMC.

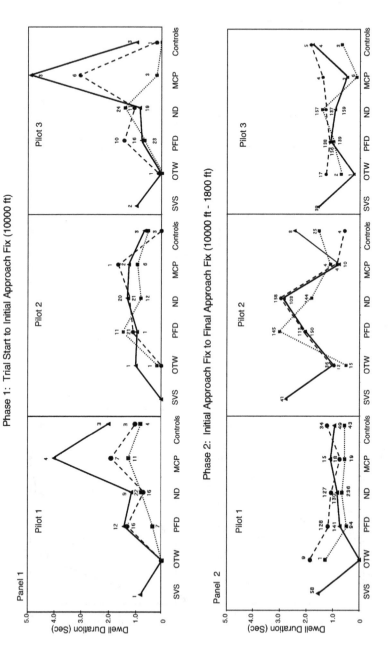

FIGURE 3.11A Pilot mean dwell durations and dwell counts across six areas of interest during approach Phases 1 and 2 of nominal landing trials. Three display/visibility configurations are compared: baseline VMC, baseline IMC, and SVS IMC.

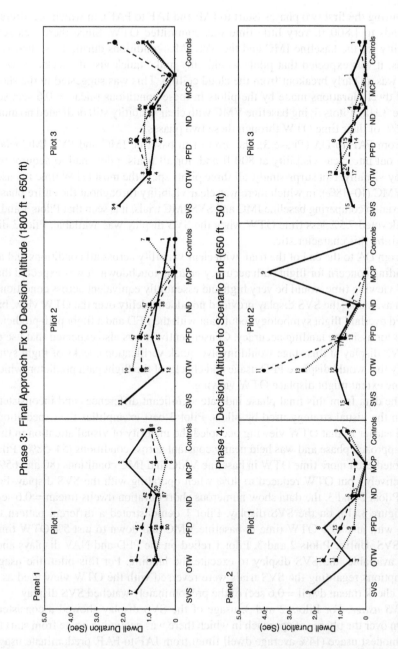

FIGURE 3.11B Pilot mean dwell durations and dwell counts across six areas of interest during approach Phases 3 and 4 of nominal landing trials. Three display/visibility configurations are compared: baseline VMC, baseline IMC, and SVS IMC.

toward landing. That said, there are wide differences in the percentage of time spent OTW among the three pilots and across the three display conditions, as will be noted.

During the first two phases (start to IAF and IAF to FAF), in which the aircraft descends to 1,800 ft, very little time was committed OTW. Since there was zero visibility for the baseline IMC and the SVS IMC conditions during these first two phases, it was expected that pilots would only make quick visual checks to see if there was an early breakout from the cloud ceiling. This was suggested in the short OTW dwell durations made by the pilots in these conditions (mean = 0.6 sec; see Figure 3.11a). Pilots flying baseline VMC with clear visibility still dedicated no more than 6% of their time OTW through these two phases.

From FAF to DA (Phase 3), the aircraft in baseline IMC and SVS IMC trials broke out into clear visibility at 800 ft and, for all trials, pilots had to acquire the runway visually. Not surprisingly, all three pilots spent the most OTW time in baseline VMC (10–18%), in which there was clear visibility throughout the entire phase. However, in comparing baseline IMC and SVS IMC trials, it is seen that Pilots 1 and 2 each devoted 7.5% less time OTW when the SVS display was available. Pilot 3 did not exhibit this characteristic.

From DA to the end of the trial, with clear visibility across all conditions and an overriding concern for flight path accuracy nearing touchdown, it was expected that OTW viewing time would be very high and essentially equivalent across conditions. When available, the SVS display provided no added fidelity over the OTW view, but offered overlaid flight symbology redundant with the PFD and a flight path predictor useful for assisting landing accuracy. Consequently, it was also expected that use of the SVS display at this stage would involve quick verification checks of flight symbology that would displace PFD usage and checks of the flight path predictor, which to some extent might displace OTW viewing.

The data from this final phase indicate significant differences and inconsistencies in the visual strategy used by pilots. Pilot 3 best exemplifies the expectations stated earlier in that OTW viewing occupied the majority of visual attention in this final approach phase and was held nearly constant across conditions (54–58%). Pilot 2 devoted even more time OTW in baseline VMC and IMC conditions (80 and 65%, respectively), but OTW reduced to 40% when operating with the SVS display. For both Pilots 2 and 3, the data show numerous short duration dwells (mean = 0.6 sec; see Figure 3.11b) on the SVS display. Pilot 1 demonstrated a different pattern of usage with just 10% OTW time in baseline VMC and down to just 5% OTW time with SVS. Unlike Pilots 2 and 3, Pilot 1 relied on the PFD and NAV displays and, when available, the SVS display to execute the landing. For this pilot, the usage assumptions regarding the SVS trials were reversed with the OTW view used as a quick check (mean dwell = 0.6 sec) of the predominately watched SVS display.

SVS usage. For Pilots 2 and 3, usage of the SVS display showed a consistent pattern over the phases of approach in which there was negligible usage from start to IAF, modest usage (14% average dwell time) from IAF to FAF, predominate usage (43% average dwell time) during the breakout and runway acquisition from FAF to DA, and then tapering back to moderate usage (20% average dwell time) during the short final activities from DA to the end of the trial. For Pilot 1, SVS usage was moderate through IAF to FAF and FAF to DA (25 and 20%, respectively), but predominated from DA to trial end (50%).

Comparing differences in dwell time percentages for display regions between SVS trials and both baseline trials indicates from where time spent viewing the SVS display was diverted. Averaged across the three pilots from IAF to FAF, the addition of the SVS reduced ND usage by 14%, PFD usage by 4.5%, and OTW usage by 0.5%. From FAF to DA, the time spent viewing the ND in the SVS condition compared to the baseline conditions was reduced by 16%, time spent on PFD was reduced by 10%, and time spent OTW was reduced by 9%. From DA to trial end, the presence of the SVS reduced viewing time on the ND by 9%, on the PFD by 1%, and, interestingly, OTW was reduced by 18%. It should be remembered that this last percentage is the result of the visual performance of Pilots 1 and 2, but not Pilot 3. That two of three pilots selected a strategy of SVS usage during the final phase of approach (actually, the final two phases per discussion previously) that significantly reduces OTW viewing is noteworthy.

PFD and ND usage. The PFD was an important and consistently accessed display with an overall mean percent dwell time across phases, pilots, and conditions of 36%. ND viewing figures most prominently from start to IAF, and IAF to FAF. When taken together with the PFD, the ND provided the primary source of spatial/navigation awareness to pilots during that period (combined average dwell time of 78%). From FAF to DA, when examining just the baseline conditions, that combined average still remained a robust 81%. But with SVS, combined usage of the PFD and ND dropped to 58%. Most of this decline was in the usage of the ND, which seemed to diminish in relative importance with the increased reliance on the SVS display and the proximity to the runway.

From DA to trial end, the PFD still remained an important source of information and was the second most attended display during this final phase. The ND, however, was used sparingly (with the exception of Pilot 1 in baseline conditions), with very short dwell durations (mean = 0.36 sec).

SUMMARY

Of particular interest in this study were observed changes in visual performance associated with the use of an SVS display. As noted in the selected analyses of the eye-tracking data, these changes included systematic reductions in the dwell times allocated to the ND, PFD, and OTW displays, as a function of the phase of approach. Despite such regularities, there were clear, localized differences in SVS usage strategies among the three study pilots. Most significantly, the data show that two of the three pilots spent less time scanning OTW during final approach when the SVS display was available, even though there was unlimited forward visibility. Such usage suggests possible over-reliance on the SVS display.

The variability among the three pilots presents a real challenge for the modeling teams, raising the important question faced by each modeling team—whether to develop and/or validate their model based on a single pilot or the average of all three. Following is a list of empirical data and qualitative information provided as a primary source of guidance for the construction and validation of the human performance models of pilot performance during approach and landing operations with and without SVS:

- Physical layout of the HITL simulation facility with the location, viewing distances, and subtended visual angle of informational elements;
- Descriptions of the symbology and functionality of displays and controls and the field of view provided by the SVS display;
- Detailed cognitive task analysis of B757 approach and landing operations with and without SVS (as prepared by Keller & Leiden, 2002). This analysis covered all aspects of B757 operational knowledge from area navigation, approach procedures, flight controls, instrumentation, displays, and ATC communications down to a detailed event/task timeline of nominal procedures and the informational requirements for their proper execution. The document also noted potential off-nominal events and their triggers; and
- HITL simulation data (Goodman et al., 2002), including: time-referenced digital data of aircraft position and state; pilot control inputs; video recordings from both an ambient room camera and eye-tracking camera with superimposed fixation cursor; eye-movement data for predefined sceneplanes that corresponded to the three displays in the cockpit (PFD, ND, SVS), the MCP, controls, and the OTW scene, including percent fixation time, percent dwell time, dwell duration, dwell counts, and transition matrices; and post-trial questionnaires assessing pilot workload and SA.

HUMAN PERFORMANCE MODELING CHALLENGES FOR SVS DESIGN AND INTEGRATION

The modelers were not required to predict the NASA HPM project data set (partially described previously); rather, they were to use these low-level data as building blocks, coupled with the structures within their model architectures, to answer higher level questions about the safe development and integration of SVS technology into aviation operations. Specifically, the NASA modeling teams were challenged to determine the extent to which human performance models can be used to:

- Assess the safety of SVS for approach and landing and identify potential safety vulnerabilities;
- Define the CONOPS associated with its integration into actual cockpits;
- Inform the design of SVS in terms of information content, symbology format, physical location in the cockpit, or other design dimensions; and
- Inform the development of SVS usage procedures.

CONCLUSION

The information provided to the modeling teams for each of the two domain problems included detailed task analysis, data from HITL simulations, and other supporting information about the cockpit and external environment. The extent of the available data was different across the two domains as was the need for modeling. For the taxi study, the data set was fairly robust and, since it was within the realm of current-day operations, the standard operating procedures, crew roles and responsibilities, and description of the cockpit and external environments were very well defined. The HITL data provided an understanding of the nature of navigation errors (how many, where they occurred, the taxiway geometry, and intracockpit activities that preceded the errors), but less was known about the underlying causes of the error. For example, the HITL study revealed three types of errors (decision, planning, and execution), and we now turn to the human performance modelers to attempt to better understand these error classes and the factors that might lead to each type of error.

The SVS study was of a different nature. At the time of the study, SVS technology was still in the concept-development phase. There were no commonly accepted crew roles, responsibilities, or procedures; the precise nature of the SVS display (i.e., where it would be located, what data it would present, and in what format) was undefined; and various CONOPS had been proposed but not agreed upon. Also, as would be typical of most projects in the conceptual development phase, robust HITL data sets did not exist. Instead, modelers were provided data from only three pilots flying a number of different scenarios that, as discussed earlier, were quite variable. The challenge of the modelers in this case was to make assumptions about how the SVS could or would be used to determine the viability of the concept, uncover potential safety concerns for further exploration in HITL simulation, and help define the CONOPS or the actual design of the SVS.

Providing HITL data to the five modeling teams was intended to foster the development of valid models and meaningful results. At the same time, it is expected that the subsequent results of these modeling efforts will guide future HITL research by identifying issues and variables of interest to be confirmed or further investigated in HITL simulations.

ACKNOWLEDGMENTS

The authors wish to acknowledge Anthony Andre, Bonny Parke (San Jose State University Research Foundation), and Dan Renfroe for their contributions to the surface operations simulation and error analysis; Lynne Martin and Allen Goodman (San Jose State University Research Foundation) for compiling the surface operations task analyses and materials provided to the modeling teams; Allen Goodman, John Wilson (San Jose State University Research Foundation), and Glenn Meyer (QSS Group, Inc.) for conducting the SVS experiment and data analyses; Tom Marlow (Silicon Valley Simulation) and Monterey Technologies, Inc. for developing the SVS simulation; Ken Leiden, John Keller, Jon French, and Ron Small (Micro Analysis and Design, Inc.) for completing the task analyses of the approach and landing task, which is presented in this chapter in an abbreviated form; and Barbara Burian, Anthony Andre, and Brian Gore (San Jose State University Research Foundation) for technical reviews of earlier versions of this chapter.

REFERENCES

Alexander, A. L., Wickens, C. D., & Hardy, T. J. (2003). Examining the effects of guidance symbology, display size, and field of view on flight performance and situation awareness. *Proceedings of the 47th Annual Meeting of the Human Factors and Ergonomics Society*. Santa Monica, CA: HFES.

Andre, A. D. (1995). Information requirements for low-visibility taxi operations: What pilots say. In R. S. Jensen & L. A. Rakovan (Eds.), *Proceedings of the Eighth International Symposium on Aviation Psychology* (pp. 484–488). Columbus, OH: The Ohio State University.

Andre, A.D., Hooey, B. L., Foyle, D., & McCann, R. (1998). Field evaluation of T-NASA: Taxi navigation and situation awareness system. *Proceedings of the IEEE/AIAA 17th Digital Avionics Systems Conference, 47:* 1–8, Seattle, WA.

Bartolone, A. P., Glaab, L. J., Hughes, M. F., & Parrish, R. V. (2005). Initial development of a metric to describe the level of safety associated with piloting an aircraft with synthetic vision systems (SVS) displays. In J. G. Verly (Ed.), *Proceedings of SPIE, Enhanced and Synthetic Vision 2005, 5802,* 112–126.

Beringer, D. B. (2000). Development of highway-in-the-sky displays for flight-path guidance: History, performance, results, guidelines. *Proceedings of the IEA 2000/HFES 2000 Congress.* Santa Monica, CA: HFES.

Billings, C. (1996). *Human-centered aviation automation: Principles and guidelines* (NASA Technical Memorandum 110381). Moffett Field, CA: NASA Ames Research Center.

Comstock, J. R., Glaab, L. J., Prinzel, L. J., & Elliot, D. M. (2001). Can effective synthetic vision system displays be implemented on limited size display space? *Proceedings of the Eleventh International Symposium on Aviation Psychology.* Columbus, OH: The Ohio State University.

FAA (Federal Aviation Administration). (2002a). *FAA runway safety report: Runway incursion trends at towered airports in the United States 1998–2001.* Washington, DC: FAA.

FAA (Federal Aviation Administration). (2002b). *Runway safety blueprint 2002–2004.* Washington, DC: FAA.

FAA (Federal Aviation Administration) Safe Flight 21. (2002). Safe flight 21 master plan 3.0 (August, 2002). The Mitre Group.

Fadden, S., Ververs, P. M., & Wickens, C. D. (1998). Costs and benefits of head-up display use: A meta-analytic approach. *Proceedings of the 42nd Annual Meeting of the Human Factors and Ergonomics Society,* 16–20. Santa Monica, CA: HFES.

Flight Safety Foundation. (1998). Killers in aviation: FSF task force presents facts about approach-and-landing and controlled-flight-into-terrain accidents. *Flight Safety Digest Special Report, 17*(11–12), 1–256.

Goodman, A. (2001). *Taxi navigation errors: Taxiway geometry analyses.* Unpublished raw data.

Goodman, A., Foyle, D. C., & Hooey, B. L. (2002). *Synthetic vision systems study data and analyses.* Unpublished report.

Goodman, A., Hooey, B. L., Foyle, D. C., & Wilson, J. R. (2003). Characterizing visual performance during approach and landing with and without a synthetic vision display: A part-task study. In D. C. Foyle, A. Goodmam, & B. L. Hooey (Eds.), *Proceedings of the 2003 NASA Aviation Safety Program Conference on Human Performance Modeling of Approach and Landing with Augmented Displays* (NASA/CP-2003-212267). Moffett Field, CA: NASA Ames Research Center.

Goodman, A., & Martin, L. (2001). *Airport surface operations cognitive tasks analysis.* Unpublished report.

Hooey, B. L., & Foyle, D. C. (2006). Pilot navigation errors on the airport surface: Identifying contributing factors and mitigating solutions. *International Journal of Aviation Psychology, 16*(1), 51–76.

Hooey, B. L., Foyle, D. C., & Andre, A. D. (2000). Integration of cockpit displays for surface operations: The final stage of a human-centered design approach. *SAE Transactions: Journal of Aerospace, 109,* 1053–1065.

Hooey, B. L. Foyle, D. C., Andre, A. D., & Parke, B. (2000). Integrating datalink and cockpit display technologies into current and future taxi operations. *Proceedings of the AIAA/IEEE/ SAE 19th Digital Avionics System Conference* (7.D.2-1–7.D.2-8). Philadelphia, PA.

Keller, J. W., & Leiden, K. (2002). *Information to support the human performance model-ing of a B757 flight crew during approach and landing SVS addendum.* (Technical report). Boulder, CO: Micro Analysis & Design.

Keller, J. W., Leiden, K., & Small, R. (2003). Cognitive task analysis of commercial jet air-craft pilots during instrument approaches for baseline and synthetic vision displays. In D. C. Foyle, A. Goodman, & B. L. Hooey (Eds.), *Proceedings of the 2003 NASA Aviation Safety Program Conference on Human Performance Modeling of Approach and Landing with Augmented Displays* (NASA/CP-2003-212267). Moffett Field, CA: NASA Ames Research Center.

Lasswell, J. W., & Wickens, C. D. (1995). *The effects of display location and dimensionality on taxi-way navigation* (Tech. Rep. ARL-95-5/NASA-95-2). Savoy, IL: University of Illinois Aviation Research Laboratory.

McCann, R. S., Andre, A. D., Begault, D. R., Foyle, D. C., & Wenzel, E. M. (1997). Enhanc-ing taxi performance under low visibility: Are moving maps enough? *Proceedings of the 41st Annual Meeting of the Human Factors and Ergonomics Society*, 37–41, Santa Monica, CA: HFES.

McCann, R. S., Foyle, D. C., Andre, A. D. & Battiste, V. (1996). Advanced navigation aids in the flight deck: Effects on ground taxi performance under low-visibility conditions. *SAE Transactions: Journal of Aerospace, 105*, 1419–1430.

McCann, R. S., Hooey, B. L., Parke, B., Foyle, D. C., Andre, A. D., & Kanki, B. (1998). An evaluation of the taxiway navigation and situation awareness (T-NASA) system in high-fidelity simulation. *SAE Transactions: Journal of Aerospace, 107*, 1612–1625.

Meyer, T., & Bradley, J. (2001). The evolution from area navigation (RNAV), required naviga-tion performance (RNP) to RNP RNAV. In *Proceedings of the ICAO Global Naviga-tion Satellite System Panel Meeting* (ICAO GNSSP IP11).

Norman, R. M. (2002). *Synthetic vision systems (SVS) concept assessment report* (NASA Contrac-tor Report NAS1-00106 Task#1002). Hampton, VA: NASA Langley Research Center.

Palmer, M. T., Abbott, T. S., & Williams, D. H. (1997). Development of workstation-based flight management simulation capabilities within NASA Langley's Flight Dynam-ics and Control Division. In R. S. Jensen & L. A. Rakovan (Eds.), *Proceedings of the Eighth International Symposium on Aviation Psychology*, 1363–1368. Columbus, OH: The Ohio State University.

Reason, J. (1990). *Human error.* New York: Cambridge University Press.

Williams, D., Waller, M., Koelling, J., Burdette, D., Capron, W., Barry, J., et al. (2001). *Concept of operations for commercial and business aircraft synthetic vision systems* (Version 1.0) (NASA TM-2001-211058). Hampton, VA: NASA Langley Research Center.

Williams, K. W. (2002). Impact of aviation highway-in-the-sky displays on pilot situation awareness. *Human Factors, 44*(1), 18–27.

Woods, D., & Dekker, S. (2000). Anticipating the effects of technological change: A new era of dynamics for human factors. *Theoretical Issues in Ergonomic Science, 1*(3), 272–282.

APPENDIX A

Task Analysis of a Nominal Landing Rollout and Taxi-to-the-Gate Procedure[a]

Task/Subtask Actions	Operator	Task Type (Input/ Output Channels)
Roll-out Segment (Median Duration = 34 sec)		
Coordinate A/C ground maneuvering		*Continuous*
Control A/C speed, direction, and braking	CA	Visual, psychomotor
Monitor ground environment		*Continuous*
Scan OTW scene for centerline tracking and possible incursions	CA	Visual
Scan OTW scene for possible incursions and exit	FO	Visual
Perform rollout		*Discrete*
Apply reverse thrust as required	CA	Visual, psychomotor
Monitor rollout progress and proper autobrake operations	CA	Visual
Disarm autobrakes, continue with manual braking	CA	Visual, psychomotor
Verify thrust levers are closed and speed brakes are up	FO	Visual
Initiate ground-speed callouts at 100 kts	FO	Visual, verbal
Call out 90, 80, 70 kts as indicated	FO	Visual, verbal
Call out 60 kts	FO	Visual, verbal
Position levers to idle forward thrust as taxi speed reached	CA	Visual, psychomotor
Navigate turnoff		*Discrete*
Visually reference chart; note current location relative to preferred exit	FO	Visual, verbal
Monitor OTW scene to identify preferred exit and coordinate turnoff	CA and FO	Visual, verbal
Disengage autopilot and begin runway turnoff	CA	Visual, psychomotor
Slow A/C onto exit, stop once clear of runway	CA	Visual, psychomotor
Communicate with ATC		*Discrete*
Contact tower—"clearing runway at ..."	FO	Verbal
Switch frequency to ground control	FO	Visual, psychomotor
Notify ground control—"clear of runway at ..."	FO	Verbal
Taxi-in Segment (Median Duration = 3 min, 34 sec)		
Monitor ground communications		*Continuous*
Listen to radio for directed communication and party-line chatter	CA and FO	Auditory
Establish taxi clearance		*Discrete*
Write down taxi route	FO	Visual, psychomotor
Read back taxi route to ground control	FO	Visual, verbal
Discuss taxi route; visually reference chart	CA and FO	Auditory, visual, verbal
Begin taxi in to gate		*Discrete*
Accelerate A/C towards first turn	CA	Visual, psychomotor
Control A/C speed, direction, and braking	CA	Visual, psychomotor
Perform navigation		*Intermittent*
Visually reference taxiway chart and review progress	FO	Visual
Confer about current location, upcoming turns, and observed traffic	CA and FO	Auditory, visual, verbal

Stop A/C during taxi to check chart, only if needed for navigation awareness	CA	Visual, psychomotor
Call out traffic	CA and FO	Visual, verbal
Call out approaching route turn	CA and FO	Visual, verbal
Complete taxi in to gate		*Discrete*
Scan to identify correct lead-in line	CA and FO	Visual, verbal
Look for marshall and follow signals	CA	Visual, psychomotor

Notes: This task analysis assumes the captain is the pilot flying, and first officer is the pilot not flying. Procedures and task times vary across airlines, aircraft, and airports. A/C = aircraft; CA = captain; Comm = communication; FO = first officer; OTW = out the window.

[a]Adapted from Goodman, A., & Martin, L. (2001). Airport surface operations cognitive tasks analysis. Unpublished report.

Task Analysis for an Area Navigation Approach and Landing for a Boeing-757 at a Nominal Airport[a,b]

Event/Task Description	Operator	Type
Receive approach clearance		
• ~2,500 ft AGL		
• 15 miles out		
• ATC communication and read back takes approximately 5 sec.		
• Once the read back is complete, the task sequence listed for this event takes approximately 10 sec for the crew to complete.		
ATC communication: "You are 10 miles from the marker, cleared for approach, slow to 160"	ATC	Discrete
Read back clearance and speed	PNF	Discrete
Check to ensure LNAV/VNAV PATH	PF	Discrete
Check airspeed	PF	Discrete
Call out "Speed 160, flaps 15"	PF	Discrete
Set speed to 160	PF	Discrete
Check speed setting	PNF	Discrete
Check speed against reference bugs	PF	Discrete
Mentally confirm flaps versus speed settings	PNF	Discrete
Set flaps 15, say "Flaps 15"	PNF	Discrete
Aircraft attitude adjustment		
• Flap deployment takes about 30–40 sec to complete.		
Hear flap lever go into detent	Both	Discrete
Feel pitch change	Both	Continuous
Scan flap indicators	PNF	Discrete
Monitor PFD	Both	Intermittent
After sufficient slowing		
• ~13 miles out		
• The task sequence listed for this event takes approximately 10 sec to complete.		
Call out, "Gear down, flaps 20, speed plus 5"	PF	Discrete
Deploy gear and say "Gear"	PNF	Discrete
Set flaps 20 and say "Flaps 20"	PNF	Discrete
Set target speed	PF	Discrete
Check speed setting	PNF	Discrete

(Continued)

Task Analysis for an Area Navigation Approach and Landing for a Boeing-757 at a Nominal Airport[a,b] (Continued)

Event/Task Description	Operator	Type
Aircraft attitude adjustment		
• Flap deployment takes between 30 and 40 sec to complete.		
Hear flap lever go into detent	Both	Discrete
Feel pitch change	Both	Continuous
Scan flap indicators	PNF	Discrete
Monitor PFD	Both	Intermittent
After further slowing		
• Speed ~ 150 kts		
• The task sequence associated with this event takes less than 10 sec to complete.		
Call out "Flaps 25"	PF	Discrete
Set flaps 25 and say "Flaps 25"	PNF	Discrete
Aircraft attitude adjustment		
• Flap deployment takes approximately 20 sec to complete.		
Hear flap lever go into detent	Both	Discrete
Feel pitch change	Both	Continuous
Scan flap indicators	PNF	Discrete
Monitor PFD	Both	Intermittent
Final flaps and landing checklist		
• Speed ~ 140 kts		
• The task sequence associated with this event takes less than 30 sec to complete.		
Call out "Flaps 30 and landing checklist"	PF	Discrete
Set flaps 30 and say "Flaps 30"	PNF	Discrete
Get list or starting from memory	PNF	Discrete
Call out "Cabin notification? ... Complete"	PNF	Discrete
Call out "Gear down?"	PNF	Discrete
Check gear lights	Both	Discrete
Call out "Down and checked" (or "down, three green lights")	PNF	Discrete
Call out "Speed brakes armed?"	PNF	Discrete
Check speed brakes (manually)	PNF	Discrete
Check speed brakes (visually)	PF	Discrete
Call out "Armed"	PNF	Discrete
Call out "Flaps?"	PNF	Discrete
Check flap settings	Both	Discrete
Call out "30 planned, 30 indicated"	PNF	Discrete
Call out "Landing checklist complete"	PNF	Discrete
Stabilization gate		
• ~11 miles out		
• 1–2 miles prior to FAF, the aircraft must be stabilized, fully configured, flaps 30 and landing checklist complete.		
• Set next altitude in MCP prior to beginning of last descent.		
Set decision altitude in MCP	PF	Discrete
Verify decision altitude setting	PNF	Discrete
Verify LNAV/VNAV PATH	PF	Discrete

Task Analysis for an Area Navigation Approach and Landing for a Boeing-757 at a Nominal Airport[a,b] (Continued)

Event/Task Description	Operator	Type
Crossing final approach fix (FAF)		
• ~9 miles out		
Call out name of FAF	PNF	Discrete
Call out, "Decision altitude is set at 600 ft barometric"	PF	Discrete
Feel descent	Both	Continuous
Stabilized on descent		
• ~2,000 ft		
• Set missed approach altitude.		
• If altitude capture occurs while resetting to the missed approach altitude, then the crew will execute a missed approach.		
Verify LNAV/VNAV PATH	PF	Discrete
Set missed approach altitude	PF	Discrete
Monitor to determine altitude capture has not occurred	Both	Continuous
Cross outer marker		
• ~5 miles out		
• 125 kts		
• The time associated with this task sequence can vary depending on how long it takes for ATC to respond with the landing clearance. It should take less than 45 sec. The call to ATC could happen anytime after starting the descent from the FAF.		
Switch to tower radio frequency	PNF	Discrete
Make ID, location, and intention call	PNF	Discrete
Tower responds with clearance to land	ATC	Discrete
Read back clearance to land	PNF	Discrete
Set taxi light	PF	Discrete
1,000- ft Callout		
• ~1,000 ft AGL		
Scan all instruments looking for error/warning flags	PNF	Discrete
If no flags, "1,000 feet, flags checked"	PNF	Discrete
Call out, "Runway XX, cleared to land"	PF	Discrete
~100 ft above decision altitude (~700 ft)		
• The pilot must make the landing determination prior to reaching the DA or the aircraft will automatically initiate a missed approach when reaching the DA due to the altitude setting in the MCP.		
• The point at which the PF begins manually flying the plane could have taken place before this whenever the runway is visible.		
• If the runway is not visible by the time DA is reached, then the pilot will execute a missed approach.		
Call out, "Approaching DA"	PNF	Discrete
If runway is in sight, call out, "Runway in sight"	PF	Discrete
Manually fly the landing		
Looking far down runway	PF	Intermittent
Hand flying A/C	PF	Continuous
Monitor instruments	PNF	Intermittent

(Continued)

Task Analysis for an Area Navigation Approach and Landing for a Boeing-757 at a Nominal Airport[a,b] (Continued)

Event/Task Description	Operator	Type

Decision altitude (600 ft)
- ~600 ft AGL
- If the pilot has not determined to land and disengaged the autopilot, the aircraft will automatically execute a missed approach.

Event/Task Description	Operator	Type
Call out "DA" or "Minimums"	PNF	Discrete
If not determined to land, call out, "Missed approach point"	PNF	Discrete
If not determined to land, call out, "Going around"	PF	Discrete

500-ft Callout
- 500 ft AGL
- This is a stabilization gate at which the speed, descent rate, and flaps must all be configured for landing.

Either automatic or call out, "500"	PNF	Discrete
State speed relative to bug, descent rate, and final flaps	PF	Discrete

100-ft Callout
- ~100 ft AGL

Call out "100 feet"	PNF	Discrete

Flare and wheel touch
- ~30 ft AGL
- ~117 kts

Flare and let A/C settle onto main landing gear	PF	Discrete

Notes: This task analysis begins as the aircraft transitions into the approach phase (approximately 15 miles from the runway threshold, flying level at 2,500 ft, at 200 kts). Procedures vary across airlines, aircraft, and airports. For a specific airport, such as SBA as used in the SVS HITL study, actual approach plate altitude values would be substituted. All altitude values are referenced to touchdown zone elevation. AC = aircraft; AGL = above ground level; ATC = air traffic control; DA = decision altitude; HITL = human-in-the-loop; PF = pilot flying; PNF = pilot not flying; MCP = mode control panel; SBA = Santa Barbara Municipal Airport; SVS = synthetic vision system.

[a] Adapted from Keller, J. W., & Leiden, K. (2002). Information to support the human performance modeling of a B757 flight crew during approach and landing SVS addendum. (Technical report). Boulder, CO: Micro Analysis & Design.

[b] This represents a nominal task analysis for an airport with a 600-ft DA.

Part 2

Application of Individual Modeling Tools to the Aviation Problems

4 Overview of Human Performance Modeling Tools

Kenneth Leiden, Michael D. Byrne,
Kevin M. Corker, Stephen E. Deutsch, Christian
Lebiere, and Christopher D. Wickens

CONTENTS

INTRODUCTION

An overview of each of the five human performance modeling tools addressed in the National Aeronautics and Space Administration Human Performance Modeling (NASA HPM) project is presented to provide the reader with initial insight into each HPM tool's theory of operation, general capabilities, programming language, and operating system requirements. The five HPM tools discussed are:

- Adaptive Control of Thought-Rational (ACT-R);
- Improved Performance Research Integration Tool (IMPRINT)/ACT-R;
- Air Man–machine Integration Design and Analysis System (Air MIDAS);
- Distributed Operator Model Architecture (D-OMAR); and
- Attention-Situation Awareness (A-SA).

In subsequent chapters, each HPM tool is discussed in greater detail, but this chapter serves as a resource for a high-level understanding of the HPM tools.

HPM TOOL DESCRIPTION

ACT-R

ACT-R (implementation discussed in chapter 5) is both a theory and a cognitive modeling tool. The theory describes how humans organize knowledge and produce intelligent behavior. The cognitive modeling tool simulates the application of this theory to the extent that is practically possible. ACT-R researchers consider ACT-R a work in progress and continuously update both the theory and modeling tool to reflect advances in psychology and neuroscience. ACT-R was originally developed at Carnegie Mellon University under sponsorship from the Office of Naval Research. The first version of ACT-R, version 2.0, was released in 1993. ACT-R perceptual-motor (ACT-R/PM) integrates previous versions of ACT-R with perceptual and motor modules inspired by, and in some cases essentially identical to, those of another HPM tool known as executive-process/interactive control (EPIC) (Meyer & Kieras, 1997). ACT-R version 5.0, the most recent version available during the time period of this HPM effort, built upon ACT-R/PM and incorporated other improvements.

With regard to the NASA HPM project, ACT-R/PM was utilized to model the surface operations effort described in chapter 3. For the synthetic vision system (SVS) modeling effort (also described in chapter 3), ACT-R 5.0 had become available and was integrated with the X-Plane flight simulator package to provide a high-fidelity environment to represent the closed-loop behavior between a pilot's actions (or inactions) and the aircraft's response.

The theory underlying ACT-R 5.0 is described in Anderson et al. (2004) and much of the overview presented here is based on that document. Including books, journal articles, and conference proceedings, there are well over 500 ACT-R publications to date, many of which are available at the ACT-R Website (http://act.psy.cmu.edu). ACT-R is open source and the associated software, models, and tutorials are available from the Website. ACT-R is implemented in the common LISP programming language as a collection of LISP functions, data structures, and subroutines. ACT-R runs on MacOS, Windows, Linux, and Unix platforms (any platform with a compatible LISP compiler). A number of tools are available for model development and execution, including a graphical user interface to author, run, and debug ACT-R models.

An overview of ACT-R 5.0 is depicted in Figure 4.1. Each block in the figure represents an independent unit that supports a particular set of cognitive processes as well as the primary region of the brain where those processes are performed. In ACT-R parlance, the three types of units are referred to as modules, buffers, and the central production system. Together, they embody a unified, or integrated, theory of cognition. In concise terms, *modules* serve to represent different types of information, such as goals and declarative memory. Each module has an associated *buffer* for holding a single declarative unit of knowledge, referred to as a *chunk,* for two-way exchange with the central production system. The *central production system,* which represents procedural memory (i.e., memory of skill or procedure), detects patterns in these buffers to determine what to do next. In this way, the central production system coordinates between itself and the modules via the buffers.

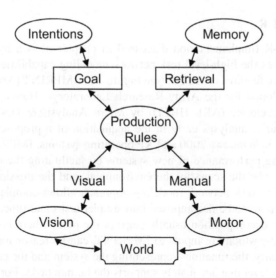

FIGURE 4.1 ACT-R 5.0 overview.

Although not all ACT-R modules and buffers are represented in Figure 4.1 (e.g., rudimentary modules for speech and audition are not depicted), the ones that are represented emphasize the primary focus of ACT-R 5.0 as follows:

- The declarative module defines the methods for retrieving information from declarative memory. The module determines the probability and latency of chunk retrieval based on past experiences with that chunk and the chunk's relevance to the current context. The retrieval buffer stores the current chunk retrieved from declarative memory.
- The visual module identifies objects in the visual field. There are two visual buffers—one to keep track of object location and one to keep track of object identity—corresponding to the "where" and "what" distinctions used by vision scientists to describe human visual processing.
- The manual module and buffer specify how the hands are controlled, detailing in particular the sequencing and timing of individual motor movements, such as key presses, while using an interface.
- The goal buffer keeps track of the current step of the goal. The goal stack from ACT-R 4.0 has been removed and replaced with the existing mechanisms for declarative memory. As a result, goal steps are tracked through the declarative module. The intentional module in ACT-R 5.0, as depicted in Figure 4.1, is currently a placeholder for future refinements.
- The central production system, represented by the matching, selection, and execution blocks in Figure 4.1, determines the "next step" of procedural memory (i.e., production rule) to execute based on the constraint that only one production rule at a time can be executed.

IMPRINT/ACT-R

IMPRINT/ACT-R (implementation discussed in chapter 6) is a hybrid modeling tool that combines the high-level task network modeling capabilities of IMPRINT with the cognitive features of ACT-R (see Figure 4.2). IMPRINT (Archer & Adkins, 1999) was developed for the Army Research Laboratory—Human Research and Engineering Directorate (ARL HRED) by Micro Analysis & Design to conduct human performance analyses early in the acquisition of a proposed new system. IMPRINT runs on Windows 2000 and XP operating systems. IMPRINT assists the user in predicting performance of new systems by facilitating the construction of models that describe the scenario, the environment, and the mission that must be accomplished. The task network modeling approach allows complex human tasks and system functions to be decomposed into simpler tasks and functions. The level of breakdown or decomposition usually stops at the point where human task time and task accuracy, which are inputs to IMPRINT, can be reasonably estimated or measured. Similarly, the functions representing the system and the environment are decomposed to a level that adequately supports the human tasks. For example, pilot tasks and activities can be supported in many situations by aircraft functions that stop short of detailed models of how the engines operate.

Executing an IMPRINT model multiple times allows the analyst to determine variability in human performance and its impact on system and mission performance. The ease of use and the task network approach of IMPRINT make it an ideal tool for incorporating the human element into trade-off studies and sensitivity

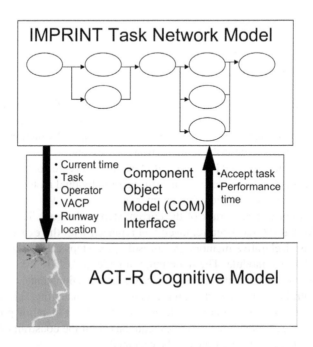

FIGURE 4.2 IMPRINT/ACT-R overview.

analyses of complex systems. However, it does not include an embedded model of cognitive or psychological processes. Rather, it relies on the modeler to specify and implement these constructs. In the IMPRINT/ACT-R approach, these constructs are provided through a dynamic link to ACT-R.

The version of ACT-R coupled with IMPRINT as implemented for this NASA HPM project refers to a hybrid of ACT-R versions 4.0 and 5.0. This variant utilizes the ACT-R 5.0 architecture for specifying the interaction of declarative and procedural knowledge and, in general, the integration of multiple buffers for simulating the synchronization of mental activities. However, the variant also incorporates many aspects of the ACT-R 4.0 hierarchical reinforcement scheme, which was based on the goal stack from ACT-R 4.0. The goal stack of ACT-R 4.0 was removed from ACT-R 5.0 due to the implausibility of perfect goal memory. This hierarchical scheme enabled a specific type of learning by allowing a top-level goal to take the success and failure of subgoals into account when determining its utility by propagating success and failure back up the goal stack. This learning mechanism was necessary to the IMPRINT/ACT-R team's goal of exploring the pilot's adaptation in scanning behavior to find rewarding information on the cockpit displays.

Air MIDAS

Development of Air MIDAS (implementation discussed in chapter 7) began in 1985 during the joint Army–NASA Aircrew/Aircraft Integration Program to explore the computational representations of human–machine performance to aid crew system designers. Air MIDAS runs on a PC using the Windows 2000 or XP operating system and requires Visual C++ 6.0, Alegro CL, and JAVA software.

Air MIDAS is an HPM tool specifically designed to assess human–system interaction in the complex, dynamic environments of aviation and air traffic control. Utilizing an integrated approach to human performance modeling, Air MIDAS gives users the ability to model the functional and physical aspects of the operator, the system, and the environment, and to bring these models together in an interactive, event-filled simulation for quantitative and visual analysis. Through its facilities for constructing, running, and analyzing simulations, Air MIDAS can provide measures of operator workload and task performance across a range of conditions. Additional metrics of human–system effectiveness can be derived by users for a given scenario, based on observed system or vehicle operating parameters during simulation. Such Air MIDAS output is intended to support the evaluation of current or future equipment, procedures, and operating paradigms. With respect to the NASA HPM project, Air MIDAS was integrated with NASA's PC-Plane flight simulator package to provide a high-fidelity environment to represent the closed-loop behavior between pilot action and aircraft response (and vice versa).

Operator behavior within the Air MIDAS simulation is driven by a set of user inputs specifying operator goals, procedures for achieving those goals, and declarative knowledge appropriate to a given simulation. These asserted knowledge structures interact with and are moderated by embedded models of perception for extracting information from the modeled world and embedded models of cognition for managing resources, memory, and actions (see Figure 4.3). In this way, Air MIDAS seeks to capture the perceptual–cognitive cycle of real-world operators who

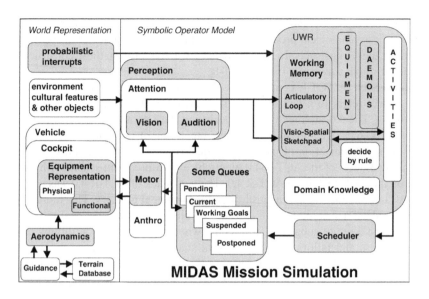

FIGURE 4.3 Air MIDAS overview.

work toward their most immediate goals, given their perception of the situation and within the limitations of their physical and cognitive capacities. In Air MIDAS, as in the real world, perceived changes in the world—new information or events—may cause changes in the adjudged context of the situation, triggering new goals or altering methods to achieve current goals. Such perceived changes may be precipitated through the behavior of other modeled entities or by user-specified time-based, condition-based, or probabilistic events set within the simulation (e.g., a system failure or the receipt of an incoming air traffic control clearance). It may, in fact, be the impact of the operator's own actions that leads to changes in context and goals.

This complex interplay between top-down and bottom-up processes means behavior flows from, and is sensitive to, the goals, knowledge, and domain expertise imparted to the simulated operator, the perceptual availability of information in the constructed environment, and the nature and timing of external events. This process is partially enabled by micromodels of human performance that feed forward and feed back to the other constituent models in the complex human/system representation. Together, this affords Air MIDAS users flexibility in selecting the focus of study as each of these parameters can be manipulated individually or in concert in order to assess the resulting effects on operator and system performance. Investigative approaches can range from single-factor sensitivity analyses to "what if" studies that explore large test matrices of interacting manipulations and conditions.

While the Air MIDAS framework simulates human behavior at a detailed level, the structure is not intended to be a "unified theory of cognition" in the terms specified by ACT-R. The Air MIDAS system should instead be viewed as an attempt at capturing and integrating many of the best practices in behavior modeling as expressed through various independent research efforts (e.g., see Card, Moran,

& Newell, 1983). In particular, the general Air MIDAS research program does not emphasize validating the overall architecture, but the individual elements of the system capture many of the elements of human cognitive processes and mechanisms (e.g., modality specific working memories, independent perception and motor processes, attentional processes). The Air MIDAS approach, while not focused on validation of a unified architecture in an ACT-R sense, is very concerned with validation of the output of the model and the research team has gone to significant effort to validate the model's predictions against operational data for air traffic control (Corker, 2005; Corker, Fleming, & Lane, 2001).

D-OMAR

D-OMAR (implementation discussed in chapter 8) was developed by BBN Technologies under sponsorship from the Air Force Research Laboratory (AFRL). Rather than a particular cognitive architecture, D-OMAR provides an event-based simulator and knowledge representation languages that are the basis for defining an architecture and constructing human performance models (Deutsch, 1998). The human performance models may operate in the simulation environment running in real time or fast time. The D-OMAR knowledge representation languages were designed to *facilitate the modeling of the human multitasking behaviors* of team members interacting with complex equipment in pursuing collaborative enterprises.

The representation languages that form the basic building blocks for the modeling effort are:

- Simple frame language (SFL) for defining the agents and objects in the simulation world; and
- Simulation core (SCORE) procedure language for defining the behaviors of agents and objects.

The D-OMAR model will run on most popular hardware/software configurations. It is built on the Franz Allegro implementation of LISP. The user interface components of D-OMAR, the graphical editors and browsers, the simulation control panel, and the timeline displays, have been built in Java. D-OMAR is open-source software available at http://omar.bbn.com. In addition, the Website has D-OMAR references and an online user manual. D-OMAR runs on Windows and Linux (and has run on Solaris in the past). External simulations can interoperate with D-OMAR via HLA, CORBA, or a straightforward socket protocol.

Figure 4.4 provides a schematic view of the current architecture as implemented in D-OMAR. At the periphery are the basic perceptual and motor components of a model. The core of the architecture is the network of procedures that emulate the capabilities of the brain's perceptual, cognitive, and motor centers. The goals that establish the network implement the model's proactive agenda and channel responses to impinging events. Basic person procedures establish capabilities such as a coordinated hand–eye action to set a lever. Domain-specific goals and procedures govern operations such as the sequence of flap settings during an approach and landing.

The model's goal and procedure network constitutes the operator's long-term procedural and declarative memory. A publish-subscribe protocol moves long-term

D-OMAR

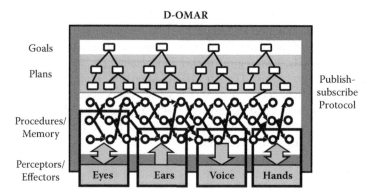

FIGURE 4.4 D-OMAR overview.

and working memory items among the nodes of the procedure network and coordinates the execution of distinct processing centers. The SCORE language provides constructs to define concurrent execution necessary at the within-task level, as in coordinated hand–eye actions and to model multitasking realistically. Activities spawned by the model's proactive agenda and those spawned by events in the environment compete for attention. Bounds on concurrent execution—the limits on multitasking—are established through contention arbitration.

Contention arbitration operates at several levels within the D-OMAR models. At the lowest level, it serves to allocate perceptual and motor resources among contending tasks. A proactive task to adjust flap settings on the approach will preempt a periodic background scan of the instrument panel by gaining access to the vision system based on the higher priority of the proactive task vis-à-vis the background task. At this level, priority-based access to perceptual and motor resources determines which tasks precede and which are delayed. The same conflict resolution strategy operates at the policy level as well. When contention between tasks is present, policy-based priorities determine the next action. A simple, clarifying example of policy-driven behavior is communication between controllers and the flight deck. Conversations between individual aircrew members of the flight deck are immediately halted when a radio message from a controller is heard. The aircrew listens to the message, even if it is directed toward another aircraft. Likewise, flight deck-initiated communication with a controller must wait until the party line is clear before the aircrew member can speak.

A-SA

The A-SA model (implementation discussed in chapter 9) predicts pilot attention allocation and the resulting situation awareness (SA). A-SA is unlike the other HPM tools discussed in this book in that it does not attempt to model complete human performance. For example, the model does not incorporate action-related phenomena (such as the tasks necessary to manipulate the flight controls of an aircraft) or low-level cognition. A-SA is typically run stand-alone as a dynamic model, but it could be embedded into complete HPM frameworks (such as ACT-R, Air MIDAS,

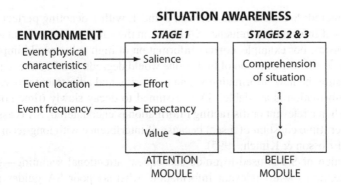

FIGURE 4.5 A-SA overview.

or D-OMAR) to predict errors of situation awareness, such as making a wrong turn on a taxiway. A-SA has thus far been tailored primarily to the aviation domain, but has also been applied to highway driving (Horrey, Wickens, & Consalus, 2006) and could be applied to other domains provided the analyst is capable of specifying the "knowledge" that comprises SA in the applicable domain.

A-SA is composed of two modules representing attention and belief updating (see Figure 4.5). The *attention module* describes the allocation of attention to events and flight deck displays within the aircraft environment, approximating Endsley's (1988) Stage 1 SA (perception). The *belief-updating module* describes SA in terms of understanding of or belief in the current and future state of the aircraft, approximating Endsley's Stages 2 and 3 SA (comprehension and projection).

Attention Module

Underlying the allocation of attention module is a static, analytic model known as SEEV (for salience, effort, expectancy, and value). SEEV predicts the probability of attending to a particular flight deck display or out-the-window (OTW) view as well as the period of neglect in which potentially important channels may be ignored. SEEV is based on the premises that:

A pilot's attention is *influenced* by two factors:
- Bottom-up capture of salient or conspicuous events (visual or auditory); and
- Top-down selection of valuable event information in expected locations at expected times.

A pilot's attention is *inhibited* by a third factor:
- Bottom-up level of effort required, which encompasses both the effort of moving attention elsewhere and the effort required to carry out concurrent cognitive activity.

Belief-Updating Module

The module for understanding the current and future state of the aircraft is based on belief updating. Situation awareness is updated any time new pieces of information are encountered; SA has a value between 0 and 1, with 1 denoting perfect SA. The new pieces of information influence SA based on the value and attentional allotment of those pieces. For example, correct information of high value would improve SA, whereas incorrect information of high value would degrade SA. In either case, SA is updated using the new information via an anchoring and adjustment process. When no new information is available, SA is assumed to decay slowly (time constant of 60 sec). When irrelevant or distracting information is encountered, SA is assumed to decay faster (time constant of 5 sec) because of interference with long-term working memory (Ericsson & Kintch, 1995).

The value of SA is used to guide subsequent attentional scanning—good SA guides attention toward relevant information, whereas poor SA guides attention toward less relevant information. In addition, SA determines the likelihood that the pilot will choose correctly when confronted with a decision point (e.g., turn or go straight when encountering a taxiway intersection). If the pilot has poor SA, default behavior provided by the analyst influences the decision (e.g., when in doubt, go straight), which may be erroneous.

REFERENCES

Anderson, J. R., Bothell, D., Byrne, M. D., Douglass, S., Lebiere, C., & Quin, Y. (2004). An integrated theory of the mind. *Psychological Review, 111*, 1036–1060.

Archer, S. G., & Adkins, R. (1999). *IMPRINT User's Guide*. Prepared for U.S. Army Research Laboratory, Human Research and Engineering Directorate.

Card, S. K., Moran, T. P., & Newell, A. (1983). *The psychology of human–computer interaction*. Hillsdale, NJ: Lawrence Erlbaum Associates.

Corker, K. (2005). Computational human performance models and air traffic management. In B. Kirwan, M. Rodgers, & D. Shafer (Eds.), *Human factors impacts in air traffic management* (pp. 317–350). London: Ashgate.

Corker, K., Fleming, K., & Lane, J. (2001). Air ground integration: Dynamics in exchange of information for control. In L. Bianco, P. Dell'Olmo, & A. Odoni (Eds.), New concepts and methods in air traffic management (Transportation Analysis) (pp. 125–142). Berlin: Springer.

Deutsch, S. (1998). Multi-disciplinary foundations of multiple-task human performance modeling in OMAR. *Proceedings of the 20th Annual Meeting of the Cognitive Science Society*.

Endsley, M. R. (1988). Design and evaluation for situation awareness enhancement. *Proceedings of the Human Factors Society 32nd Annual Meeting* (pp. 97–101). Santa Monica, CA: HFES.

Ericsson, K. A., & Kintsch, W. (1995). Long-term working memory. *Psychological Review, 2002*, 211–245.

Horrey, W. J., Wickens, C. D., & Consalus, K. P. (2006). Modeling driver's visual attention allocation while interacting with in-vehicle technologies. *Journal of Experimental Psychology: Applied, 12*, 67–78.

Meyer, D. E., & Kieras, D. E. (1997). A computational theory of executive cognitive processes and multiple-task performance: I. Basic mechanisms. *Psychological Review, 104*, 3–65.

5 An ACT-R Approach to Closing the Loop on Computational Cognitive Modeling
Describing Dynamics of Interactive Decision Making and Attention Allocation

Michael D. Byrne, Alex Kirlik,
and Michael D. Fleetwood

CONTENTS

INTRODUCTION

It has now been over 25 years since prominent human factors researchers such as Sheridan (1976) and Rasmussen (1976) called for a shift of focus in modeling human performance in technological systems. Prior to this call, human manual control behavior, such as hand flying an aircraft or steering a ship, occupied the majority of modelers' attention (see Jagacinski & Flach, 2002, for a recent overview and integration). But as researchers such as Sheridan and Rasmussen noted, many, if not all, of the fastest control loops in many human–machine systems are now available to be controlled by automation of one sort or another (e.g., autopilots). Additionally, a system operator's perceptual access to the controlled system is no longer direct but instead mediated by technological interfaces presenting processed and designed information rather than raw data (e.g., Kirlik, 2006). As such, the most pressing and important knowledge gap to be filled has changed. The central new questions have shifted to how a system operator monitors, manages, or otherwise interacts with highly automated systems—that is, "supervisory control" (Sheridan)—and also how an operator uses the highly processed information locally available from a techno-logical interface to diagnose and respond to system faults and environmental distur-bances occurring in a remote location—that is, "behind" the interface, so to speak (Rasmussen).

The supervisory control and fault diagnosis demands spawned by increasing levels of automation have a common psychological origin. Both emerged due to auto-mation and interface technology playing an increasingly *mediating* role, in standing between the operator and the controlled system and its environment. As such, an ever increasing need existed for the operator to go "beyond the information given" (Bruner, 1957) in the performance of these supervisory control and fault manage-ment tasks. As one of the fathers of the "cognitive revolution," Bruner pointed out that both the need and ability to perform this "going beyond" effectively are one of the hallmarks of a process that was once controversial in psychology, but is now taken as commonplace: *cognition*. Regardless of the exact terms in which Rasmus-sen and Sheridan originally framed their charge to the human performance model-ing community, we can now safely say that their charge amounted to an appeal to extend modeling into the cognitive realm.

As a result and due to this shift of emphasis, one can find many if not most authors, nearly all of them engineers, writing in the 1976 landmark volume *Monitor-ing Behavior and Supervisory Control* (Sheridan & Johannsen, 1976) grappling with the issues of how to incorporate cognitive activities such as decision making, atten-tion, memory, planning, and problem solving into their quantitative human perfor-mance modeling approaches. This trend, demonstrating an optimistic (some might say, "engineering") outlook—that cognitive-level performance could largely be for-mulated in quantitative terms—held sway in the "man–machine systems" tradition throughout the late 1970s and early 1980s, as revealed by the contents of the first volume in the well known "Advances in Man–Machine Systems Research" series, edited by the prominent human–machine systems engineer William Rouse (1984).

But history has demonstrated that modeling cognition in complex, operational task environments would not quickly yield to quantitative techniques such as Bayesian

decision or estimation theories, optimal control theory, or even early artificial intelligence (AI)-inspired models such as expert systems and "intelligent" planners or
problem solvers (Kirlik & Bisantz, 1998). Thus, to the extent that modeling has still
proven useful and influential in human factors of the last 20 years or so, it has been
aimed at a highly broad, abstract, and qualitative level, perhaps best illustrated by
Rasmussen's (1983) "skills, rules, and knowledge" framework and Klein's (1989)
"recognition-primed decision" model, or else aimed at a relatively micro level, either
with respect to one particular psychological "module" or one particular type of task.
As noted by Kirlik (2003), one likely explanation for the slow progress in developing
both more general *and* more formal cognitive performance models has been the difficulty in discovering or identifying sufficiently context-free abstractions of complex
operational contexts; these abstractions are the hallmark of engineering modeling
techniques.

But times are changing. As Byrne and Gray (2003) have observed, perhaps the
time is ripe for formal—that is, computational or quantitative—approaches to once
again provide a useful role in the engineering of complex human–machine systems.
The present chapter and, indeed, this volume itself hopefully provide evidence for at
least some merit to this claim. During this past 20 years or so, another community of
scientists, working within the traditions of cognitive science and human–computer
interaction, have been developing, iteratively refining, and validating modeling techniques that we believe hold promise in eventually becoming useful engineering tools
for the analysis, design, and evaluation of human–machine systems.

Broadly speaking, these "cognitive architectures" (see Byrne, 2003, for an
introduction and overview) represent an attempt to provide general frameworks for
cognitive modeling in which the findings from scores of experimental studies and
the latest psychological theory are implicitly embedded in the design of the architecture itself. As such, they can be viewed as computational artifacts that effectively
embody much state-of-the-art theory and empirical knowledge in cognitive psychology, thus allowing the informed users of these artifacts to benefit by this knowledge to construct plausible cognitive models of particular cases of human–machine
interaction. This "artifact as knowledge" view highlights another important feature,
or way of looking at, these modeling techniques—that is, the cognitive modeling
architecture as the repository of cumulative scientific knowledge. To illustrate, none
of the current authors would call himself an expert on the retrieval of information
from long-term memory. Yet, the models described here allow us to leverage exactly
this expertise as it is in a very real sense "cached" directly into the design of the
cognitive architecture we have chosen to model the NASA taxiing and approach and
landing scenarios.

Yet, as will be made clear, we believe that these architectures still have some
way to go before they can be used as efficiently and usefully as, say, Fitts's law (Fitts,
1954) for the analysis, design, and evaluation of human–machine system performance. This point needs to be articulated further to illustrate the rationale underlying the presentation of much of the research to follow. In particular, we would like to
pause to first explain why a considerable amount of time and effort had to be devoted
to "applying" the particular modeling architecture with which we have chosen to
work to the NASA scenarios and, importantly, why we do *not* think that this time

and effort bodes poorly for the future prospects of computational cognitive modeling becoming a useful engineering technique in human factors.

MODEL VERSUS APPLICATION

While an ideal modeling tool for the analysis and design of human–machine systems would require no special expertise on the part of the modeler and would not require the modeler to engage in a protracted and potentially effortful exercise in applying the model, we know of no engineering discipline in which models are used in this way. Were this the case, there would be no need to hire structural engineers to design bridges, no need to hire electrical engineers to design circuits, and no need to hire aeronautical engineers to design airplane wings, even though every one of these professionals is equipped with a rich toolbox of models. Especially as human–machine systems increase in technological complexity, we think that it is simply naive to believe that the profession of cognitive engineering or human factors engineering will proceed any differently.

Next, we would like to point out that even in the simplest cases in human factors, the apparent simplicity of a model can belie the complexity of the tasks involved in its application. Take the aforementioned Fitts's law, which is a simple log-linear mathematical model that is often said to be able to "predict" human discrete movement times for various tasks. But Fitts's law, in and of itself, predicts nothing of the sort. Making such predictions requires performing work to parameterize the equation according to the details of the encounter between a particular human and a particular environment. The values of both the slope and intercept terms of Fitts's law hardly have the status of physical constants, as they vary across the degrees of freedom of the motor systems involved in the movement in question and also across individuals. We know of no truly reliable published tables for these values; thus, to make any point "prediction" using Fitts's law one will first have to study the performer of interest in order to estimate these human-centered parameters of the equation from behavioral data. Therefore, we do not find it defensible to suggest that the need for cognitive modelers to use data associated with the prior behavior of performers in order to parameterize models in order to make analogous point predictions somehow reveals a deficiency of cognitive modeling methods.

Finally, consider the environmentally-centered variables in Fitts's law: namely, target distance (or movement amplitude) and target width (or the need for accuracy). Estimates of the values of these variables must come from a detailed analysis of the task environment; when those movements are of greater than one dimension, or if the target is irregularly shaped, making such measurements becomes notoriously difficult (MacKenzie, 2003). What is notable is that this task can even be nontrivial in the case of Fitts's law, where the functionally relevant variables of the task environment (distance and target width) are known. This problem is much more complicated in the case of applying a cognitive architecture of any kind because, in the vast majority of cases, they do not speak to the issue of specifying the environmental variables that play a functional role in cognition and behavior. This is not to say that the design of the architecture itself may not reflect a deep concern with environmental structure and thus contain mechanisms designed in an environmentally adaptive fashion (see Anderson, 1990).

It is the case, however, that the relative lack of resources contained in these architectures to support environmental modeling is, admittedly, an arguable weakness of many such approaches (e.g., in comparison to a model such as Fitts's law), and it owes to the origin of most cognitive architectures in the view that specifying functionally relevant environmental variables was outside the scope or purview of cognitive modeling per se. This issue is especially crucial for human factors applications, due to the following long-standing recognition:

> Human behavior, either cognitive or psychomotor, is too diverse to model unless it is sufficiently constrained by the situation or environment; however, when these environmental constraints exist, to model behavior adequately, one must include a model for that environment. (Baron, 1984, p. 6)

Contrasting views on what should be the appropriate scope of cognitive modeling do exist, of course, but a discussion of this issue is beyond the scope of the present chapter (e.g., see Hammond & Stewart, 2001; Hutchins, 1995; Kirlik, Miller, & Jagacinski, 1993; Tolman & Brunswik, 1935). Suffice it to say that modeling a performer in a realistically complex task environment using a cognitive architecture not only requires a knowledge engineering component, but also can require a parallel effort to discover the functionally relevant environmental variables of interest. Additionally, it may even require, as Baron suggests, actually including these variables in a model in its own right, or perhaps "hooking up" the computational cognitive model to an existing simulation of the task environment—tasks that we discuss in some detail in the following sections. In fact, we take one contribution of the research presented in this chapter to be demonstrating how, in at least the two scenarios of interest, the task of supplementing an architecturally inspired cognitive model with models or simulations of the controlled system (aircraft) and the more encompassing task environment can at least be approached.

In a larger context, however, this was but one of the many challenges we faced in bringing what to this point had largely been a cognitive architecture applied and validated primarily in laboratory tasks to bear on realistically complex, operational scenarios (for another example, see Salvucci's 2001 work in the domain of automobile driving). In the following descriptions of the two modeling efforts we performed in taxi navigation and approach and landing with and without synthetic vision system (SVS) displays, we share what we consider to be not only our successes, but also our failures, compromises, and shortfalls, as we believe all of this information is potentially relevant to moving cognitive modeling research forward.

ACT-R MODELING: A GENERAL OVERVIEW

ACT-R (Anderson et al., 2004; see also Anderson & Lebiere, 1998) is a computational architecture designed to support modeling of human cognition and performance at a detailed temporal grain size. Figure 5.1 depicts the general system architecture. ACT-R allows for modeling of the human in the loop as the output of the system is a time-stamped stream of behaviors at a very low level, such as individual shifts of visual attention, keystrokes, and primitive mental operations such as retrieval of a simple fact. In order to produce this, ACT-R must be provided two things: knowledge

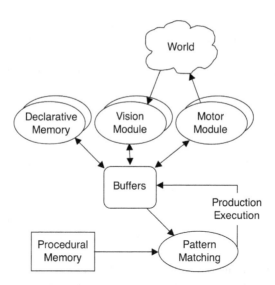

FIGURE 5.1 The ACT-R cognitive architecture. The less used auditory and vocal modules are visible behind the other perceptual–motor modules.

and a world or environment (usually simulated) in which to operate. The environment must dynamically respond to ACT-R's outputs and thus must also often be simulated at a high degree of fidelity. The knowledge that must be provided to ACT-R to complete a model of a person in an environment is essentially of two types: declarative and procedural. Declarative knowledge, such as "George Washington was the first president of the United States" or "'IAH' stands for Bush Intercontinental Airport in Houston," is represented in symbolic structures known as *chunks*.

Procedural (sometimes referred to as "how-to") knowledge, such as the knowledge of how to lower the landing gear in a 747, is stored in symbolic structures known as *production rules* or simply *productions*. These consist of IF–THEN pairs; IF a certain set of conditions holds, THEN perform one or more actions. In addition, both chunks and productions contain quantitative information that represents the statistical history of that particular piece of knowledge. For example, each chunk has associated with it a quantity called *activation* that is based on the frequency and recency of access to that particular chunk, as well as its relationship to the current context. Because the actual statistics are often not known, in many cases these values are left at system defaults or estimated by the modeler, though in principle ACT-R can learn them as well.

The basic operation of ACT-R is as follows. The state of the system is represented in a set of buffers. The IF sides of all productions are matched against the contents of those buffers. If multiple productions match, a procedure called *conflict resolution* is used to determine which production is allowed to fire, or apply its THEN actions. This generally changes the state of at least one buffer, and then this cycle is repeated every 50 ms of simulated time. In addition, a buffer can change without a production explicitly changing it. For example, there is a buffer that represents the visual object currently in the focus of visual attention. If that object changes or disappears, this buffer will change as a result. That is, the various perceptual and motor processes (and declarative memory as well) act in parallel with each other and with the central cognitive production cycle. These processes are modeled at varying levels of fidelity. For example, ACT-R does not contain any advanced machine vision component that would allow it to recognize objects from analog light input. Rather, ACT-R needs to be given an explicit description of the visual object to which it is attending.

A COMPUTATIONAL COGNITIVE MODEL OF TAXI NAVIGATION

One of the first decisions that had to be made in modeling the NASA taxi navigation simulation and scenarios (chapter 3) was a decision about scope. To limit the scope of the project, we chose to model only the captain in ACT-R, and treated both the ground controller and the first officer (FO) as items in the environment. We felt this decision balanced tractability and relevance, since the captain made the final decisions and the captain also controlled the aircraft.

A second important aspect of scoping model coverage was to select the focal psychological activities to model. Our research team was one of many teams also creating cognitive models of the same NASA surface operations simulation data (as described in chapter 1). In this light, we considered both the strengths and weaknesses of our ACT-R approach with the alternative approaches taken by other research teams, with the goal of providing a unique contribution to the overall research effort. For example, we ruled out focusing on multitasking and situation awareness (SA) issues (such as losing track of one's location on the airport surface), as these were the focal points for other approaches.

All things considered, including our own previous experience in human performance modeling (e.g., Kirlik et al., 1993; Kirlik, 2006), we decided to focus on the interactive, dynamic decision-making aspects of the task in its closed-loop context. As a result, we focused on those contributions to error that may result from the interaction of the structure of a task environment and the need often to make rapid decisions on the basis of imperfect information, resulting from decay of clearance information from memory, low visibility, and sluggish aircraft dynamics. Our focus on decision making, which assumed pilots had accurate knowledge of their current location, was complemented by another modeling team's focus on a particular class of SA errors associated with losing track of one's location on the airport surface (chapter 9).

THE MODEL'S ENVIRONMENT

We created an ACT-R model of one human pilot, but this pilot model still had to be situated in an accurate environment. Thus, three external entities were modeled to describe the environment: the simulated aircraft controlled by the pilot model, the simulated visual information available to the pilot model, and the simulated runway and taxiway environment through which the aircraft traveled. Each of these three environmental entities was modeled computationally and then integrated with the cognitive components of the pilot model to create an overall representation of the interactive human–aircraft–environment system.

It would have been extremely difficult to interface ACT-R with the code for the vehicle dynamics that was used to drive the actual NASA flight simulator (chapter 3) in which behavioral data were collected. We therefore had to create a simplified vehicle model with which the pilot model could interact. Although we were not interested in control issues per se, the dynamics of the aircraft played an important role in determining decision time horizons, a key factor in the cognitive representation of the pilot's activities. Given vehicle size, mass, and dynamics, however, we still did require a somewhat reasonable approximation to the actual aircraft dynamics used in the experiments in order to be able to get a handle on timing issues. The aircraft model we constructed assumed that the pilot controlled the vehicle in three ways: applying engine power, braking, and steering. For the purposes of modeling an aircraft during taxiing, these three forms of control are sufficient. Based on Cheng, Sharma, and Foyle's (2001) analysis of the NASA simulated aircraft dynamics, we proceeded with a model in which it was reasonable to assume that throttle and braking inputs generated applied forces that were linearly related with aircraft speed.

Steering, however, was another matter. After consideration of the functional role that steering inputs played in the NASA taxi simulation scenario (chapter 3), we decided that we could finesse the problem of steering dynamics by assuming that the manual control aspects of the steering problem did not play a significant role in the navigation errors that were observed. That is, we assumed that making an appropriate turn was purely a decision-making problem, and that no turn errors resulted from correct turn decisions that were erroneously executed. Note that this assumption does not completely decouple the manual and cognitive aspects of the modeling, however. It was still the case that manual control of the acceleration and braking aspects of the model did play a role in determining the aircraft's position relative to an impending turn and, importantly, placed a hard constraint on the aircraft's maximum speed of approach to each turn.

The maximum aircraft speeds for the various types of turns required in the NASA simulation were calculated under the constraint that lateral acceleration be limited to 0.25 *g* for passenger comfort (Cheng et al., 2001) and also the field data reported in Cassell, Smith, and Hicok (1999). These maximum speeds partially determined the time available to make a turn decision, and, as will be seen, as this time is reduced there was a greater probability of an incorrect decision. Our simplification regarding steering merely boiled down to the fact that once the model had made its decision about which turn to take, that turn was executed without error.

To implement this aspect of the model, we decided to model the Chicago O'Hare (KORD) airport taxiway as a set of interconnected "rails" upon which travel of the simulated aircraft was constrained. Taxiway decision making in this scheme, then, boiled down to the selection of the appropriate rail to take at each taxiway intersection. In this manner, we did not have to model the dynamics of the aircraft while turning: We simply moved the aircraft along each turn rail at the specified, turn-radius-specific speed.

The model used to represent the visual information available to our ACT-R pilot model was obtained from the actual NASA flight simulator (chapter 3) in the form of a polygon database. This database consisted of location-coded objects (e.g., taxiways, signage) present on the KORD airport surface, or at least those objects presented to flight crews during NASA experimentation. Distant objects became "visible" to the pilot model at distances similar to those at which these same objects became visible to human pilots in the NASA simulation.

TASK ANALYSIS AND KNOWLEDGE ENGINEERING

The task-specific information required to construct the model was obtained by study-ing various task analyses of taxiing (e.g., Cassell et al., 1999) and through extensive consultation with two subject matter experts (SMEs) who were experienced airline pilots. We first discovered that, in many cases, pilots have multiple tasks in which to engage while taxiing. Based on this finding, our ACT-R model only concerned itself with navigation decision making when such a decision was pending. In the interim, the model iterated through four tasks deemed central to the safety of the aircraft.

These four tasks included monitoring the visual scene for incursions, particu-larly objects like ground vehicles, which are difficult to detect in poor visibility; maintaining the speed of the aircraft, since the dynamics of a commercial jetliner require relatively frequent adjustments of throttle and/or brake to maintain a constant speed; listening for hold instructions from the ground controller; and maintaining an updated representation of the current position of the aircraft on the taxi surface and the location of the destination. While these tasks often have little direct impact on navigation, they do take time to execute, and time is the key limited resource in making navigation decisions in our integrated pilot–aircraft–environment system model. Essentially, the presence of these tasks served to reduce the time available for the model to make navigation decisions.

With respect to those navigation decisions, we found that decision making is highly local. That is, the planning horizon is very short; flight crews are quite busy in the time after landing and thus, in situations like KORD in poor visibility, report they do not have the time to "plan ahead" and consider turns or intersections other than the immediately pending one. Second, the decision process tends to be hierarchical: Pilots first decide if the next intersection requires a turn, and if it does, then decide which turn to make. We found that in the error corpus available to us (chapter 3; Goodman, 2001), errors in the first decision (whether to turn or not) were rare (which was also consistent with our SME reports), so we concentrated our efforts on under-standing how pilots made the second decision.

The first issue to be addressed was: What kinds of knowledge and strategies are actually brought to bear by real pilots in the kinds of conditions experienced by the pilots in the NASA study? Largely through interviews with SMEs, we discovered a number of key strategies employed by pilots, and also discovered that some of these strategies would *not* have been available to our model, either because they were not available to the pilots in the NASA simulation or due to technical limitations of the simplified model. We thus excluded these strategies from consideration. At the end of both our task analyses and SME interviews, we had identified five primary deci-sion strategies available for making turn decisions:

1. Remember the correct clearance. While fast, this strategy is increasingly inaccurate as time lapses between obtaining the list of turns described in the clearance and the time at which turn execution is actually required.

This is, in some sense, the normative strategy because this is what pilots are generally expected to do.
2. Make turns toward the gate. While somewhat slower than the first strategy, this strategy has a reasonable level of accuracy at many airports.
3. Turn in the direction that reduces the larger of the X or Y (cockpit-oriented) distance between the aircraft and the gate. We deemed this strategy to be moderately fast, like Strategy 2, but with a potentially higher accuracy than Strategy 2, since more information is taken into account.
4. Derive from map/spatial knowledge. This is the slowest strategy available, with high accuracy possible only from a highly experienced (at a given airport) flight crew.
5. Guess randomly. This is a very fast strategy, although it is unlikely to be very accurate, especially at multiturn intersections. However, we did include it as a possible heuristic in the model for two reasons: (a) It may be the only strategy available, given the decision time available in some cases; and (b) it provides insights into chance performance levels.

The next modeling issue to be dealt with was how to choose between strategies when faced with a time-constrained decision horizon. This type of metadecision is well modeled by the conflict resolution mechanism ACT-R uses to arbitrate between multiple productions matching the current situation. The accuracy of strategies 1 (recall the clearance) and 4 (derive from map knowledge) is primarily a function of the accuracy of the primitive cognitive operations required of these tasks, moderated by factors such as ACT-R's memory decay and effectively constrained working memory. However, the accuracy of Strategies 2, 3, and 5 is less cognitively constrained and instead critically dependent on the geometry of actual clearances and taxiways. Thus, we employed an SME as a participant in a study to provide data for an analysis of the heuristic decision Strategies 2 and 3 (the accuracy of Strategy 5, random guessing, was determined by the taxiway geometry itself). We would perhaps never have thought of performing this study if the ACT-R model had not required us to provide it with high-level (i.e., airport-neutral) strategies that pilots might use in deciding what turns to make during taxi operations, along with their associated costs (times required) and benefits (accuracy).

IDENTIFYING TAXI DECISION HEURISTICS

To obtain this information, which was required to inform modeling, we provided our SME (a working B-767 pilot for a major U.S. carrier) with Jeppesen charts (taxi diagrams) for most major U.S. airports, and then asked him to select charts for those airports for which he had significant experience of typical taxi routes and clearances. He selected nine airports (DFW, LAX, SFO, ATL, JFK, DEN, SEA, MIA, ORD). The SME was asked to draw, using a highlighter on the charts themselves, the likely or expected taxi routes at each airport from touchdown to his company's gate area. A total of 284 routes were generated in this way.

Our goal at this point was to identify whether any of the heuristic strategies identified during task analysis and knowledge engineering would be likely to yield acceptable levels of decision accuracy. We obtained an estimate of the accuracy of heuristic Strategies 2 (turn toward the company's gates), and 3 (turn in the direction that minimizes the largest of the X or Y distance between the current location and the gates) by comparing the predictions these heuristics would make with the data provided by the SME for the nine airports studied. We recognize that these accuracy estimates may be specific to the (major) carrier for whom the SME flew, since other carriers' gates may be located in areas at these nine airports such that their pilots are provided more or less complex, or geometrically intuitive, clearances than those providing the basis of our SME's experience. However, we do believe that this study resulted in enlightening results regarding the surprisingly high level of accuracy of simple, "fast and frugal" decision heuristics (Gigerenzer & Goldstein, 1996) in this complex, operational environment.

Figure 5.2 presents the results of an analysis of the effectiveness of these two heuristic strategies. Note that the XY heuristic is quite good (nearly perfect performance) across the board, and the even simpler "toward terminal" heuristic is reasonably accurate at many major U.S. airports. Thus, we created the turn decision-making components of the pilot model to make decisions according to the set of

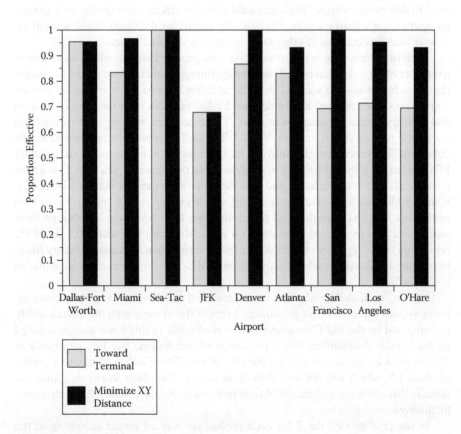

FIGURE 5.2 Accuracy of "toward terminal" and "minimize greater of XY distance" heuristics.

five strategies described previously, including the two surprisingly frugal and robust "toward terminal" and "XY" heuristics portrayed in Figure 5.2. One can think of these five strategies as being hierarchically organized in terms of their costs (time requirements) and benefits (accuracies). The decision components of the cognitive model worked by choosing the strategy that achieved the highest accuracy given the decision time available.

Detailed Description of Dynamic Decision Modeling

From a time-horizon (cost) perspective, the selection of decision strategies was informed by a procedure for estimating the time remaining before a decision had to be made. Time remaining was based on the aircraft's distance to an intersection and the amount of slowing necessary to make whatever turns were available, which was thus dependent on aircraft dynamics. Recall that we had an algorithm available to calculate the maximum speed with which a turn of a given type could be negotiated. Thus, the computation of time remaining assumed a worst-case scenario for each specific intersection. That is, the time horizon for decision making was determined by the intersection distance combined with knowledge of aircraft dynamics, used to determine whether braking could slow the aircraft sufficiently to negotiate an intersection's sharpest turn. Horizons for particular decisions could actually be shorter than this because of other tasks occupying the pilot's time.

Each turn-related decision strategy was one production rule, which was allowed to enter conflict resolution only if the average time it would take the model to execute the procedure was <0.5 sec less than the decision horizon. This somewhat conservative approach was used to compensate for the fact that both the time estimation and strategy execution times were noisy. Those productions meeting this criterion competed in a modified version of ACT-R's conflict resolution procedure. In the default ACT-R procedure, the utility of each production is estimated by the quantity PG-C, where P is the probability of success if that production is selected, G is a time constant (20 sec is the default), and C is the time taken until an outcome is reached if that production fires. Because time cost was irrelevant in this application as long as the cost was less than the time remaining, this term was removed, though there was a 1-sec penalty applied to productions whose time cost was within 0.5 sec of the remaining time—again, a conservative move to ensure that a decision strategy likely to be completed will be selected (one of our SMEs indicated a conservative bias in this direction).

The utility of each production is also assumed in ACT-R to be a noisy quantity, so the system was not always guaranteed to select the strategy with the highest utility as computed by the PG-C measure. (Amount of noise in this computation is based on the logistic distribution. The s parameter of that logistic is a free parameter in ACT-R models and was set to 1 for our simulations. This yields a standard deviation of about 1.8, which was not varied to fit the data.) Thus, there were two sources of noise in this situation: estimation of time remaining and the utilities of the strategies themselves.

In the pilot model, the P for each production was estimated according to the actual probability of success of each of the decision strategies. Thus, P for the production

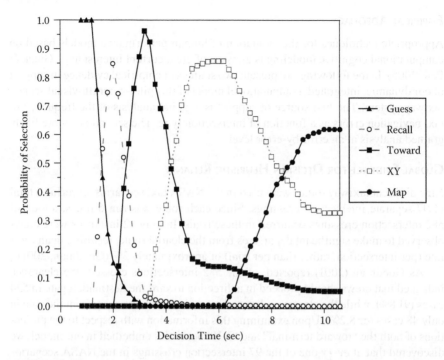

FIGURE 5.3 Selection probability for each decision strategy as a function of decision time horizon.

initiating the "turn toward the gate" production was 80.7% since that was the success rate for that strategy as determined by the SME study. P values for the other two decision heuristics (3 and 5 previously) were calculated in an analogous fashion, and P values for Strategies 1 (recall the actual clearance) and 4 (derive from the map) were determined by the boundedly rational cognitive mechanisms inherent in the ACT-R cognitive architecture. With the entire model in place, we then ran a Monte Carlo simulation (300 repetitions at each of 50 time horizons) to determine the probability of selection for each strategy as a function of decision time available. These simulation results are presented in Figure 5.3.

As is clear from Figure 5.3, as the decision horizon decreases, so does the likelihood that the pilot model will select a less accurate strategy. In fact, in the time window of from about 2.5 to about 8 sec, the environmentally derived heuristics dominate alternative strategies, and, based on our viewing of the NASA-supplied scenario videos (chapter 3), this time horizon was not unrepresentative of many actual decision horizons observed. However, this can be viewed as adaptive since a fast and frugal strategy that can run to completion can frequently outperform an analytically superior decision strategy that must be truncated due to time constraints (Gigerenzer & Goldstein, 1996).

EMPIRICAL ADEQUACY

Appropriate techniques for the validation of human performance models based on computational cognitive modeling is an issue of great current interest (e.g., Gluck & Pew, 2005). In the following, we present two sources of empirical evidence in support of our dynamic, integrated, computational model of this pilot–aircraft–visual scene–taxiway system. The first source of support is a global analysis of the frequency of taxi navigation errors as a function of intersection type. The second is a more finely grained analysis at an error-by-error level.

GLOBAL EVIDENCE FOR DECISION HEURISTIC RELIANCE

Nine different taxiway routes were used in the NASA taxi scenarios, covering a total of 97 separate intersection crossings. Since each route was run 6 times, a total of 582 intersection crossings occurred in these trials. In only 12 instances were crews observed to make significant departures from the cleared route, resulting in an error rate (per intersection, rather than per trial) of approximately 2% (Goodman, 2001).

As Goodman (2001) reported, of the 582 intersections crossed, the clearance indicated that crews should proceed in a direction toward the destination gate in 534 cases (91.8%), while the clearance directed crews in directions away from the gate in only 48 cases (or 8.2%). Upon examining this information with respect to the predictions of both the "toward terminal" and "XY" heuristics embodied in our model, we discovered that at *every* one of the 97 intersection crossings in the NASA scenarios at which the cleared route conflicted with *both* these two heuristics, at least one taxi error was made. These accounted for 7 of the 12 taxi errors observed.

In addition, as will be discussed in the following section, 4 of the 12 taxi errors were attributed not to decision making, but rather to a loss of SA (i.e., losing track of one's position on the airport surface; see Goodman, 2001, and Hooey & Foyle, 2006), a cognitive phenomenon beyond the scope of the present modeling. Our modeling approach assumed that the primary source of errors was time-stressed decision making combined with what might be called "counterintuitive" intersection and clearance pairs—that is, those at which both the "toward terminal" and "XY" heuristics failed due to atypical geometry or to clearances.

LOCAL EVIDENCE OF DECISION HEURISTIC RELIANCE

The Goodman (2001) report provided a detailed analysis of each of the 12 taxi errors observed in the baseline conditions of the NASA taxi simulation. In Byrne and Kirlik (2005), we analyzed each error individually. In the present context, we will merely summarize these results by saying that 4 of the 12 errors were unambiguously classified by Goodman as owing to a loss of situation awareness (e.g., crews indicating their uncertainty about their location on the airport surface). Of the eight remaining errors, which we consider to be decision-making errors, every one involved either an incorrect or premature turn toward the destination gate.

SUMMARY

The crux of the interpretation of taxi errors in the NASA taxi simulation is that pilots had multiple methods for handling individual turn decisions, and used the most accurate strategy possible given the time available (cf. Payne & Bettman, 2001). When time was short as a function of poor visibility, workload, and aircraft dynamics, the model assumed that the pilot tended to rely on computationally cheaper, but less specific, information gained from experience with the wider class of situations of which the current decision was an instance. In the case of the NASA taxi simulation, this more general information pertained to the typical taxi routes and clearances that would be expected from touchdown to gate at major U.S. airports. This interpretation is also consistent with the fact that the suite of display aids used in the high-technology conditions of the NASA simulation experimentation, by providing improved information to support local decision making, effectively eliminated taxi errors (Hooey & Foyle, 2006; Hooey, Foyle, & Andre, 2000). We believe that the ACT-R model we constructed not only received a reasonable degree of validation in terms of agreement with the empirical data, but also, perhaps even more importantly, prompted us to ask questions (e.g., about pilots' decision heuristics) that we likely would never have asked if we had not attempted to accomplish the task of constructing a running computational cognitive model of this behavioral situation.

A COMPUTATIONAL COGNITIVE MODEL OF THE IMPACT OF SYNTHETIC VISION SYSTEMS

Chapter 3 provides a detailed account of the experiments performed by NASA to examine the effects of introducing synthetic vision systems' (SVSs) display technology into a cockpit during a simulated approach and landing scenario at Santa Barbara municipal airport (SBA). Due to the relatively low number of participants (three) and the limited amount of overt behavioral data available due to flying a largely automated approach, we decided early to focus our modeling efforts on trying to assess the influence of the SVS on pilot's visual attention allocation, for which much data were available through eye tracking. As such, our first step in this modeling exercise was to conduct a fairly encompassing analysis of visual attention allocation, based largely on an examination of fixation frequencies at the various areas of interest (AOIs) within the cockpit (including the out-the-window [OTW] scene). A detailed analysis of these data is presented in Byrne and Kirlik (2004). The manner in which these analyses guided model construction and validation will be described in following sections.

Additionally, very early in this exercise, strongly influenced by our experience in modeling the taxi scenarios, we realized the importance of having a fairly veridical simulation of the cockpit, aircraft, and task environment that we could couple with our ACT-R model to enable truly closed-loop simulation. To accomplish this, we needed a dynamic and accurate model of the pilot's environment with which ACT-R could interact. For the aircraft/flight simulation, we selected X-Plane for this purpose (note that X-Plane has been certified by the FAA for training; see http://www. x-plane.com/FTD.html) and linked ACT-R to X-Plane via a user datagram protocol (UDP) network interface that we constructed from the ground up. X-Plane natively supports sending certain kinds of information such as altitude and heading via the network interface, but other things cannot be sent, including the visual scene. This is a problem since the ACT-R model needs something to "see." Most of this problem

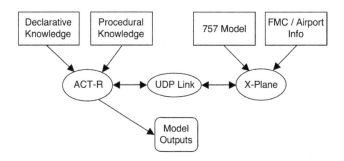

FIGURE 5.4 A model of the closed-loop, interactive system.

was solved by mocking up the primary displays (navigation display [ND]; primary flight display [PFD]; mode control panel [MCP], etc.) in the language of ACT-R so that the cognitive model could directly "view" those pieces of the display. These displays were updated based on data sent from X-Plane over the network connection.

In addition, we had to supply X-Plane with the aircraft specifications (a Boeing 757) and the appropriate approach/navigation and flight management computer (FMC) programming (e.g., approach waypoints) for SBA. Fortunately, the Boeing-757 specifications and the airport and geography for SBA were freely available and could simply be plugged in. Figure 5.4 depicts the overall interactive system.

MODELING AND TASK ANALYSIS

After detailed eye-movement data analysis and the creation of an interactive, closed-loop simulation capability, our first order of business was to try to understand the task at a detailed level. We relied on three primary sources of information: the task analysis information collected and supplied by NASA Ames Research Center (Keller, Leiden, & Small, 2003), related work in the human factors of aviation, and conversations with our SME. We synthesized these into the ACT-R formalism. An example of some of the resulting control structure appears in Figure 5.5.

The first insight from the knowledge engineering process is that the bulk of the task, particularly for the first two phases of flight, consists primarily of monitoring the state of the aircraft and maintaining an up-to-date representation of that state. Additionally, we learned that pilots very actively check for a number of events and conditions that do not occur in the scenarios, such as late changes of wind direction that might lead to wind shear. Thus, for the bulk of the experimental trials, there is little activity in the form of overt control actions despite the high engagement of the pilots. This is fairly realistic for most landings, which are, in fact, routine. However, pilots do have to monitor for nonroutine conditions. In order to simulate the true workload accurately, we have included checks for many of these things in the model even though they do not occur in the experimental scenarios.

The ACT-R architecture provided a great deal of constraint. Working within the parameters of the architecture sets certain boundaries and delimits scope. In particular, it meant that we were modeling the task at a highly detailed level of analysis.

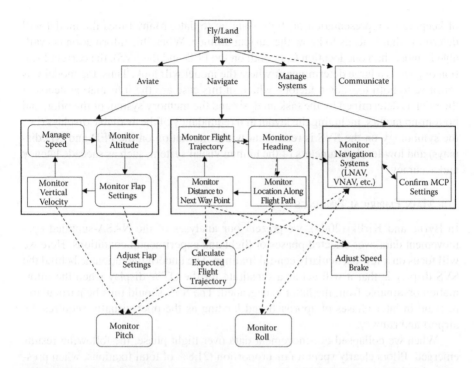

FIGURE 5.5 Flow of control resulting from task analyses. Dashed lines represent conditional subgoals that may not be executed every time; rounded boxes indicate that some information required by that subgoal may be found on the SVS, if present.

ACT-R provides end-to-end modeling of the human operator side of the human in the loop, from basic visual and auditory attentional operators to complex cognition and back down to basic motor movements. This impacts the strategies that are even possible and the way in which knowledge about dynamic state has to be updated in order to be maintained accurately.

Because the eye-movement data were the primary focus of the modeling effort, we examined other data and models in the "allocation of attention" domain in the human factors literature (e.g., Senders, 1964; Wickens, 2002). These are high-level (relative to ACT-R) accounts of how operators choose which objects to sample visually and at what frequency. The basic findings are that the rate at which particular displays are sampled depends jointly on the task importance of the displayed information as well as the rate of change of the information. As one might expect, more important information is sampled more often, and more dynamic information is sampled more often. We believe that these accounts provide a useful high-level starting point; one of our ultimate aims is to provide the explanation for how these high-level phenomena emerge from a combination of task and environmental constraints and relatively low-level cognitive–perceptual capabilities.

The final ACT-R model is driven primarily from the top down by the goal structure, and seeks information from the environment in order to fulfill the requirements of keeping its representation of flight state up to date. Many times the model will determine that it needs to know the current altitude. When this information is available in more than one location (e.g., both on the PFD and the SVS), the current location of gaze is a large determiner of where the model will look. Thus, the model was sensitive to both top-down factors, which in this case are the information needs of the pilot as determined by the task analysis and the memory system of the pilot, and bottom-up factors, including the layout and redundancy of the available displays (i.e., the symbology on the SVS is redundant with information on the PFD and ND displays) and low-level parameters of the human visual system, such as saccade latency and accuracy.

THE MAIN EMPIRICAL FINDINGS

In Byrne and Kirlik (2004), we present our analyses of the NASA-supplied eye-movement data over selected phases of flight and experimental conditions. Here we will focus on a few particularly central findings. The underlying rationale behind the SVS display is that it will act as a substitute for the OTW display when the information obtainable from the latter is degraded. The SVS should thus be particularly relevant in later phases of approach and landing as the pilot visually acquires the airport and runway.

When we collapsed eye-movement data over flight phase, the following results emerged. Pilots clearly spent a fair proportion (21.8% of total fixations, when present) of their gaze on the SVS when it was present (see Figure 5.6). An interesting question we wanted to address is from where they "stole" these gazes. Namely, they

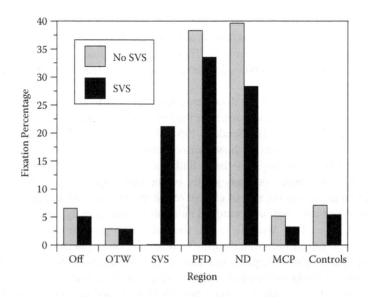

FIGURE 5.6 Percentage of all fixations falling on each scene plane across all phases.

must have reduced some portion of fixations associated with other areas of interest. A natural suspect for the location from which fixations would be stolen was the OTW display, due to the redundancy between the perceptual information gained from the SVS and OTW displays.

Surprisingly, however, the fixation dwell time and frequencies on different areas of interest under SVS versus baseline (no SVS) conditions revealed no obvious difference in the amount of gaze directed out the window. Instead, it appeared the SVS was associated with a reduction in the amount of gaze directed at the PFD and ND displays. Thus, the SVS display was drawing attention away from other sources of information from within the cockpit; it was not acting as a "substitute" for the information provided by the OTW display. Instead, our data analyses indicated that it did not act as a substitute source of environmental information (at least for these pilots). As a possibly unintended result of the presence of the SVS, less attention was paid to other displays.

We found this result both counterintuitive and quite interesting. Subsequent analyses, separated by phase of flight (approach toward landing), indicated that this tendency only increased with flight phase, and that pilots were not using the SVS as merely a proxy for OTW information, but mainly using it as a proxy for the PFD and ND displays, due to redundantly displayed information on both by virtue of the symbology overlaid on the SVS. The informal hypothesis that the symbology is responsible for the high proportion of attention allocated to the SVS is exactly the kind of hypothesis that can be formalized and tested in an ACT-R model, which is what we did.

EVALUATING AND VALIDATING THE ACT-R MODEL

One evaluation and validation approach that could have been taken would have been to perform split-half validation based on the eye-movement data provided. Split-half validation methods are popular with researchers who use regression methods and collect data from large numbers of individuals, such as questionnaire data. However, with only three pilots (who clearly demonstrated individual differences) it was not clear to us that this was a wise idea. It also raised the issue for us of how the data might be split; we felt we could not randomly sample fixations because that destroys the sequential nature of the data.

In the end, we decided to capitalize on the strengths of the supplied data. While there were few subjects, there was a great deal of data that could be considered at multiple levels. We chose to evaluate the model on the basis of fit to data at one level of abstraction and then validate against more difficult, lower level criteria. That is, we decided to attempt to fit the more global attention-allocation data at the level of models like SEEV (Wickens, 2002); this is the question of the percentage of time that the model looked at each display versus the percentage of time that the human pilots did. We validated by examining the performance of the same model at the more fine-grained level of transitions. That is, we asked how well the model, with parameters selected to fit the more global data, fits the more detailed data in the transition matrices.

In a critical sense, the first two phases of flight are the key phases because anything that changes in these phases as a function of the presence of the SVS is, if not unintended, then not easily anticipated. The primary goal of the SVS was not to change these phases of flight. However, as seen in the previous section, fairly dramatic changes did occur. The primary question, then, is to what degree the model captures those changes.

There were a few parameters that could be tweaked in this model that affect how well it captures the data. These generally involved strategy selection (conflict resolution) parameters that affected the model's behavior in two key circumstances: first, when choosing which high-level task (e.g., aviate, navigate) to pursue next and, second, when choosing which display to look at for a particular datum (e.g., look for altitude on the SVS or on the PFD?). Again, the option to find such information on the SVS is due to the overlay on the SVS. Many of these parameters could be set on an a priori basis without searching for best-fitting values, and thus made to be nonfree parameters. The parameters that control the second type of decision are of this type, as the values could be estimated by considering the distance to be traveled by the saccade, which affects how long it takes on average to complete successfully. Note that these parameters have a large impact on how often the model refixates in the same region, for this cost difference drives the model to prefer to refixate in the same display if the needed datum is available in multiple locations. However, we did not change these parameters in order to achieve a better fit.

Thus, we hand-optimized (purely through trial and error) only the parameters that control how often the model selects among the available high-level (e.g., aviate, navigate) goals. These two parameters were then kept constant through all simulations. Values were selected to produce a good fit to the fixations in each region for the initial approach fix to final approach fix (IAF to FAF) phase of flight with SVS. Thus, the behavior of the model in all other conditions, and on the transitions, is essentially a parameter-free prediction. At the most qualitative level of fit assessment, watching the model do the task does have the same general feel as watching video of the actual pilots doing the task (though without the eye-tracker-induced noise). However, we believe the bar should be somewhat higher than that.

Since this was the primary criterion and the basis for parameter tuning, the model's best fit is to the total fixations in each region for IAF to FAF with SVS present, as depicted in Figure 5.7. The model-to-data r-squared here is 0.978. While the model very slightly underpredicts the PFD and MCP proportions and overpredicts the SVS and controls proportions, overall the model does a good job of capturing the pilots' attention-allocation performance. Indeed, it appears that the presence of the SVS symbology is enough not just to drive pilots' attention to the SVS some of the time but to provide the basis for a quantitative fit. So, while the model captures the performance with the SVS, how does it fare when the SVS is not present? This is a prediction without parameter manipulation and is shown in Figure 5.8. The model is somewhat off, in that it ended up allocating slightly too many of the SVS fixations to the PFD and slightly too few to ND. However, the prediction is by no means poor, with an r-squared of 0.932.

The next question is how the model did with the initial phase, from start to the initial approach fix ("start to IAF"). This phase is probably in general somewhat less important than IAF to FAF because it is somewhat less realistic (pilots in the real world do not begin a flight at that point) and also fairly short. However, it is a

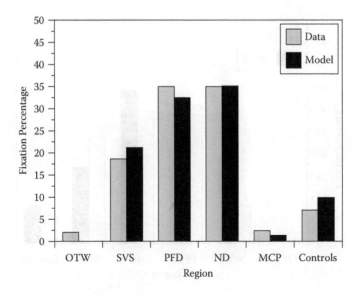

FIGURE 5.7 Model versus data overall attention allocation for IAF to FAF with SVS present.

useful test to see if the model can capture the differences between the two phases. Figure 5.9 presents the model-data comparison for the SVS condition for "start to IAF." This is not a great fit but at least the general trends are captured, explaining almost 70% of the variance (r-squared of 0.678). The model is a little too focused on

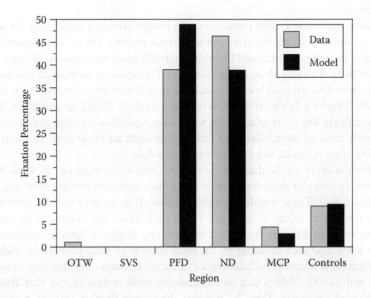

FIGURE 5.8 Model versus data overall attention allocation for IAF to FAF without SVS.

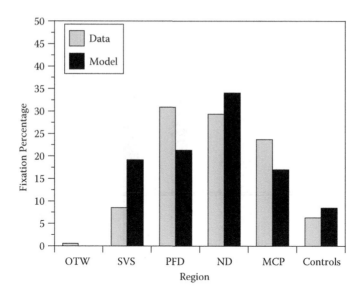

FIGURE 5.9 Model versus data overall attention allocation for "start to IAF" with SVS present.

navigating at this point and does not spend enough fixations assessing the state of the FMC (which is displayed on the MCP).

However, the situation is somewhat better in the no-SVS condition. The model again does not spend enough of its fixations on the MCP and still overpredicts ND (and PFD as well), but the fit is somewhat better, r-squared of 0.849. The fit is shown in Figure 5.10.

Taken across these four conditions, the model averages explaining about 85% of the variance in allocating attention to various regions. Given the relatively small number of participants here and the high interindividual variance, we believe this is about as well as can be done on this data set. However, as mentioned before, there exists a more fine-grained level at which the model may be validated: the transition matrices. This is a fairly stringent test of the model's ability to match the human pilots at a fairly fine grain of analysis; we believe a process-oriented cognitive model is the only kind of model likely to fare well in such an evaluation, as many other modeling approaches do not even produce such data.

Presentation of all the data here would be somewhat laborious, so we will illustrate with a particular selection of them (for a more complete presentation, see Byrne & Kirlik, 2004). These are difficult data to visualize, so they will be presented in tabular form. The values in each cell of all such tables are computed by counting each transition of fixation, which can be from one display to another or from a display to itself, and then dividing by the total number of such transitions, yielding a proportion (or probability) of occurrence of each transition (thus, the sum of all table values will be 1.0). Values that would generate table entries of less than 0.01 have been omitted for clarity. Table 5.1 presents the human fixation-transition matrix for IAF to FAF without SVS, while Table 5.2 presents the model's matrix under the

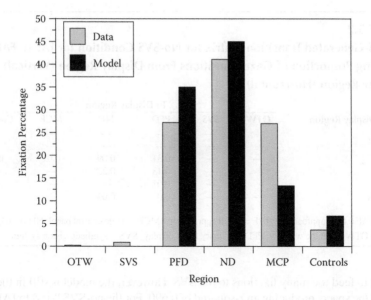

FIGURE 5.10 Model versus data overall attention allocation for "start to IAF" without SVS.

same conditions. The model explains nearly 80% of the variance in the data with an r-squared of 0.772.

For IAF to FAF with SVS, the model does not do quite as well. The ND and PFD displays were well fit, but the model was not as accurate with the SVS. In particular, the SVS tended to send too many fixations off to other displays and the other displays

TABLE 5.1

Human Transition Matrix for No-SVS Condition for IAF to FAF, Showing Proportion of Gaze Transitions From Display Region (Vertical) to Display Region (Horizontal)

From Display Region	To Display Region							
	Off	OTW	SVS	PFD	ND	MCP	Controls	Overlap
Off	0.03	—	—	0.01	0.02	0.01	0.01	—
OTW	—	—	—	—	—	—	—	—
SVS	—	—	—	—	—	—	—	—
PFD	0.01	—	—	0.22	0.11	—	—	—
ND	0.02	—	—	0.11	0.32	—	0.01	—
MCP	—	—	—	—	—	0.03	—	—
Controls	0.01	—	—	0.01	0.01	—	0.05	—
Overlap	—	—	—	—	—	—	—	—

Notes: FAF = final approach fix; IAF = initial approach fix; MCP = mode control panel; ND = navigation display; OTW = out the window; PFD = primary flight display; SVS = synthetic vision system.

TABLE 5.2

Model-Generated Transition Matrix for No-SVS Condition for IAF to FAF, Showing Proportion of Gaze Transitions From Display Region (Vertical) to a Display Region (Horizontal)

From Display Region	To Display Region					
	OTW	SVS	PFD	ND	MCP	Controls
OTW	—	—	—	—	—	—
SVS	—	—	—	—	—	—
PFD	—	—	0.31	0.09	0.01	0.11
ND	—	—	0.13	0.22	—	—
MCP	—	—	0.01	—	—	—
Controls	—	—	0.07	0.04	—	—

Notes: FAF = final approach fix; IAF = initial approach fix; MCP = mode control panel; ND = navigation display; OTW = out the window; PFD = primary flight display; SVS = synthetic vision system.

tended to feed too many fixations to the SVS. However, the model is still in the right part of the space, producing an r-squared of 0.690. For the no-SVS "start to IAF," the match to empirical data is fairly reasonable. The model also favors the diagonal and matches the ND and PFD entries quite well. The model explains almost 80% of the variance in the transition matrix, r-squared 0.792. Just as with the overall attention allocation, the model does not do quite as well in the SVS condition with the "start to IAF" phase. The model does not even explain half the variance in the data, r-squared 0.419. Given the other results obtained, we suspect this is the result of either an aberrant model run or an aberrant pilot performance, though it could instead indicate a flaw in the model. Overall, though, the model's ability to predict pilots' attention allocation at both the molar level and the detailed level is good, though there is certainly still room for improvement.

Discussion of Model Results

In general, model results were satisfactory. While the fits are not perfect, the majority of the variance is explained not only in overall attention allocation, but also in terms of the transitions that underlie the more global behaviors. This was done with a bare minimum of numerical parameter fitting, meaning these fits have credibility as predictions. Just as importantly, the model provides some explanation for why the data are as they are. Pilots use the SVS at the rate they do, at least in the early phases of flight, entirely as a result of the symbology overlaid on the SVS. The symbology is the *sole* reason the model looks at the SVS. This leads to what we believe is one of our most important take-home messages for SVS design: The overlaid symbology matters, and may matter a great deal.

Ultimately, someone will have to decide for SVS systems deployed in future aircraft what to overlay. We believe it would be egregiously bad for the content and visual properties of the overlay to be lightly considered, as it can have a tremendous impact on pilots' attention allocation. We spent some time exploring the effects of varying the symbology on the model's allocation performance, but it was difficult to know in advance which pieces of information would have the most impact. To our surprise, the addition or removal of single indicators (e.g., altitude) did not greatly impact the model's performance. The degree to which models such as ours could be used to inform decisions about which indicators should go where is not yet clear, but we believe we have taken the first steps in that direction.

More generally, we believe we have developed a model that is dynamically sensitive to both top-down (i.e., task structure and information-seeking goals) and bottom-up (i.e., parameters of visual attention and structure of the information environment) constraints. This is an important insight in and of itself; one cannot predict performance by looking just at a system interface or by considering just the task structure. Integrated, closed-loop computational cognitive modeling seems necessary to reach such goals.

CONCLUSIONS

Along with our colleagues reporting their modeling research in this volume, we hope we have demonstrated that computational cognitive modeling, suitably extended and amended to handle the complexities of operational environments, has a promising future as an engineering tool for the analysis, design, and evaluation of human–machine systems. Just as importantly, our chapter should also indicate how far we have to go before this sort of modeling can reliably be put in the hands of professional practitioners. Compromises had to be made, extensions had to be improvised, and some ad hoc assumptions made simply to close the loop on a computational modeling approach that has, until very recently, been applied primarily to static and isolable tasks characteristic of the experimental psychology laboratory. As these modeling techniques mature, we certainly should expect even more rigor in model validation than could be provided in this effort, motivated largely to push the boundaries and test the limits of this style of modeling.

That said, no science has yet sustained itself without spawning a socially relevant and valuable branch of engineering or technology, and we expect that cognitive science will be no different in this regard. As such, we invite those cognitive scientists interested in furthering the development of cognitive architectures or modeling techniques to join us in this venture of taking, head on, the complexities of modeling human cognition and behavior in contexts that are at once scientifically challenging and also of enormous practical consequence.

ACKNOWLEDGMENTS

The authors would like to acknowledge the following people who played an essential role in conducting the research reported here. They include David Huss and Chris Fick of Rice University; and Sarah Miller, Alex Kosorukoff, Ruei-Sung Lin, Sam Zheng, and Piotr Adamczyk of the University of Illinois at Urbana-Champaign. We could never have performed this research without their hard work and dedication to this project in the realms of ACT-R modeling, computer networking and data communications, laboratory task design and programming, reviewing the literature, experimental design, data collection, eye-movement data coding and analysis, statistical modeling and analysis, and decision modeling. The authors would also like to thank the editors of this volume for their generous support and assistance throughout this research.

REFERENCES

Anderson, J. R. (1990). *The adaptive character of thought*. Hillsdale, NJ: Lawrence Erlbaum Associates.

Anderson, J. R., Bothell, D., Byrne, M. D., Douglass, S., Lebiere, C., & Quin, Y. (2004). An integrated theory of the mind. *Psychological Review, 111*, 1036–1060.

Anderson, J. R., & Lebiere, C. (1998). *The atomic components of thought*. Mahwah, NJ: Lawrence Erlbaum Associates.

Baron, S. (1984). A control theoretic approach to modeling human supervisory control of dynamic systems. In W. B. Rouse (Ed.), *Advances in man–machine systems research* (Vol. 2, pp. 1–47). Greenwich, CT: JAI Press.

Bruner, J. S. (1957). Going beyond the information given. In J. S. Bruner, E. Brunswik, L. Festinger, F. Heider, K. F. Muenzinger, C. E. Osgood, & D. Rapaport (Eds.), *Contemporary approaches to cognition* (pp. 41–69). Cambridge, MA: Harvard University Press. Reprinted in Bruner, J. S. (1973). *Beyond the information given* (pp. 218–238). New York: Norton.

Byrne, M. D. (2003). Cognitive architecture. In J. A. Jacko & A. Sears (Eds.), *The human–computer interaction handbook: Fundamentals, evolving technologies and emerging applications* (pp. 97–117). Mahwah, NJ: Lawrence Erlbaum Associates.

Byrne, M. D., & Gray, W. D. (2003). Returning human factors to an engineering discipline: Expanding the science base through a new generation of quantitative methods— Preface to the special section. *Human Factors, 45*, 1–4.

Byrne, M. D., & Kirlik, A. (2004). Integrated modeling of cognition and the information environment: A multilevel investigation (process and product modeling) of attention allocation to visual displays (Tech. rep. AHFD-04-14/NASA-04-4). Savoy: University of Illinois, Human Factors Division.

Byrne, M. D., & Kirlik, A. (2005). Using computational cognitive modeling to diagnose possible sources of aviation error. *International Journal of Aviation Psychology, 15*, 135–155.

Cassell, R., Smith, A., & Hicok, D. (1999). *Development of airport surface required navigation performance (RNP)* (NASA/CR-1999-20910).

Cheng, V. H. L., Sharma, V., & Foyle, D. C. (2001). A study of aircraft taxi performance for enhancing airport surface traffic control. *IEEE Transactions on Intelligent Transportation Systems, 2*(2), 39–54.

Fitts, P. M. (1954). The information capacity of the human motor system in controlling the amplitude of movement. *Journal of Experimental Psychology, 47*, 381–391.

Gigerenzer, G., & Goldstein, D. G. (1996). Reasoning the fast and frugal way: Models of bounded rationality. *Psychological Review, 103*, 650–669.

Gluck, K. A., & Pew, R. W. (2005). *Modeling human behavior with integrated cognitive architectures: Comparison, evaluation*. Mahwah, NJ: Lawrence Erlbaum Associates.

Goodman, A. (2001). *Taxi navigation errors: Taxiway geometry analyses*. Unpublished report.

Hammond, K. R., & Stewart, T. R. (2001). *The essential Brunswik: Beginnings, explications, applications*. New York: Oxford University Press.

Hooey, B. L., & Foyle, D. C. (2006). Pilot navigation errors on the airport surface: Identifying contributing factors and mitigating solutions. *International Journal of Aviation Psychology, 16*, 51–76.

Hooey, B. L., Foyle, D. C., & Andre, A. D. (2000). Integration of cockpit displays for surface operations: The final stage of a human-centered design approach. *SAE Transactions: Journal of Aerospace, 109*, 1053–1065.

Hutchins, E. (1995). *Cognition in the wild*. Cambridge: MIT Press.

Jagacinski, R., & Flach, J. (2002) *Control theory for humans*. Mahwah, NJ: Lawrence Erlbaum Associates.

Keller, J. W., Leiden, K., & Small, R. (2003). Cognitive task analysis of commercial jet aircraft pilots during instrument approaches for baseline and synthetic vision displays. In D. C. Foyle, A. Goodman, & B. L. Hooey (Eds.), *Proceedings of the 2003 NASA Aviation Safety Program Conference on Human Performance Modeling of Approach and Landing with Augmented Displays* (NASA/CP-2003-212267). Moffett Field, CA: NASA Ames Research Center.

Kirlik, A. (2003). Human factors distributes its workload. Review of E. Salas (Ed.), *Advances in human performance and cognitive engineering research,* Vol. 1. *Contemporary Psychology, 48*(6), 766–769.

Kirlik, A. (2006). *Adaptive perspectives on human–technology interaction: Methods and models for cognitive engineering and human–computer interaction*. New York: Oxford University Press.

Kirlik, A., & Bisantz, A. (1998). Cognition in human–machine systems: Experiential and environmental components of adaptation. In P. Hancock (Ed.), *Human performance and ergonomics* (pp. 47–68). New York: Academic Press.

Kirlik, A., Miller, R. A., & Jagacinski, R. J. (1993). Supervisory control in a dynamic uncertain environment: A process model of skilled human–environment interaction. *IEEE Transactions on Systems, Man, and Cybernetics, 23*(4), 929–952.

Klein, G. (1989). Recognition-primed decisions. In W. B. Rouse (Ed.), *Advances in man–machine systems research* (Vol. 5, pp. 47–92). Greenwich, CT: JAI Press.

MacKenzie, I. S. (2003). Motor behavior models for human–computer interaction. In J. M. Carroll (Ed.), *Toward a multidisciplinary science of human–computer interaction* (pp. 27–54). San Francisco, CA: Morgan Kaufmann.

Payne, J. W., & Bettman, J. (2001). Preferential choice and adaptive strategy use. In G. Gigerenzer & R. Selten (Eds.), *Bounded rationality: The adaptive toolbox* (pp. 123–146). Cambridge, MA: MIT Press.

Rasmussen, J. (1976). Outlines of a hybrid model of the process plant operator. In T. B. Sheridan & G. Johannsen (Eds.), *Monitoring behavior and supervisory control* (pp. 371–383). New York: Plenum Press.

Rasmussen, J. (1983). Skills, rules, and knowledge: Signals, signs, and symbols, and other distinctions in human performance models. *IEEE Transactions on Systems, Man, and Cybernetics, SMC-13,* 257–266.

Rouse, W. B. (1984), *Advances in man–machine systems research* (Vol. 2). Greenwich, CT: JAI Press.

Salvucci, D. D. (2001). Predicting the effects of in-car interface use on driver performance: An integrated model approach. *International Journal of Human–Computer Studies, 55,* 85–107.

Senders, J. (1964). The human operator as a monitor and controller of multidegree of freedom systems. *IEEE Transactions on Human Factors in Electronics, HFE-5,* 2–6.

Sheridan, T. B. (1976). Toward a general model of supervisory control. In T. B. Sheridan & G. Johansen (Eds.), *Monitoring behavior and supervisory control* (pp. 271–282). New York: Plenum Press.

Sheridan, T. B., & Johannsen, G. (1976). *Monitoring behavior and supervisory control*. New York: Plenum Press.

Tolman, E. C., & Brunswik, E. (1935). The organism and the causal texture of the environment. *Psychological Review, 42,* 43–77.

Wickens, C. D. (2002). Multiple resources and performance prediction. *Theoretical Issues in Ergonomic Science, 3*(2), 159–177.

6 Modeling Pilot Performance with an Integrated Task Network and Cognitive Architecture Approach

Christian Lebiere, Rick Archer, Brad Best, and Dan Schunk

CONTENTS

INTRODUCTION

This chapter describes the approach for modeling pilot error in the National Aeronautic and Space Administration Human Performance Modeling (NASA HPM) project adopted by the Micro Analysis and Design (MAAD) team. The approach consists of using an integration of a task network modeling architecture and a cognitive architecture to model the state of the environment and the actions of the pilots as they operate the commercial aircraft. The primary benefit of this effort is in taking advantage of the strengths of each tool in order to increase the fidelity of human performance models, without unnecessarily burdening the model developer. An additional benefit is in introducing a clean practical and conceptual separation between models of the environment and of the pilots preventing unwanted and erroneous dependencies and emphasizing flows of information. Additionally, this project exploits the synergy between the computer science and cognitive science communities. This synergy will promote the advancement of human performance modeling approaches and tools through the selective application of computational cognitive modeling technology.

Our modeling strategy throughout this project is to emphasize the predictive ability of our model, in particular the ACT-R cognitive model. This strategy manifests itself in a number of ways. First, we follow a path of starting with the simplest, most constrained model possible and only incrementally add complexity when clearly necessary. This enables us to understand clearly which part or mechanism of the model is responsible for which aspect of the behavioral data. Second, we did not resort to parameter fitting but instead use default parameter values and leverage architectural constraints and mechanisms to try to predict the data rather than hardwire it into the model. Third, we only included in the knowledge constituting the model information that was clearly available to the pilots, such as specified directly in procedural lists or flight manuals. The result is a modeling approach that attempts to explain, understand, and predict human performance rather than replicate it.

IMPRINT

IMPRINT (Improved Performance Research Integration Tool; Archer & Adkins, 1999) is a task network modeling tool that was developed for the Army Research Laboratory–Human Research and Engineering Directorate (ARL HRED) by MAAD to conduct human performance analyses very early in the acquisition of a proposed new system. It consists of a set of automated aids to assist researchers in conducting human performance analyses.

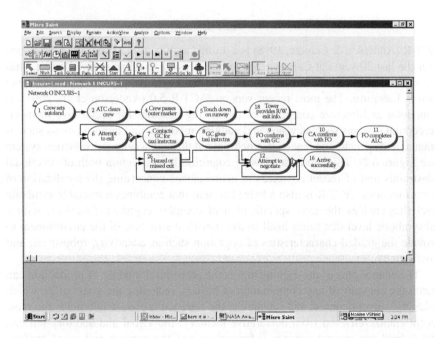

FIGURE 6.1 IMPRINT task network.

IMPRINT assists a user in estimating the likely performance of a new system by facilitating the construction of flow models that describe the scenario, the environment, and the mission that must be accomplished (see Figure 6.1). Since it is typically easier to describe the mission by breaking it into smaller "sub" functions than trying to describe the mission as a whole, users build these models by breaking down the mission into a network of functions, reflecting the decomposition of complex tasks into increasingly simpler ones. Each of the functions is then further broken down into a network consisting of other functions and tasks. Then, a user estimates the time it will take to perform each task and the likelihood that it will be performed accurately. Finally, by executing a simulation represented by the flow model of the mission multiple times, the user can study the range of results that occur. A description of the variability of each element can be obtained for further analysis. Additionally, at the completion of the simulation, IMPRINT can compare the minimum acceptable mission performance time and accuracy to the predicted performance. This will determine whether the mission met its performance requirements.

IMPRINT has been used successfully to predict human performance in complex and dynamic operational environments (Archer & Allender, 2001; Archer & LaVine, 2000). It has been shown to be easy to use and fairly quick to apply. However, it does not include an embedded model of cognitive or psychological processes. Rather, it relies on the modeler to specify and implement these constructs. In our approach for this effort, these constructs will be provided through a link to the Adaptive Control of Thought-Rational (ACT-R) cognitive architecture.

ACT-R

ACT-R (Anderson & Lebiere, 1998) is a unified architecture of cognition developed over the last 30 years at Carnegie Mellon University. At a fine-grained scale, it has accounted for hundreds of phenomena from the cognitive psychology and human factors literature. The most recent version, ACT-R 5.0 (Anderson et al., 2004), is a modular architecture composed of interacting modules for declarative memory, perceptual systems such as vision and audition modules, and motor systems such as manual and speech modules—all synchronized through a central production system (see Figure 6.2). This modular view of cognition is a reflection both of functional constraints and of recent advances in neuroscience concerning the localization of brain functions. ACT-R is also a hybrid system that combines a tractable symbolic level that enables the easy specification of complex cognitive functions, with a subsymbolic level that tunes itself to the statistical structure of the environment to provide the graded characteristics of cognition such as adaptivity, robustness, and stochasticity.

The central part of the architecture is the procedural module. A production can match the contents of any combination of buffers, including the goal buffer, which holds the current context and intentions; the retrieval buffer, which holds the most recent chunk retrieved from declarative memory; the visual and auditory buffers, which hold the current sensory information; and the manual and vocal buffers, which hold the current state of the motor and speech module. The highest rated

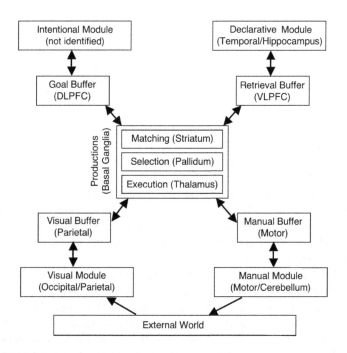

FIGURE 6.2 The overall flow of control in ACT-R.

matching production is selected to effect a change in one or more buffers, which in turn trigger an action in the corresponding modules. This can be an external action (e.g., movement) or an internal action (e.g., requesting information from memory). Retrieval from memory is initiated by a production specifying a pattern for matching in declarative memory. Each chunk competes for retrieval, with the most active chunk being selected and returned in the retrieval buffer. The activation of a chunk is a function of its past frequency and recency of use, the degree to which it matches the requested pattern, plus stochastic noise. Those factors confer memory retrievals and behavior in general, desirable "soft" properties such as adaptivity to changing circumstances, generalization to similar situations, and variability (Anderson & Lebiere, 1998).

The current goal is a central concept in ACT-R, which as a result provides strong support for goal-directed behavior. However, the most recent version of the architecture is less goal-focused than its predecessors by allowing productions to match to any source of information, including the current goal, information retrieved from declarative memory, objects in the focus of attention of the perceptual modules, and the state of the action modules. The content of many of those buffers, especially the perceptual buffers, might have changed—not as a function of an internal request, but rather as a result of an external event happening, perhaps unexpectedly, in the outside world. This emphasis on asynchronous pattern matching of a wide variety of information sources better enables ACT-R to operate and react efficiently in a dynamic, fast-changing world through flexible, goal-directed behavior that gives equal weight to internal and external sources of information.

There are three main distinctions in the ACT-R architecture. First, there is the procedural-declarative distinction that specifies two types of knowledge structures: chunks for representing declarative knowledge and productions for representing procedural knowledge. Second, there is the symbolic level, which contains the declarative and procedural knowledge, and the subsymbolic level of neural activation processes that determine the speed and success of access to chunks and productions. Finally, there is a distinction between the performance processes by which the symbolic and subsymbolic layers map onto behavior and the learning processes by which these layers change with experience.

Human cognition can be characterized as having two principal components: (1) the knowledge and procedures codified through specific training within the domain, and (2) the natural cognitive abilities that manifest themselves in tasks as diverse as memory, reasoning, planning, and learning. The fundamental advantage of an integrated architecture like ACT-R is that it provides a framework for modeling basic human cognition and integrating it with specific symbolic domain knowledge of the type specified by domain experts (e.g., rules specifying what to do in a given condition, a type of knowledge particularly well suited for representation as production rules).

However, performance described by symbolic knowledge is mediated by parameters at the subsymbolic level that determine the availability and applicability of symbolic knowledge. Those parameters underlie ACT-R's theory of memory—providing effects such as decay, priming, and strengthening—and make cognition adaptive, stochastic, and approximate, capable of generalization to new situations and robustness in the face of uncertainty. They also can account for the limitations of human performance, such as latencies to perform tasks and errors that can originate from a number of sources. Finally, they provide a basis for representing individual differences such as those in working memory capacity, attentional focus, motivation, and psychomotor speed as well as the impact of external behavior moderators such as fatigue (Lovett, Reder, & Lebiere, 1999; Taatgen, 2001) through continuous variations of those subsymbolic architectural parameters that affect performance in complex tasks.

The subsymbolic level in which continuously varying quantities are processed, often in parallel, to produce much of the qualitative structure of human cognition also determines the quantitative predictions of performance. These subsymbolic quantities participate in neural-like activation processes that determine the speed and success of access to chunks in declarative memory as well as the conflict resolution among production rules. ACT-R also has a set of learning processes that can modify these subsymbolic quantities. Similar computations are at work in other modules, such as the perceptual–motor modules. Especially important are the parameters controlling the time course of processing as one attempts to execute a complex action, or as one shifts visual attention to encode a new stimuli (Byrne & Anderson, 2001). ACT-R can predict direct quantitative measures of performance such as latency and probability of errors; from the same mechanistic basis can also arise more global, indirect measures of performance such as cognitive workload (Lebiere, 2001).

IMPRINT AND ACT-R INTEGRATION

Because ACT-R and IMPRINT were targeted at different behavioral levels, they perfectly complement each other. IMPRINT is focused on the task level—how high-level functions break down into smaller-scale tasks, and the logic by which those tasks follow each other to accomplish those functions. ACT-R is targeted at the "atomic" level of thought—the individual cognitive, perceptual and motor acts that take place at the subsecond level. As seen in Figure 6.2, ACT-R is centered on the concept of the current goal. At each cycle, a production will be chosen that best applies to the goal and other buffers such as the contents of memory retrieval or visual perception, then it performs a number of internal or external actions such as memory retrieval or manual movement, and almost always modifies the current goal. Those cycles will repeat until the current goal is solved, at which point it is removed and another one is selected.

The ACT-R theory specifies in detail the performance and learning that takes place at each cycle within a specific goal, but has comparatively little to say about the selection of those goals. Since goals in ACT-R closely correspond to tasks in IMPRINT, that weakness matches perfectly IMPRINT's strength. Conversely, since IMPRINT requires the characteristics of each task to be specified as part of the model, ACT-R can be used to generate those detailed characteristics in a psychologically plausible way without requiring extensive data collection.

An IMPRINT model specifies the network of tasks used to accomplish the functions targeted by the model (e.g., landing a plane and taxiing safely to the gate). The network specifies how higher order functions are decomposed into tasks and the logic by which these tasks are composed together. As input, it takes the distribution of times to complete the task and the accuracy with which the task is completed. It can also take as input the workload generated by each task. Additional inputs include events generated by the simulation environment. Finally, a number of additional general parameters such as personnel characteristics, level of training or familiarity, and environmental stressors can be specified. IMPRINT specifies the performance function by which these parameters modulate human performance. The outputs include mission performance data such as time and accuracy as well as aggregate workload data.

An ACT-R model specifies the knowledge structures such as declarative chunks and production rules that constitute the user knowledge relevant to the tasks targeted by the model. It also specifies the goal structures reflecting the task structure and the architectural and prior knowledge parameters that modulate the model's performance. For each goal on which ACT-R is focused, it generates a series of subsecond cognitive, perceptual, and motor actions. The result of those actions is the total time to accomplish the goal, as well as how the goal was accomplished, including any error that might result. Errors in ACT-R originate from a broad range of sources. They include memory failures (including the failure to retrieve a needed piece of information or the retrieval of the wrong piece of information), choice failures (including the selection of the wrong production rule), and attentional failures (such as the failure to detect the salient piece of information by the perceptual modules).

While those errors could arise because of faulty symbolic knowledge (either declarative or procedural), this is often not the case, especially in domains that involve highly trained crews, such as airline pilots. More often, those errors occur because the subsymbolic parameters associated with chunks or productions do not allow the model to access them reliably or quickly enough to be deployed in the proper situation. Moreover, because those parameters vary stochastically and their effect is amplified by the interaction with a dynamic environment, those times and errors will not be deterministic but will vary with each execution, as is the case for human operators. Thus, the ACT-R model for a particular goal can be run whenever IMPRINT selects the corresponding task to generate the time and error distribution for that task in a manner that reflects the myriad cognitive, perceptual, and motor factors that enter into the actual performance of the task. ACT-R can also generate workload estimates for each task that reflect the cognitive demands of the actions taken to perform the task (Lebiere, 2001, 2005).

HUMAN PERFORMANCE MODEL
OF PILOT NAVIGATION WHILE TAXIING

As a practical application of the IMPRINT and ACT-R integration, a complex and dynamic task such as the development of models of pilot navigation while taxiing from a runway to a gate seemed to be a good candidate because it is a task that consists of numerous procedures and memory requirements that make it somewhat prone to errors, thus involving aspects of both paradigms. Research on pilot surface operations has shown that pilots can commit numerous errors during taxi procedures (Goodman et al., 2003; Hooey & Foyle, 2006). NASA was hoping to reduce the number and scope of pilot errors during surface operations by using information displays that would improve the pilots' overall situation awareness (SA). Our goal for this effort was to try to determine what sensitivities to error in the taxiing task could be uncovered by the various modules in the ACT-R architecture when coupled to the IMPRINT model.

As described in chapter 3, NASA researchers provided the modeling teams with data describing pilot procedures during prelanding and surface (taxi) operations. The IMPRINT and ACT-R modeling team used the scaled map of Chicago's O'Hare Airport (KORD), described in chapter 3, to estimate the time between taxiway turns. IMPRINT handled the higher level, task-oriented parts of the taxiing and landing operations (i.e., turning, talking on radio, looking at instrumentation), while ACT-R handled the more cognitive and decision-making parts of the task (i.e., remembering where to turn and remembering the taxi route). By using the scaled map of the airport, we were able to determine the amount of time between each taxi turn (based on speed estimated from the video tape data) and then use these data to estimate the decay rate for the list of memory elements (i.e., runway names) that the pilot would have to remember. It appeared from the tapes that the pilots did not write down the clearance.

Using this integrated architecture allowed us to be able to represent a complex, dynamic task by using the strengths of each architecture. The IMPRINT model allowed us to represent with a minimum of complexity the key aspects of the environment and the decomposition of the mission into its component tasks. The ACT-R model, which can be complicated to create and time-consuming to debug and fine-tune, could then focus on the cognitively essential parts of the mission. The various assumptions in each model were segregated by the separation between the task network and cognitive model, and the flow of information between them was precisely specified.

THE IMPRINT MODEL

The IMPRINT task network model consists of the tasks that the captain and the first officer (FO) perform from the time that the airplane approaches the airport from about 12 nautical miles (nmi) out until the airplane either commits a taxi navigation error or arrives at the correct terminal gate without committing an error. The tasks in the model are grouped into three general segments. Prior to the execution of the first segment, an initialization task communicates all of the correct information about where the airplane will be directed to land and the taxi route information to the gate destination.

The first segment is the approach, which begins with the tasks that the captain and FO must perform in preparation for landing. It is during this approach segment that air traffic control (ATC) communicates runway landing information to the crew.

Next, in the "rollout" segment, the crew lands the aircraft and proceeds down the runway until the preferred runway turnoff has been reached. After this first turnoff is performed, the taxi route and gate information is communicated to the crew by the control tower. However, it is often given during the runway rollout segment.

As the crewmembers observe their taxi chart and communicate with one another, this information is shared with the ACT-R model via shared variables using the Microsoft component object model (COM) functionality. As the aircraft approaches each potential runway turnoff, information about whether runway signage has been noticed or communicated is also passed to the ACT-R model. In turn, ACT-R passes information back to IMPRINT about whether to make a turn and in which direction to turn.

After turning off the runway the "taxi in" segment takes the aircraft to the gate. Information about taxiway signage, the crew's use of displays, and their communication is passed to the ACT-R model. ACT-R again passes information back to IMPRINT as to whether a turn should be made, whether the aircraft should stop and wait, or whether it should proceed to the next taxiway intersection.

Aside from the normal procedures that are being performed by the crew, other events such as radio communications or taxiway traffic can occur. If the pilot makes an error in which turn to make or in which direction to turn, the simulation model is terminated. The model was executed repeatedily to predict the likelihood of navigation errors.

THE ACT-R MODEL

In the spirit of concentrating on the areas where cognitive accuracy is most critical, the ACT-R model focused on the task of memorizing and recalling the list of taxiways to follow after landing. This task depends fundamentally upon the detailed workings of the declarative memory module, and we will expand here in more detail upon the subsymbolic processes that underlie memory performance.

A central concept in the storage and retrieval of chunks of information in memory is that of activation. Formally, activation reflects the log posterior odds that a chunk is relevant in a particular situation. The activation A_i of a chunk i is computed as the sum of its base-level activation B_i plus its context activation:

$$A_i = B_i + \sum_j W_j S_{ji} \qquad \text{Activation equation}$$

In determining the context activation, W_j designates the attentional weight given the focus element j. An element j is in the focus, or in context, if it is part of the current goal chunk (i.e., the value of one of the goal chunk's slots). S_{ji} stands for the strength of association from element j to a chunk i. ACT-R assumes that there is a limited capacity of source activation and that each goal element emits an equal amount of activation. Source activation capacity is typically assumed to be 1 (i.e., if there are n source elements in the current focus, each receives a source activation of $1/n$). The associative strength S_{ji} between an activation source j and a chunk i is a measure of how often i was needed (i.e., retrieved in a production) when chunk j was in the context. Associative strengths provide an estimate of the log likelihood ratio measure of how much the presence of a cue j in a goal slot increases the probability that a particular chunk i is needed for retrieval to instantiate a production. The base level activation of a chunk is learned by an architectural mechanism to reflect the past history of use of a chunk i:

$$B_i = \ln \sum_{j=1}^{n} t_j^{-d} \approx \ln \frac{nL^{-d}}{1-d} \qquad \text{Base-level learning equation}$$

In the preceding formula, t_j stands for the time elapsed since the jth reference to chunk i, while d is the memory decay rate and L denotes the lifetime of a chunk (i.e., the time since its creation). As Anderson and Schooler (1991) have shown, this equation produces the power law of forgetting (Rubin & Wenzel, 1990) as well as the power law of learning (Newell & Rosenbloom, 1981). When retrieving a chunk to instantiate a production, ACT-R selects the chunk with the highest activation A_i. However, some stochasticity is introduced in the system by adding Gaussian noise of mean 0 and standard deviation σ to the activation A_i of each chunk. In order to be retrieved, the activation of a chunk needs to reach a fixed retrieval threshold σ that limits the accessibility of declarative elements. If the Gaussian noise is approximated with a sigmoid distribution, the probability P of chunk i to be retrieved by a production is:

$$P = \frac{1}{1+e^{-\frac{A_i-\tau}{s}}} \qquad \text{Retrieval probability equation}$$

where $s = \sqrt{3}\sigma/\pi$. The activation of a chunk i is directly related to the latency of its retrieval by a production p. Formally, retrieval time T_{ip} is an exponentially decreasing function of the chunk's activation A_i:

$$T_{ip} = Fe^{-fA_i} \qquad \text{Retrieval time equation}$$

where F is a time scaling factor.

In addition to the latencies for chunk retrieval as given by the retrieval time equation, the total time of selecting and applying a production is determined by executing the actions of a production's action part, whereby a value of 50 ms is typically assumed for elementary internal actions. External actions, such as pressing a key, usually have a longer latency as determined by the perceptual–motor modules. In summary, subsymbolic activation processes in ACT-R make a chunk active to the degree that past experience and the present context (as given by the current goal) indicate that it is useful at this particular moment.

This task of learning and retrieving a sequence of taxiway turns is similar to the cognitive psychology task of list learning, for which an ACT-R model had already been developed (Anderson et al., 1998). We adapted that model to the task at hand while preserving its fundamental representation and parameters, thus eliminating degrees of freedom and inheriting that model's empirical validation. The taxiway turns were represented by two chunks each, one indicating the name of the taxiway and the other holding the direction to turn. While the taxi clearances did not contain the direction information, we assumed that the pilot disambiguated the direction from a map of the airport based on subsequent turns. This cognitive requirement at a critical time could be avoided by having the ATC communicate the turn direction directly rather than implicitly. The ACT-R model was called for the initial memorization of the list of taxiways and then each time the aircraft approached a taxiway intersection.

RESULTS

The model could reproduce a wide range of errors observed in human pilots. Omission errors occurred when a chunk holding a turn could not be recalled because of time-based decay or activation noise. The resulting error would be a missed turn. Two kinds of commission errors could occur. The first kind would result in the wrong chunk recalled because of interference, similarity-based partial matching, priming, or activation noise. This would cause the model to schedule a turn on the wrong taxiway. The second kind of commission error would happen when the wrong direction chunk was retrieved, again for reasons of interference or noise. This would result in a turn on the correct taxiway but in the wrong direction. A small sample of five model runs produced a diversity of outcomes. Two runs produced all the correct turns; two runs ended with a missed turn and one run ended with a turn in the wrong direction. While a larger sample of runs could be collected and analyzed, this indicates that our model can capture the range of real-world outcomes.

Clearly, we focused here on only one possible source of errors—specifically, memory errors that might be present in the human-in-the-loop data. Other possible sources of errors include perceptual inaccuracies (especially in limited-visibility environments), loss of SA, and pilot distraction. Most or all of those could be modeled in a principled approach in the ACT-R architecture if enough data were available. However, as one tries to capture more factors, the model grows more complex and degrees of freedom start multiplying, unless experimental conditions provide enough accuracy to constrain the model. To provide a maximally constrained model, our approach in this case focused on one key cognitive component of the task—the memorization and recall of taxiing directions—and leveraged existing, validated models to provide a priori predictions of task behavior.

HUMAN PERFORMANCE MODEL OF APPROACH AND LANDING

The focus of the second part of the project is on commercial aircraft approach and landing scenarios and the tasks the pilots must perform. The modeling compares pilot procedures using current cockpit displays for the Boeing 757 with procedures using augmented displays such as a synthetic vision system (SVS).

Data to characterize the performance of pilots during four different approach and landing scenarios were collected by NASA via pilot participation in a part-task simulation study (as presented in chapter 3). The four scenarios were: (1) a vectored approach, (2) a late reassignment of runways for the landing, (3) a missed approach, and (4) a mismatch of the terrain shown in the SVS and the out-the-window (OTW) view from the cockpit.

The method that was used to perform the approach and landing modeling was a similar integration of IMPRINT and ACT-R as was used for the pilot taxi model. There were several sources of data for the modeling effort. The primary data source was a cognitive task analysis (CTA) that was conducted as a separate effort in support of all of the modeling teams (see chapter 3). The CTA was supplemented by a number of published papers and other background documents. In addition, video-tapes of pilots flying approach and landing tasks in a NASA part-task simulator were provided.

The result of this phase is a simulation model of an aircraft making its final approach and landing into an airport. The model built represents an aircraft and its environment as well as the pilots operating the aircraft. For the simulation of the aircraft, IMPRINT represents the autopilot as well as the physics of the aircraft. These aspects include the aircraft's location in time and space, its deceleration, its descent rate, and all physical changes in the aircraft, including its landing gear, flap settings, and air brakes. The model also includes the controls and displays of the aircraft, including all autopilot functions. Represented in the model are the mode control panels (MCP), the primary flight display (PFD), the navigation display (ND), and an OTW view. The model also handles all communication between the aircraft and ATC. With these controls and displays, the model is able to simulate how a plane will react in its environment when these controls and displays are manipulated. Currently, the simulation is set up to work with a vertical navigation (VNAV) path auto-pilot setting, but the model is capable of utilizing the other types of autopilot (e.g., glideslope and localizer) for future analysis. The simulation model will terminate when the pilot switches off the autopilot for the manual portion of the landing.

In the simulation, a model of the pilot was developed using the ACT-R cognitive architecture. Following the practice of decomposing complex behavior into a set of unit tasks, the ACT-R model is composed of a set of goals, together with the proce-dural and declarative knowledge necessary to meet those goals. The top-level goal is essentially a monitoring loop that repeatedly sets subgoals to check the settings of the various controls. Each of these subgoals typically requires acquiring the value of one or more environmental variables (e.g., speed, altitude, etc.) by reading the instru-ments or looking OTW. A decision is then made as to what the desired control value is, given those readings. If that value is different from the current control value, the appropriate action is performed to change that value.

Decisions are made using either declarative or procedural means. For procedural control, a production rule is applied that supplies the control value given the envi-ronmental readings. This type of decision best captures crisp, symbolic decisions relying on precise values provided by instruments (e.g., "set flaps to 15 when speed is 200 knots"). For declarative control, instances are defined in declarative memory linking environmental readings to control values. Given a particular condition, the most relevant instance is retrieved from memory using a similarity-based partial matching mechanism, and the control value is extracted from it. Multiple memory instances can also be retrieved using a mechanism called blending (Lebiere, 1999) and a consensus control value extracted that best satisfies the set of instances. This control is similar to that provided by neural networks and best describes approxi-mate, iterative adjustments as practiced in OTW flying.

THE IMPRINT MODEL

The backbone of the IMPRINT simulation is a series of networks representing the different aspects of the simulation (see Figure 6.3). When the IMPRINT simulation is started, the model starts in network 2, the start/end landing network. All of the

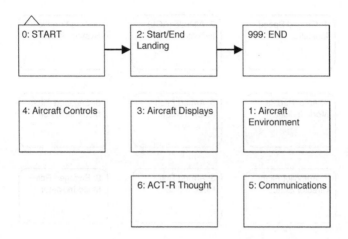

FIGURE 6.3 Main IMPRINT network.

initial communication takes place between ACT-R and IMPRINT in this network. From the start/end landing network, the simulation starts the ACT-R thought network, the aircraft environment subnetwork, and the communication network.

When the communication network is started, IMPRINT will schedule random communication tasks between the captain and FO as well as continually check for scheduled communication between the pilots and ATC. Inside the ACT-R thought network, the main communication between ACT-R and IMPRINT takes place. With this communication, ACT-R decides what controls and displays to look at or manipulate. The aircraft controls and aircraft displays networks represent the various controls and displays of the aircraft. In the aircraft environment network, IMPRINT updates the current state of the aircraft, its controls, its displays, and the environment itself every 0.1 sec. This is a continual action until the end of the simulation.

An aircraft's displays are further broken down in the aircraft displays network (see Figure 6.4). In these subnetworks, the displays are separated into the main displays of the aircraft, specifically the PFD, the ND, and the OTW. It is important to point out that while there is no physical display specifically for OTW, the pilot has the ability to look OTW in order to perform such actions as seeing whether the ground is in sight or what the flap settings currently are. The lowest level of the aircraft displays network is the actual displays represented by tasks in the IMPRINT simulation. Figure 6.5 shows how each part of the PFD is represented by a single task. During the IMPRINT simulation, ACT-R will communicate to IMPRINT

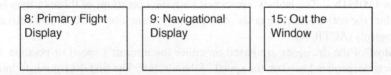

FIGURE 6.4 Aircraft displays subnetwork.

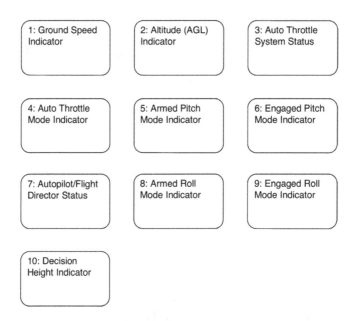

FIGURE 6.5 Primary flight display subnetwork.

the display for which it would like the value. IMPRINT will then start the task for the appropriate display and return that display's current value. The information is then communicated back to the ACT-R thought subnetwork located in the main IMPRINT network (see Figure 6.3).

THE ACT-R MODEL

The ACT-R model cycles through six decisions: setting the flaps, altitude, speed, lowering the landing gear, setting speed brakes, and engaging and disengaging the autopilot. The model keeps cycling through these decisions until it either turns off the autopilot or misses the runway.

The model uses two special-purpose buffers, which mediate between ACT-R and the world (IMPRINT) in a manner similar to ACT-R's perceptual–motor modules. The first buffer is LOOK. If ACT-R wants the value of some instrument or the setting of a dial, it uses the LOOK buffer to pose a query. The ACTION buffer is used to change the setting of a dial. The contents of both of these buffers are communicated to IMPRINT, which then updates the LOOK buffer with the value stored within IMPRINT. The latency associated with the operations of those buffers is set to reflect the real-world time costs of acquiring information (LOOK) and activating the controls (ACTION).

Most of the decisions are based on either the aircraft's speed or position. The set-flaps decision is based on the speed of the aircraft. The first decision is to look at the speed indicator. ACT-R then matches this speed against its speed flaps value in memory and then looks at the current flap setting. If it is the same, it just goes to the next decision—in this case, dial altitude. If it is different, it uses the ACTION buffer to update the flap setting.

The altitude settings depend on which waypoint is next in the approach phase. Once again, ACT-R uses the LOOK buffer to "read" the waypoint indicator. It then "remembers" the correct altitude for the waypoint and, if it is not already set, it communicates to IMPRINT to dial in the new altitude. If it is working correctly, the altitude should only be changed once for a waypoint position.

The speed setting depends on the distance to the airfield. However, there is not a single reading that directly gives this datum. Instead, ACT-R looks at the waypoint and then looks at the distance to waypoint. ACT-R then recalls the range of this waypoint to the airfield and computes the distance to the airfield. After this more complex set of productions, ACT-R then retrieves and, if needed, sets the speed in the same style as the previous decisions.

The other decisions are done in a slightly different style. These are more like rule-based productions. Instead of being based on a set of memorized relations, the productions directly encode the decision point. For example, the abort-landing production checks to see if the distance to the runway is zero. If it is, it signals an abort and clears all of the goals.

The decide-gear production checks if the distance to the runway is less than or equal to 15.0 nmi. If it is, it lowers the landing gear. There is currently no production for raising the gear.

The decision for the speed brakes is to compare the current speed to 145 kts. If the speed is less, the brakes are turned off. If the speed is greater than 145 kts, the brakes are turned on. Like the other decisions, before performing the action a production checks the current setting and performs the action if the decision is not equal to the current setting. Unlike the landing gear, the current model could turn off the speed brakes, but, given the current scenarios, this never happens.

The decision to disengage the autopilot depends on the visibility of the runway. ACT-R uses the OTW values to determine if the runway is visible. If it is, then the next production will disengage the autopilot.

When control returns to the top-level decision cycle, if the autopilot has been disengaged, ACT-R ends the simulation and signals success. If the autopilot is engaged, ACT-R cycles back to the set-flap decision.

A separate production responds to communications. The only effect it has is to take time away from the top-level control loop. This can cause the model to "miss" the runway.

COMMUNICATIONS PROTOCOL

In order to have successful communications between IMPRINT and ACT-R, a protocol was set up so that each program responded correctly to the other program's request. In this project IMPRINT acts as the aircraft and the environment affecting the aircraft. ACT-R acts as the pilot of the aircraft and decides which displays to look at and which controls to manipulate. The first step in creating this protocol was to develop a simple generic protocol that all IMPRINT and ACT-R links will use as a basis for developing intercommunications with each other. This generic protocol has three main areas: starting a model, ACT-R actions to be performed that take time, and ending a model.

When an IMPRINT model that requires communication with an ACT-R model begins executing, an initial communication must take place between ACT-R and IMPRINT. This is done to ensure that both programs are synchronized in all areas of the model. This allows ACT-R to reset its current data as well as check to make sure that both IMPRINT and ACT-R are running the same version of the model.

When a pilot is flying an aircraft, communications between the pilot and ATC will occur at various intervals. Communications fall into three different categories:

- Communications to which the pilot should listen and respond;
- Communications to which the pilot should listen and not respond; and
- Communications that the pilot should ignore.

During the simulation execution, various communications take place between ATC and the pilot. When a communication event occurs, IMPRINT will interact with ACT-R, informing ACT-R that an ATC communication has occurred as well as the type of the communication. ACT-R then makes the decision during the course of the simulation as to if or when it should reply back to ATC in the form of an action.

FINDINGS AND IMPLICATIONS

The model has many potential parameters, but we can aggregate them into five main ones:

- Visual speed;
- Manual speed;
- Communications;
- Decision consistency; and
- Procedural speed.

Each of these parameters corresponds to a separate module in the ACT-R architecture. Variations in the parameters can correspond to individual differences as well as changes in equipment or procedures. Four of the parameters are latencies (in seconds), which represent the time for the pilot to perform procedural, visual, motor, and auditory actions. The other is ACT-R's activation noise value, which is a measure of the stochasticity of the model's decision-making. This noise captures the human decision-making phenomenon of occasionally sampling from lower probability outcomes to maximize eventual learning and keep up with a changing environment. Without activation noise, the model would prematurely settle into erroneous or suboptimal behavior and never explore other possibilities (see Lebiere, 1998).

The first speed parameter is "visual speed," which represents the mean time for the pilot to look at an instrument and perceive its value. The next is "manual speed," which is the mean time it takes to perform an action such as dialing in a new setting. The next is the "procedural speed" parameter that represents the time for a production rule to be selected and fire. The last is "aural speed," which represents the time that the pilot spends listening to a communication before replying. The visual, manual, and aural speed parameters are scaling factors that vary depending on the

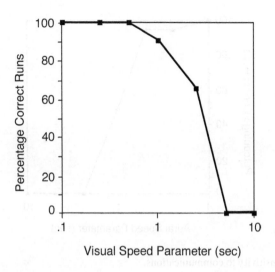

FIGURE 6.6 Sensitivity to visual speed.

precise perceptual, motor, or communication act. These values specify a random distribution to represent adequate between- and within-subjects variability.

To test the model's sensitivity to these parameters, their values were varied over a range of possible quantities. Since all five parameters had similar values, we used the same test range for each parameter, except for procedural speed, corresponding to a module that is generally about an order of magnitude faster than the others, due to its central role in the architecture.

Figures 6.6–6.10 show the influence of the different pilot psychometric parameters on performance and their implications for cockpit design and procedures.

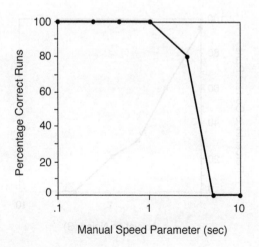

FIGURE 6.7 Sensitivity to manual speed.

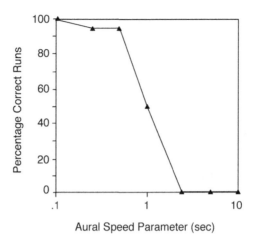

FIGURE 6.8 Sensitivity to communications.

The model used for these analyses was the baseline late reassignment scenario described in chapter 3 without the SVS. The performance measure is percentage correct landings as estimated by Monte Carlo runs of 100 samples for each parameter combination. The percentage of landings without an error for default parameters was about 90%.

Figure 6.6 shows that pilot performance is very sensitive to the speed of visual shifts and instrument acquisition. A visual shift occurs when the pilot needs to shift his visual attention from one display to another to gain information. Attention management in a complex cockpit with multiple displays is a key component of efficient cognition. The default value of this parameter is 1 sec, which corresponds to 90%

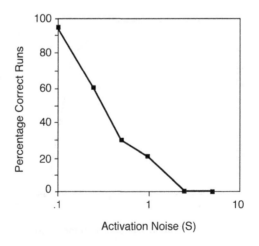

FIGURE 6.9 Sensitivity to decision consistency.

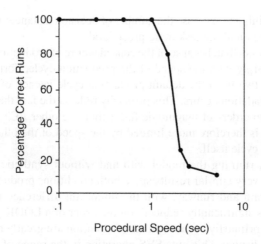

FIGURE 6.10 Sensitivity to procedural speed.

correct landings. Improving the visual speed parameter by a factor of two yields perfect performance, which emphasizes the benefits of visual aids such as SVS that can improve perceptual performance by combining multiple instruments onto a single display, thereby reducing the time taken by attention shifts between widely separated instruments or areas such as the OTW scene. Conversely, impairments in visual speed lead to a rapid deterioration in performance.

Figure 6.7 shows that pilot performance is also very sensitive to the speed of manual operations. The default manual speed is 2.5 sec per action. Performance drops sharply when actions take significantly longer to perform, leading to total failure for a doubling of the action time. This factor supports the concept of division of labor between the pilot flying (PF) the aircraft and the pilot not flying (PNF); one is in charge of monitoring and decision making and the other is in charge of actually performing the actions.

Figure 6.8 shows that performance is highly sensitive to the overhead of communications. The default time to listen to a communication and decide whether it is relevant to performance is 0.5 sec, which is quite small. However, even when reduced to still smaller values, a small risk of failure remains. The impact of communications is hard to eliminate because of the random nature of their temporal distributions. Communications can occur at any time, including at critical moments in the control loop when even a small temporary disruption in attention can lead to failure. Increases in the number of communications or in the average duration of each communication (one can view the aural speed parameter as reflecting either) lead to a very rapid deterioration in performance.

Figure 6.9 shows that performance degrades gradually with the activation noise that controls the stochasticity in the memory retrieval of decision instances. Interestingly, further reducing activation noise from its default value of 0.1 does not lead in the elimination of errors because activation noise has been shown to serve a useful purpose in preventing systematic errors. The most promising remediation for this degradation is training to increase the number of decision instances (practice) and/or making the decision-making task more procedural.

Figure 6.10 shows that, because of the central nature of productions, performance degrades very sharply with the speed of the production cycle. Fortunately, there is a safety factor of two from the default production cycle latency of 50 msec before performance degradation occurs. This primarily reflects the fact that the production cycle is one to two orders of magnitude faster than the other ACT-R modules, and that performance is therefore more limited by the speed of the other modules than by the production cycle itself.

At this point, running the model with and without synthetic vision technology will produce very similar results since both conditions produce clear daytime vision of the terrain and runway, with the important difference that the average perceptual time is significantly reduced (on the order of a LOOK parameter value of about 0.5 sec), primarily because shifts of attention are greatly reduced because of the integrated display. This puts SVS operation in the range of safe, successful performance rather than the break point pictured earlier for conventional systems. We still need to more finely model additional savings achieved from the heads-up SVS display.

EXTENDING THE APPROACH AND LANDING MODELS

This phase of the project focused on further developing the approach and landing model along two principal lines. We attempted to (1) increase the generality of each model as a way of improving the range of applicability, and (2) bring more constraints to bear on the verification and validation of the models. The most straightforward generalization was the IMPRINT model. We needed to add the other scenarios to the original one to create the full set of scenario conditions encountered in the human data collection described in chapter 3 of this book.

We then needed to generalize the ACT-R model to handle the new conditions arising in the scenarios other than the one for which it had been defined. This generalization proceeded along two distinct paths: One extended the model by developing a general method for attending to sources of information based on their learned utility (described in this section), and one extended the model by incorporating a deeper task analysis based on human performance data collected for this task (described in the validation section).

GENERALIZING THE ACT-R APPROACH MODEL THROUGH LEARNED UTILITY OF INFORMATION SOURCES

We needed to handle the choice between the primary instrument panel and the SVS, both of which offer similar information about the world, in a more principled manner. The motivation for a more principled treatment was twofold. The first goal was to examine more closely the benefits offered by the SVS. The second goal was to explain some of the data analyses performed in the first phase, especially the result showing that some crews learned to rely more on the SVS while other crews mostly stuck with the traditional instrument panel. Both types of crews primarily focused their attention on one panel while occasionally glancing at the other one.

The first model had been written, its knowledge engineered in the form of production rules and chunks of declarative knowledge, to perform a particular scenario in the default aircraft configuration. We could have extended the same model to the new conditions encountered in the other scenarios. Production system models can usually be generalized quite easily by simply adding more production rules and/or knowledge chunks to handle new conditions without disrupting their existing functionality. (At least theoretically—as a practical matter, the new knowledge sometimes interferes with the existing one and can lead to difficult debugging and revalidation cycles.) However, this would not have served either of our goals outlined previously: It would not have brought new constraints to bear on the model and its validation, and it would not have explained the preponderance of reliance on either source of information regarding the world and the aircraft.

Therefore, we decided to generalize the ACT-R model through learning rather than exclusively through knowledge engineering. Learning in cognitive architectures such as ACT-R provides two principal advantages. First, it brings more constraints to bear on the model because the learning processes contain many fewer degrees of freedom than a knowledge engineering process performed by a human modeler. Second, learning allows the modeling of the fundamental human processes of encountering a new situation and adapting to it. That is certainly what the crews encountered when they were asked to perform their usual tasks in an environment in which a crucial new piece of equipment had been added, specifically the SVS. More precisely, learning would allow us to explain why some crews came to rely primarily on the new system and why others mostly chose to focus more on the conventional displays. This modeling approach provides the added practical benefit of supporting not only the design of the new system but also its real-world adoption.

LEARNING AT MULTIPLE LEVELS OF DECOMPOSED TASKS

The original ACT-R model had been designed as a two-level goal decomposition process. At the top level was the goal to decide which control to monitor next. This level was implemented as a round-robin loop that tested the previous control and scheduled the next one. At the lower level was the goal to gather needed information, decide the desired state of the control, and, if necessary, activate the change. Which source to gather the information from (instrument panel or SVS) was also prespecified.

To enable learning to proceed, the hardwired symbolic constraints on production selection at both levels were removed to allow the architectural learning processes to take their place in tuning subsymbolic parameters to perform the decision instead in an adaptive manner. Specifically, the hardwired monitoring loop at the top level was removed to allow it instead to learn which controls to monitor. Each production specifying a different control to monitor can now compete with all others in the conflict resolution process that determines which rule is selected to fire. Similarly, at the lower level two production rules are created for each goal to gather information about a specific instrument: One production rule will gather that information from the traditional instrument panel while the other will obtain it from the SVS display.

How are the competing productions selected and how is learning reflected in the conflict resolution process? The process of selecting which production to fire at each cycle, known as conflict resolution, is determined by subsymbolic quantities called utility that are associated with each production. The utility, or expected gain, E, of a production is defined as:

$$E = P \bullet G - C \qquad \text{Expected gain equation}$$

where G is the value of the goal to which the production applies, and P and C are estimates of the goal's probability of being successfully completed and the expected cost in time until that completion, respectively, after this production fires.

Conflict resolution is a stochastic process through the injection of noise of mean θ and standard deviation sigma in each production's utility, leading to a probability of selecting a production i given by:

$$p(i) = \frac{e^{\frac{E_i}{t}}}{\sum_j e^{\frac{E_j}{t}}} \qquad \text{Conflict resolution equation}$$

where $t = \sqrt{6}\sigma/\pi$.

A production's probability of success and cost are learned to reflect the past history of use of that production—specifically, the past number of times that that production led to success or failure of the goal to which it applied, and the subsequent cost that resulted, as specified by:

$$P = \frac{\sum\limits_{successes} t_i^{-d}}{\sum\limits_{successes} t_i^{-d} + \sum\limits_{failures} t_j^{-d}} \qquad \text{Probability learning equation}$$

$$C = \frac{\sum\limits_{successes \& failures} t_k^{-d} Effort_k}{\sum\limits_{successes} t_i^{-d} + \sum\limits_{failures} t_j^{-d}} \qquad \text{Cost learning equation}$$

Costs are defined in terms of the time (effort) to lead to a resolution of the current goal. Thus, the more or less successful a production is in leading to a solution to the goal and the more or less efficient that solution is, the more or less likely that production is to be selected in the future. The terms *efforts, successes,* and *failures* will decay over time according to a power law process. Thus, a recent experience will initially have more weight in the sums and ratios of the learning equations than an older one, but that weight will gradually decrease as time passes.

The learning process for production utilities at both levels thus depends fundamentally upon the defining of successes and failures at the end of each goal and the accumulations of costs (in terms of time spent) to reach the end of the goal. At the top level, a goal to choose the next control to monitor (and set a subgoal to accomplish that task) is deemed successful if the monitoring results in some action being taken. Otherwise, the monitoring of that particular control was deemed in vain and the goal resulted in failure since it wasted time and perhaps the opportunity to monitor some more pressing control. At that level, the learning will accomplish a similar but more flexible version of the initial round-robin selection.

When a control task is selected and results in an adjustment, its utility goes up because its estimate of success is also increased. It is then likely to be selected again shortly because of that high utility, but then will likely result in failure because the underlying situation has not changed enough to justify a new adjustment to that control. The utility of that production is then downgraded because of the failure, meaning that it will not be selected again until the failure has decayed and the productions associated with the other tasks have also incurred their own failures, at which point the cycle repeats. This process is quite similar to the hardwired round robin, but is considerably more flexible.

Since utility is stochastic, conflict resolution is probabilistic, leading to the constant exploration of new combinations. More fundamentally, the learning process will be sensitive through successes and failures to the frequency at which the underlying situation affecting a given control changes. This will result in productions associated with controls that require more frequent adjustments having higher utilities than others, and thus being selected more frequently. However, if that frequency changes over time, the conflict resolution process will again adapt through the time decay of the learning terms and change its frequency of selection to reflect the new situation.

At the lower level, we assume that a goal to capture information about underlying state variables, decide what to do with the given control, and (optionally) perform the action is always successful. What varies is the cost in terms of time spent performing these actions, in particular the capture of information from display panels. It takes much less time to capture information from the same source (traditional instrument panel or SVS) once one's visual attention is already there because switching attention to another display panel requires significant time. Therefore, given a low-level production that decides (and acts upon) which source to capture information from, the cost of that production will primarily reflect where a shift of attention between display panels is required, or whether a much more minor and faster shift of attention between displays on the same panel is all that is needed.

Note that that information is not represented symbolically at the production level (i.e., productions to capture information simply specify a source) and is not whether it involves switching between panels. It is simply used to learn the production utilities. What results is a gradual convergence to either display panel. Initially, productions have similar utilities and, given their stochastic selection, the model will cull information from either display with equal likelihood. But as one source starts being preferred through sheer randomness, the utilities for productions associated with that source will decrease while those associated with the other source will increase, leading to a winner-take-all process that will result in the model preferring one source over the other consistently, except for occasional glances to the other driven by conflict resolution stochasticity. This pattern reproduces the analysis of human data that we set out to emulate at the outset—not from our engineering the result into the model, but instead by turning on the architectural learning mechanisms, driven by environmental constraints.

One rather remarkable fact of this learning process is that while it takes the form of a single mechanism operating uniformly across two goal levels, it results in fundamentally opposite patterns at each level. At the top level, a round-robin selection process emerges that makes sure all controls are attended to at some regular intervals. At the bottom level, a winner-take-all process emerges that gradually chooses to rely upon either of the two main sources of information to the almost complete exclusion of the other.

FURTHER INVESTIGATIONS SUGGESTED BY LEARNED UTILITY OF INFORMATION SOURCES

In this model we have assumed that extracting information from the SVS was a process neither more nor less efficient than that from the conventional instrument panel. If an assumption was made based upon finer distinctions of the cost of harvesting information from the respective displays that one panel was more efficient than the others, this model would make predictions about the probability of adoption of the more efficient display over the other. Nothing in the model would need to be changed other than the costs of the various perceptual operations in each display. As new technologies emerge, their efficiency and rate of adoption could be easily studied by simply representing their costs rather than their entire operation.

Another process that can be studied in detail is the speed at which trained pilots adapt from one category of display to another. The process is basically one of overcoming well-trained productions with new ones with presumably higher utilities. The speed at which the skill at using the new displays will lead to the replacement of the skills associated with the established ones can also be studied under various assumptions of efficiency and errors. This can lead to the design of training procedures designed to maximize speed of adoption and reduce errors associated with the training phase during which skills conflict and might result in incorrect decisions and actions.

The IMPRINT model is designed to simulate 10 different scenarios, shown in Table 6.1. Selection of the various scenarios will change the aircraft's environment

TABLE 6.1
Scenarios Simulated in IMPRINT

Scenario	ATC Communications	Visibility
VMC, baseline vectored approach	Normal	Runway visible at 800 ft AGL
VMC, baseline late reassignment	Late reassignment request at 1,000 ft AGL	Runway visible at 800 ft AGL
VMC, baseline missed approach	Normal	Runway visible at 800 ft AGL (traffic on runway)
IMC, baseline vectored approach	Normal	Runway visible at 650 ft AGL
IMC, baseline missed approach	Normal	Runway never visible
IMC, baseline terrain mismatch	Normal	Runway visible at 650 ft AGL (runway 200 ft misaligned)
IMC, SVS vectored approach	Normal	Runway visible at 650 ft AGL
IMC, SVS late reassignment	Late reassignment request at 1,000 ft AGL	Runway visible at 650 ft AGL
IMC, SVS missed approach	Normal	Runway never visible
IMC, SVS terrain mismatch	Normal	Runway visible at 650 ft AGL (runway 200 ft misaligned)

Notes: AGL = above ground level; ATC = air traffic control; IMC = instrument meteorological conditions; SVS = synthetic vision system; VMC = visual meteorological conditions.

and which displays are available for the pilot to use. The environmental changes include the altitude where the runway can be seen, the condition of the runway (i.e., misaligned or occupied by aircraft or other obstacles), and what ATC communications take place. If the scenario uses the SVS technology, the SVS display will be available for the pilot. Since the IMPRINT model is packaged as source code, a user is able to run any scenario available and is also able to add new technologies to the aircraft for future testing. In addition, existing controls can be extended to improve their fidelity, modify their function, or otherwise alter them to better suit the simulation goals.

MODEL VALIDATION

Model validation is an extremely complicated subject; for a model to be completely validated, it would have to reproduce exactly the performance of the human subjects. Given the variable nature of human performance, this task is nearly impossible unless a task is so constrained as to make the model nearly meaningless. For any task of reasonable complexity and level of dynamic interaction, a broad range of human behavior can be expected, and the task of validation is to ensure that the model performance falls within the bounds determined by human performance.

The ACT-R theory, as an instance of first-principle modeling (Laughery, Lebiere, & Archer, 2006), provides some validation by only supporting processes that have been empirically determined to fall within the realm of human capability. That is, architecturally, ACT-R prohibits many actions that an unconstrained computer program (such as one written in C#) could accomplish. In this sense, there is a level of validity obtained simply by using ACT-R since the operations completed by the model are, in principle, operations that could be accomplished by a human performer of the task. However, these architectural limitations do not ensure that the operations undertaken by the model are the same operations undertaken by the human performer. Validating an ACT-R model of task performance requires that the constraints imposed by human performance on the task be used to inform the actual operations undertaken by the model to produce a validated model.

Validation of the IMPRINT/ACT-R model against human performance will be examined here at three distinct levels, from broad to fine grain. The first level involves achieving the criterion of successful task completion (to the extent that human performers succeed). The second level involves finer grained validation of the individual actions taken by the model to determine if they correspond to the individual actions taken by human performers, and it is largely qualitative (i.e., establishing that the model's actions have face validity). The third, final level involves examining behavioral data at an extremely fine grain and attempting to establish close quantitative correspondence between data captured from an instrumented model and data captured from human performance.

VALIDATION LEVEL 1: SUCCESSFUL TASK COMPLETION

The first level, the most basic level of model validation, is a demonstration that the model can, in fact, accomplish the task. If the model cannot complete the task and human pilots can, there is no need to proceed further along the validation path.

TABLE 6.2
Successful Task Completion

Scenario	Scenario Description	HITL Outcome	HPM Outcome
1	Baseline, VMC, normal approach	Disengage autopilot and prepare to land	Same
2	Baseline, VMC, late runway reassignment	Accept runway reassignment; land on parallel runway	Same
3	Baseline, VMC, missed approach (traffic on runway)	Initiate go-around	Same
4	Baseline, IMC, normal approach	Disengage autopilot and prepare to land	Same
5	Baseline, IMC, missed approach	Initiate go-around	Same
6	Baseline, IMC, terrain mismatch	Detect terrain mismatch; initiate go-around	Same
7	SVS, IMC, normal approach	Disengage autopilot and prepare to land	Same
8	SVS, IMC, late runway reassignment	Accept runway reassignment; land on parallel runway	Same
9	SVS, IMC, missed approach	Initiate go-around	Same
10	SVS, IMC, terrain mismatch	Detect terrain mismatch; initiate go-around	Same

Notes: ATC = air traffic control; HITL = human in the loop; HPM = human performance model; IMC = instrument meteorological conditions; SVS = synthetic vision system; VMC = visual meteorological conditions.

Our model is a model of the PF during runway approach. We have abstracted the PNF since, in the NASA human-in-the-loop (HITL) simulation, the PNF was a confederate and only operated controls upon request of the PF, whereas our primary interest is in the cognitive operations of the PF. In the case of this approach model, the model achieves the same outcome as human pilots flying the scenarios in the simulator in each case. These scenarios and the identical outcomes (scenario conclusions) reached by both the HITL simulation pilots and the model are detailed in Table 6.2.

It is not a given that the model would successfully complete the task. The simulation environment requires that the model approach the runway at the proper altitude (through controlling the simulation) and see the runway where it is expected to be. There are several ways things can go wrong. If the simulation results in the aircraft being above the altitude where the runway can be seen, the result is that the aircraft would overfly the runway. On the other hand, if decision altitude (DA) is reached prematurely and the runway is not in sight, the model would choose to go around. Correct performance is predicated upon arriving at the DA at the appropriate distance from the airfield, and therefore critically upon controlling the rate of descent and the speed of forward motion, and configuring the aircraft to allow a landing to occur. The preceding outcomes, achieved by both the HITL simulation pilots and the ACT-R model, represent the fundamental validation on which more extensive validation depends: The cognitive model successfully completes the task in agreement with the human pilots.

Validation Level 2: Assessing Subtask Correspondence

Given the first level of validation, we will proceed to the second validation level, a qualitative assessment of the correspondence between model actions and human pilot actions. This aspect of validation can be approached rigorously by encoding the actions taken by the human pilots and lining it up with a model trace. This approach is based on the techniques of protocol analysis as described by Ericsson and Simon (1993). In addition, the correspondence between the model's use of the SVS display and the actual pilots' use of the display was addressed in earlier development of this model as discussed previously, and is incorporated within the version reported here (as well as being briefly addressed later).

Following this approach, the videotaped scenarios described in chapter 3 were transcribed by recording those events that were audible and recording the corresponding timestamp present on the video. Aggregating these performances and performing a protocol analysis resulted in the emergence of a general task decomposition that was common among the scenarios.

A high-level ACT-R trace is included later. Unlike low-level traces that can be output to examine in detail every aspect of the model operation, this high-level trace is focused on the operation of each module, similarly to our parameter sensitivity analysis in the previous section. ACT-R actions that execute subtasks within a task are coded as "PROCEDURAL" in the model runs. The ACT-R model traces also produce actions that are coded as "VISUAL," indicating looking at something in particular; "MANUAL," indicating a physical action performed; "AURAL," indicating something listened to; and "VOCAL," indicating a speech act. Because we have access to its inner workings, the ACT-R model can thus provide much more fine-grained data on the operations it is performing at any time. This is a general truth of cognitive models, making the validation task more difficult since we only have access to sparser information about the detailed cognitive processes the human participants are involved in. (The model must do much more than the human pilot says he or she is doing just as the human pilot must do much more than the pilot says that he or she is doing to complete the task.)

The performance of approach tasks can be broken down into the main tasks and subtasks listed in Table 6.3. These tasks do not necessarily correspond to other previous categorizations of approach tasks; instead, they focus on the cognitive phases as demonstrated by human performance in the simulation environment as demonstrated on the videotaped scenario runs.

The main tasks can be separated into two distinct categories: (1) schematic scripts (simulation initiation, and landing checklist), and (2) reactive monitoring tasks (descent from 3,000 ft and final approach). The preceding tasks occur in the order presented. The schematic tasks are characterized by rapid performance of the subtasks while the monitoring tasks take up the time between the schematic tasks.

This sequence of tasks is worth examining in further detail. Upon simulation initiation (within 30 sec of simulation start), regardless of the scenario, the PF rapidly performs the subtasks listed previously under the main task "simulation initiation" in a scripted fashion (the pilot sets LNAV/VNAV, sets autopilots to "engaged," sets flaps to setting 1, etc.). This is followed by a period of roughly 7 min of monitoring the descent, occasionally gradually adjusting controls. At this point, the PF calls out

TABLE 6.3

Model Tasks and Subtasks

1. Simulation initiation
   a. Encode and respond to ATC instructions for approach
   b. Set aircraft configuration for approach
      i. Set speed brakes
      ii. Set LNAV/VNAV
      iii. Set autopilots on
      iv. Set flaps to 1 or 4
      v. Check altitude

2. Descent from 3,000 ft
   a. Check/adjust speed
   b. Check/adjust flaps
   c. Check/adjust altitude dial
   d. Check/adjust map scale
   e. Check altitude
   f. Check/adjust decision altitude
   g. Check distance to next waypoint
   h. Check/adjust air brakes

3. Landing checklist
   a. Set gear down
   b. Set flaps 15
   c. Set speed brakes armed
   d. Set speed 135
   e. Notify cabin of landing

4. Final approach (nearing decision altitude)
   a. Check/adjust flaps (set full flaps on)
   b. Make landing decision at decision altitude
      i. Check altitude
      ii. Check runway (out the window)
      iii. Decide to land or go around

the landing checklist—again, a scripted set of actions including lowering the landing gear, arming the speed brakes, etc.—that ensures that the aircraft is configured for landing (typically read off a card, though sometimes recalled from memory). The landing checklist takes approximately 10 sec to complete.

This is followed by a period of roughly 3 min while the final approach is monitored, during which flaps are set to their final setting, and the pilot monitors their altitude relative to the DA while looking for the runway OTW. These tasks form the macrostructure of all scenario performances and are therefore shared by all of the individual task models. (In fact, one ACT-R model performs all of the scenarios; this model is not parameterized for any scenario and is not provided with the current scenario beyond the presence or absence of the SVS.) These tasks also directly form the macrostructure of the ACT-R model itself, which is both an aid in development and validation (the model maps directly onto the cognitive constructs).

The performance of actions during these phases (for the baseline IMC, non-SVS condition) is demonstrated by protocol excerpts from scenarios for Pilot 3 and the corresponding model trace. Excerpts were selected from scenario runs that provided the most detailed information for validation. The situations described had corresponding episodes in each of the scenario runs; the most complete verbal/visual protocols available were used.

Validating Task 1: Simulation Initiation

The human performance protocol (abstracted from the HITL simulation and implemented in IMPRINT) during this phase is as follows:

	TIME	OPERANDS
PROCEDURAL	40:03:00	Preparing for approach: setting LNAV
PROCEDURAL	40:04:00	Preparing for approach: speed brakes full
PROCEDURAL	40:05:00	Autopilots engaged
PROCEDURAL	40:06:00	Preparing for approach: set flaps 1

This section involves setting the flaps, autopilot, LNAV, and VNAV just after model initiation (from 2.4 sec into the simulation to 8.9 sec into the simulation). The ACT-R model trace for this phase is as follows:

	TIME	OPERANDS
PROCEDURAL	2.363	Preparing for approach: setting flaps 1
MANUAL	2.363	Flaps set 1
PROCEDURAL	3.435	Preparing for approach: setting VNAV
VISUAL	3.435	Waypoint Value 1
PROCEDURAL	6.476	Preparing for approach: engaging speed brakes full
MANUAL	6.476	Air brakes on/off 0
PROCEDURAL	7.649	Preparing for approach: setting LNAV
VISUAL	7.649	Distance-next Value 1
PROCEDURAL	8.899	Preparing for approach: engaging autopilot
VISUAL	8.899	Autopilots value up

Note that the actions do not occur in the same order in the abstracted human protocol and the ACT-R model trace. This is typical of human performance during simulation initiation, where the same set of actions occurs, but the order is not specified. The model produces the same stochastic effect where different runs will produce a different ordering of the subtasks involved in initiating the simulation. Also note that this particular process does not have an explicit checklist (like landing does), and the pilot has NOT in this case ensured that the VNAV was set (though this pilot explicitly does so in almost every other scenario). In contrast, our model is using an explicit checklist at this point and does not miss the item.

Validating Task 2: Descent From 3,000 ft

The human protocol (abstracted from the HITL simulation and implemented in IMPRINT) during the descent from 3,000 ft is somewhat sparser than the ACT-R model protocol (the ACT-R model reports everything it is looking at while the pilot represented in IMPRINT reports only changes to controls being made). The following explicit actions take place (though there is also constant monitoring of various controls and displays as evidenced by the eye-tracking data) in the HITL human performance data:

	TIME	OPERANDS
PROCEDURAL	43:40:00	Retract speed brakes
MANUAL	43:42:00	Set speed brakes retracted
PROCEDURAL	46:30:00	Lower gear
MANUAL	46:31:00	Set gear down
PROCEDURAL	46:32:00	Flaps 25
MANUAL	46:33:00	Set flaps 25

The following segment is a section of the ACT-R model trace that illustrates actions during the descent from 3,000 ft. This segment of the trace involves monitoring the status of various systems. The pilot first checks the status of the autopilot and looks OTW at 47 sec. The pilot then checks the status of the speed brakes at 51 sec and, now that the plane speed has decreased, chooses to disengage them. The pilot then checks that the DA is set at 54 sec (which it already is). At 57 sec, the pilot checks the value of the speed dial, which is 160 kts, and decreases it to 140 kts. The pilot then looks at the altitude and notes that the current altitude is 2,500 ft. This is followed by lowering the landing gear at 65 sec, and adjusting the flaps from 15 to 25 at 69 sec. The model trace follows:

	TIME	OPERANDS
PROCEDURAL	47.436	Subgoal checking autopilot
VISUAL	48.498	OTW_runway value out of sight
VISUAL	49.65	Autopilots value up
VISUAL	50.626	Altitude value 2,500

**Goal37
    isa SET-AUTOPILOT
    visibility Out-Of-Sight peek 2500
    decision EngagedGOAL37 DIAL DIAL T Ignore-Autopilot

	TIME	OPERANDS
PROCEDURAL	51.327	Subgoal checking speed brakes
VISUAL	52.596	Speed value 170
VISUAL	53.478	Air brakes value off

**Goal41
    isa SPEED-BRAKES
    speed 170
    decision OffGOAL41 DIAL DIAL T Decide-Speed-Brakes

	TIME	OPERANDS
PROCEDURAL	54.078	Subgoal checking decision altitude
VISUAL	55.277	Waypoint value 2
VISUAL	56.261	Distance-next value 2
RETRIEVAL	56.361	Waypoint 2 next 2 distance 13

**Goal44
    isa SET-DECISION-ALTITUDE
    waypoint 2
    distance-next 2
    distance 13
    old 600
    decision 600GOAL44 DIAL DIAL T Ignore-Decision-Altitude

	TIME	OPERANDS
PROCEDURAL	57.017	Subgoal checking speed
VISUAL	57.906	Waypoint value 2
VISUAL	58.834	Distance-next value 1
RETRIEVAL	58.934	Waypoint 2 next 1 distance 12
RETRIEVAL	59.034	Speed
VISUAL	60.278	Dial-speed value 160

MANUAL	63.031	Dial-speed set 140

**Goal47
    isa SET-SPEED
    waypoint 2
    distance-next 1
    distance 12
    speed 140GOAL47 DIAL DIAL T Dial-Speed

PROCEDURAL	63.641	Subgoal checking altitude
VISUAL	64.715	Altitude value 2500

**Goal52
    isa CHECK-ALTITUDE
    altitude 2500
    previous 2500GOAL52 DIAL DIAL T Ignore-Altitude

PROCEDURAL	65.372	Subgoal checking landing gear
VISUAL	66.437	Waypoint value 2
VISUAL	67.335	Distance-next value 1
RETRIEVAL	67.436	Waypoint 2 next 1 distance 12
VISUAL	68.252	Landing-gear value down

**Goal54
    isa MOVE-GEAR
    waypoint 2
    distance-next 1
    distance 12
    decision DownGOAL54 DIAL DIAL T Decide-Gear

PROCEDURAL	68.852	Subgoal checking flaps
VISUAL	69.918	Speed value 159
RETRIEVAL	70.068	Flaps 25
VISUAL	71.025	Flaps value 15
MANUAL	73.94	Flaps set 25

**Goal58
    isa SET-FLAP
    speed 159
    flap 25GOAL58 DIAL DIAL T Set-Flaps

Again, the trace indicates many of the same actions are completed, though the order is not specified. This captures the variability of the human data as well, where this segment of flight is characterized by monitoring of various systems. Much of the trace from this segment shows the pilot alternately checking controls that already have the desired value. The segment selected previously shows a period of transition where several control values actually are altered, as they are in the human protocol.

Validating Task 3: Landing Checklist

The landing checklist is completed by notifying the cabin to prepare for landing, lowering the landing gear, setting the speed to 135 kts, arming the speed brakes, and setting the flaps to their final setting (40). These actions take approximately 10 sec to complete for pilots and roughly the same for the model (from 140 to 153 sec in this case). Here, the abstracted human protocol (Pilot 3, scenario 9, SVS, IMC, missed approach) and the model protocol are extremely similar (exactly as we would expect for the more highly scripted sections of performance).

After the pilot announces it is time to perform the landing checklist, in rapid succession (8 sec in this case) the PF and PNF check that the gear are down, set the flaps to 15, arm the speed brakes for automatic deployment on landing, and set the speed dial to 135 kts. The human protocol trace (abstracted from the HITL simulation and implemented in IMPRINT) is:

	TIME	OPERANDS
PROCEDURAL	3:05:02	Landing checklist
PROCEDURAL	3:05:03	Gear down
VISUAL	3:05:04	Gear value down
PROCEDURAL	3:05:05	Flaps 15
MANUAL	3:05:06	Set flaps 15
PROCEDURAL	3:05:07	Speed brakes armed
MANUAL	3:05:08	Set speed brakes armed
PROCEDURAL	3:05:09	Speed 135
MANUAL	3:05:10	Set speed 135

Similarly, upon deciding to execute the landing checklist, the ACT-R model quickly announces to the cabin to prepare for landing (this step was skipped by the pilot in the simulator in this run but was included by this pilot in other runs), checks that the gear are down, sets the speed to 135 kts, sets the speed brakes to armed, and sets the flaps to 15. The ACT-R model trace shows:

	TIME	OPERANDS
PROCEDURAL	139.975	Subgoal completing landing checklist
PROCEDURAL	142.313	Landing checklist: preparing cabin for approach
VOCAL	142.313	Nothing communication 2007
PROCEDURAL	145.058	Landing checklist: setting gear down 1
MANUAL	145.058	Landing-gear up/down 0
PROCEDURAL	147.611	Landing checklist: setting speed to 135
MANUAL	147.611	Speed set 135
VISUAL	147.611	Speed value 142
PROCEDURAL	151.038	Landing checklist: setting speed brakes to armed
MANUAL	151.038	Air brakes on/off 0
PROCEDURAL	153.097	Landing checklist: setting flaps 15
MANUAL	153.097	Flaps set 15
VOCAL	153.147	Landing checklist complete

Validating Task 4: Final Approach

The final approach is relatively simple, though critical. As this task is initiated, flaps are set to the final setting (40) and the pilot enters a cycle of monitoring the altitude relative to the DA, and looking for the runway. If a clear runway is visually identified by the DA, the pilot decides to land. Otherwise, the pilot calls a missed approach and initiates a go-around. The following sequence (abstracted from Pilot 3, baseline IMC missed approach) shows the operations performed by the pilot just prior to landing:

	TIME	OPERANDS
PROCEDURAL	1:33:24	Full flaps
MANUAL	1:33:25	Set flaps 40
PROCEDURAL	1:35:16	Check altitude
VISUAL	1:35:17	Altitude value 1000

PROCEDURAL	1:35:37	Check altitude
VISUAL	1:35:38	Altitude value 650—approaching decision
PROCEDURAL	1:35:42	Check altitude
VISUAL	1:35:43	Altitude value 600—decision height
PROCEDURAL	1:35:44	Check out the window
VISUAL	1:35:45	Field not in sight
PROCEDURAL	1:35:47	Missed approach—go around

Similarly, the ACT-R model supports the range of cues and decisions available to a pilot. In this case, for the vectored approach scenario, the ACT-R model checks the altitude and decides to land based on a visible, clear, aligned runway. The ACT-R model trace shows:

	TIME	OPERANDS
PROCEDURAL	287.970	Subgoal checking altitude
VISUAL	288.797	Altitude value 623
**Goal278		
isa CHECK-ALTITUDE		
altitude 623		
previous 713GOAL278 DIAL DIAL T Ignore-Altitude		
PROCEDURAL	289.455	Subgoal checking autopilot
VISUAL	290.566	OTW_runway value in-sight
VISUAL	291.751	Autopilots value up
PROCEDURAL	291.801	Disengage autopilot and land
MANUAL	294.559	Autopilots up/down 0
**Goal280		
isa SET-AUTOPILOT		
visibility In-Sight		
peek nil		
decision DisengagedGOAL280 DIAL DIAL T Disengage-Autopilot		

Several possible actions are possible at the conclusion of the final approach. The section on Task 1 presents these actions as taken by the model and by the human pilots. Only the outcome is different for both different model runs and different pilot runs in the simulator, and these outcomes are in perfect 1:1 correspondence.

Validation of Learning the Utility of Sources of Information

The previous modeling effort on this project focused on the learning aspects of progress monitoring during the descent from 3,000 ft and the final approach by guiding where to look for information based on the learned utility of looking during online task performance. A result of using this utility model that has previously been demonstrated is that the model will adapt to use primarily the SVS or rely on the PFD view, but will prefer not to alternate frequently between the views because of the cost of attention switching. The current extension of the previous model incorporates these utility learning mechanisms with the standard schematic behavior of quickly performing both the simulation initiation and the landing checklist.

This learning model provides a principled description of monitoring activity that occurs in the absence of a strongly schematic approach to task completion. This task, in particular, is extremely interesting because it depends on highly scripted task performance and on loose and adaptive reactive monitoring between the scripted episodes. The model presented here provides a model of this task and also a model of how reactive and highly goal-directed behaviors can coexist within the same architecture.

VALIDATION LEVEL 3: QUANTITATIVE CORRESPONDENCE OF BEHAVIORAL DATA WITH MODEL PERFORMANCE

The third level of validation, a quantitative assessment of correspondence between behavioral data exhibited by human task performance and data produced by the model, depends on a rich body of data on both sides to align. For this project, eye-tracking data were provided for the HITL simulator runs (see chapter 3). To establish correspondence with ACT-R model performance, the ACT-R model was instrumented to record the amount of time spent in various subtasks, with the assumption that the task performer would look at what he or she was doing. We do, however, expect one systematic variation of the model from the human data: Since the PNF operated the controls during the HITL simulation but the ACT-R model operates these directly during model runs, the ACT-R model should spend a correspondingly larger amount of time viewing the controls.

Quantitative assessment typically depends on having a large enough sample of data such that any findings can be truly expected to generalize. In the case of this project, we have data for only three pilots completing the task, all running only one trial per scenario. Given these limitations of the data, we have decided to use this as an opportunity to explore data validation methods. However, any findings in this section will be subject to the caveat that the data set used is very limited, and the findings are therefore suggestive.

The experimental data included eye fixation data for the PFs as they performed the tasks. Much of these data were at too fine a level of detail for incorporation into the models we developed here. However, the individual pilot × task data also included summary tables of eye dwells in various areas of interest, where the tables included both mean dwell time and the total number of dwells for each particular area. This allows the derivation of an exact percentage of fixation time for each area of interest during task performance. For Pilot 3, across all scenarios, summarizing in this way results in the data shown in Table 6.4. Because the model did not include the early phases of the flight and focused only on the more interesting approach and landing phases, the data summarized in Table 6.4 as well as the subsequent comparison data use only the latter two of the four phases of flight that are represented in the HITL data.

There are several points worth mentioning in this table. First, the PF spends an extremely small amount of the total time actually looking at the controls. The primary explanation for this is that, as part of this experiment, a confederate of the experiment played the role of the PNF and performed most of the control manipulations. The PF requested that the PNF set the values, but had little need to look at the controls themselves. The pilot depended on the SVS a fair amount when it was available (45% of all dwell time when available; Scenarios 7–10). In the VMC conditions, the pilot also spent 20% of the time looking OTW (Scenarios 1–3). During IMC conditions, however, the pilot relied primarily on the ND and the PFD, though

TABLE 6.4
Pilot 3 Dwell Time in Area of Interest as Percentage of Total Time (Data Weighted for Phase 3 + 4 Durations)

Scenario	MCP	ND	PFD	SVS	Controls	OTW	Off
1	0.001	0.195	0.433	0.000	0.013	0.187	0.048
2	0.008	0.276	0.315	0.000	0.021	0.257	0.027
3	0.016	0.205	0.453	0.002	0.001	0.157	0.122
4	0.000	0.447	0.425	0.000	0.071	0.020	0.028
5	0.000	0.371	0.570	0.000	0.026	0.025	0.008
6	0.011	0.363	0.493	0.000	0.048	0.031	0.047
7	0.003	0.140	0.230	0.415	0.017	0.124	0.027
8	0.011	0.146	0.294	0.422	0.021	0.051	0.018
9	0.000	0.225	0.271	0.473	0.004	0.007	0.019
10	0.001	0.189	0.229	0.500	0.001	0.008	0.054
Average	0.005	0.256	0.371	0.181	0.022	0.087	0.040

Notes: MCP = mode control panel; ND = navigation display; OTW = out the window; PFD = primary flight display; SVS = synthetic vision system.

he still looked at the controls and OTW occasionally. However, the low percentage of time the pilot sampled OTW during IMC conditions suggests that peripheral vision played a large role in determining whether or not to look there.

A priori, we expected the ACT-R model of the task discussed here to depart from these values systematically in several ways. First, we chose not to model the whole simulated run, which included substantial amounts of time where nothing of interest was happening; instead, we focused on the second half of the approach period, especially final approach, resulting in modeled runs of approximately 280 sec (vs. approximately a total of 600 sec for the pilot runs). To account for this, the pilot data used were weighted averages (with respect to duration) of Phases 3 and 4 only. Further, lacking peripheral vision, the ACT-R model cannot know there is nothing to see out the window unless it actually looks there. That is, the ACT-R model does not represent the well-known covert shifts of attention that allow a person to attend to an area without shifting gaze (and the results from the previous IMC conditions indicate strongly that this is occurring). Second, since the ACT-R model does not have a copilot, it actually looks at the controls as it sets them. Thus, the ACT-R model should have a substantially higher percentage of dwell time associated with the controls. The dwell percentages as predicted by the ACT-R model are presented in Table 6.5.

The correlations between the model predictions for dwell times and the performance of Pilot 3 for each of the 10 scenarios are shown in Table 6.6. Although there is a lot of variability between scenarios, several of the correlations are very high with an average $r = 0.63$, which means that the model accounts for a fair amount of variance in dwell time ($r^2 = 0.47$). Similar individual analyses were conducted for Pilots 1 and 2, yielding average correlations of $r = 0.78$ and $r = 0.68$, respectively, accounting for the majority of variance in these cases as well ($r^2 = 0.58$ and $r^2 = 0.48$, respectively).

Given that there were no discernible strategy differences among the pilots, it also makes sense to collapse the data and compare the model predictions against the

TABLE 6.5
Model Dwell Time in Area of Interest as Percentage of Total Time

Scenario	MCP	ND	PFD	SVS	Controls	OTW	Off
1	0.08	0.40	0.30	0.00	0.07	0.08	0.07
2	0.07	0.43	0.29	0.00	0.06	0.07	0.07
3	0.11	0.36	0.30	0.00	0.06	0.07	0.11
4	0.10	0.38	0.30	0.00	0.06	0.07	0.10
5	0.10	0.38	0.29	0.00	0.06	0.06	0.11
6	0.10	0.37	0.30	0.00	0.07	0.08	0.08
7	0.11	0.33	0.29	0.06	0.06	0.07	0.09
8	0.10	0.30	0.29	0.10	0.06	0.07	0.08
9	0.10	0.26	0.30	0.12	0.06	0.07	0.09
10	0.11	0.27	0.27	0.09	0.10	0.11	0.04
Average	0.10	0.35	0.29	0.04	0.07	0.08	0.08

Notes: MCP = mode control panel; ND = navigation display; OTW = out the window; PFD = primary flight display; SVS = synthetic vision system.

aggregate subject performance. Table 6.7 presents the averaged dwell percentages of Pilots 1, 2, and 3:

The overall fit of the ACT-R model to the aggregate subject data is $r = 0.74$ ($r^2 = 0.55$). As predicted, because the ACT-R model does not explicitly represent the PNF, the model spends more time focusing on the controls. Also as predicted, because the ACT-R model does not account for peripheral vision, the model must spend more time checking OTW than the human pilots do to see if something has changed.

The ACT-R model also spends less time using the information on the SVS than the human pilots do. This may be due to a bias introduced in the experiment simply through the presence of the SVS: The human pilots were confronted with a familiar cockpit with an unfamiliar device (the SVS) and may have responded to the demand characteristics of the experiment by focusing on the SVS more than was useful. The ACT-R model considers the SVS from a purely information-theoretic perspective

TABLE 6.6
Pilot 3 Model Correlation Results

Scenario	r	r^2
1	0.72	0.52
2	0.76	0.58
3	0.76	0.58
4	0.96	0.92
5	0.87	0.76
6	0.92	0.85
7	0.12	0.02
8	0.40	0.16
9	0.53	0.28
10	0.22	0.05
Average	0.63	0.47

TABLE 6.7
Pilot Average Dwell Time in Area of Interest as Percentage of Total Time

Scenario	MCP	ND	PFD	SVS	Controls	OTW	Off
1	0.010	0.202	0.427	0.002	0.023	0.236	0.032
2	0.010	0.259	0.362	0.000	0.048	0.221	0.024
3	0.012	0.235	0.465	0.001	0.018	0.162	0.055
4	0.054	0.342	0.419	0.000	0.064	0.092	0.027
5	0.008	0.294	0.612	0.000	0.026	0.041	0.020
6	0.013	0.394	0.314	0.000	0.069	0.084	0.122
7	0.058	0.158	0.303	0.303	0.030	0.084	0.046
8	0.012	0.172	0.316	0.335	0.031	0.085	0.028
9	0.005	0.228	0.398	0.312	0.019	0.012	0.022
10	0.005	0.196	0.399	0.265	0.019	0.058	0.043
Average	0.019	0.248	0.401	0.122	0.035	0.107	0.042

Notes: MCP = mode control panel; ND = navigation display; OTW = out the window; PFD = primary flight display; SVS = synthetic vision system.

and seeks to find information without incurring costs. It is straightforwardly possible to add a bias such as this, but the model is in many ways more interesting as it stands, since the dwell times are more purely a function of seeking information in the environment to perform necessary tasks.

To restate the point, the scan pattern as used by the model emerges from the need to complete tasks. It is nowhere represented explicitly in the model, and the model is not parameterized in any way to follow a particular scan pattern. Rather, the model uses a task decomposition to drive its activities, and as those activities require looking for information in certain locations, a scan pattern is produced through the completion of those tasks.

DISCUSSION

The modeling approach chosen here—separately modeling the cognitive agent and the simulation environment—has promise for producing validated models that can generalize to other domains and tasks. This is a result of keeping the model free of many of the entanglements that result from embedding a cognitive model within an application or embedding a simulation within a cognitive modeling environment.

We have explored model development along two separate paths. The first path involved exploring the use of architecturally supported performance and learning mechanisms to determine the ability to base pilot decision-making on generalization of recalled information from memory and the adaptive selection of external information sources based on the sources that are most often useful (i.e., provide novel information). Besides freeing a cognitive modeler from having to carefully hand-craft knowledge and arbitration rules for choosing between competing knowledge, these mechanisms have the added benefit of making it possible to expose shortcomings in a simulation environment. In the latter case, the mechanism learned to look at the displays that changed the most often.

Perhaps the most interesting finding of this modeling effort is that the dwell pattern as produced by the pilots also emerges from the model as a result of this mechanism despite not being encoded explicitly. Rather, the task demands drive information seeking; given that the model and the pilots are performing the same tasks, the model and the pilots allocate their attention to the various visual regions of interest similarly. The combination of task-driven looking and novelty-driven monitoring captures the majority of the variation in the dwell pattern of the pilots ($r^2 = 0.55$) without resorting to parameter fitting and model tweaking.

The actual definition of novelty is an interesting topic to pursue with future research. In this study, the OTW view only changed when the runway came into view. Thus, a continuously varying aspect of the environment is abstracted into categories that allow the straightforward determination of novelty. This is a situation that is true of the IMPRINT simulation environment we constructed, but that is not necessarily true of either the HITL simulation or the actual environment. For example, in both of these, the visual display is constantly updated, except in IMC conditions. It is likely that the eventual solution to this problem will be to focus on rate of change or relative amount of change as the driving factor in novelty-driven monitoring of the environment, or base the novelty detection not on the change in the environment per se, but rather on the model's inability to predict the change that actually occurs. For instance, slow, predictable change would not actually be classified as providing new information if the model can accurately predict its effect on state variables.

The second model development path pressed on the issue of the contrast between scripted tasks and actions and reactive tasks and actions. The overall task analysis that resulted from protocol analysis of pilot performance called for a high-level structure that consisted of two scripted sets of actions (one explicitly so: the landing checklist; and the other implicitly so: placing the simulator in the pilot's desired configuration during the simulation initiation) and two reactive sets of actions (monitoring flight performance and monitoring the final approach). Our initial model of scripted actions used in this approach model is simple and not prone to errors, though we have experience with more complex models of scripted actions within the ACT-R framework that are subject to intrusions and skipping steps. These more complex models do suggest that increased pilot training is not the answer (though it will help); rather, the use of explicit physical checklists and the control of interruptions during the execution of those checklists may help reduce errors. There are fundamental limits to human cognitive abilities, and cognitive modeling can help determine which external aids will best help in improving pilot performance.

This contrast between scripted and reactive behaviors is also related to the procedural errors found in the transcripts of pilot protocols from the HITL simulations. Although it is common practice to use an actual printed landing checklist, the pilots in the scenarios did not use one, but rather recalled the items from memory. As an example of the kind of error that results from this, Pilot 3 in scenario 6 neither notifies the cabin nor arms the speed brakes. The first is an oversight and understandable given that the pilots know there is no cabin to notify in the simulation. Despite being a standard item on landing checklists, this item was actually skipped fairly often. The second, however, is potentially more important. This omission occurred immediately after a communication from the tower interrupted the execution of the landing checklist. This suggests that one safety procedure might be to restart a landing checklist from scratch if it is interrupted in any way. In practice, the pilots in these scenarios appeared to compensate by issuing multiple exemplars of the critical commands in the landing checklist. In particular, pilots often lowered the landing gear and set the flaps to 15 as a preamble to the landing checklist, where within 5 sec they would then issue the command to lower the landing gear and set the flaps to 15. This redundancy indicates that the pilots may have accumulated some experience that encouraged them to rely on some extra insurance.

Despite strategies such as this, mistakes still creep in. Pilot 3, in scenario 10, after lowering the gear and setting the flaps to 15 (starting at offset of 7:01), starts to execute the landing checklist immediately afterwards (offset 7:03, just seconds later). The first item executed, however, is "speed brakes full," rather than "speed brakes armed." This confusion persists until, several minutes later and approaching 1,000 ft, the PNF asks if "speed brakes full" is really what the PF wants. The pilot then recovers and corrects the mistake, retracting the speed brakes and arming them for automatic deployment on landing. It may be that the simulation environment lacks the audible and tactile feedback that a real plane provides, and perhaps this mistake might never occur in real flight, but it is suggestive of the impact that confusing similar terms can have in an environment where many things are happening at the same time.

Despite very carefully controlled environments and precisely laid out procedures, both the limitations and the adaptivity of human cognition make for very difficult predictions. Computational cognitive architectures and integrated simulation environments have provided us with a framework to model and study the interaction between human cognition and its increasingly information-dense environment. Much work remains to be done in validating and scaling up those techniques, but the trend toward increased complexity in our environment on the one hand and our relatively fixed abilities on the other makes the work of fitting one to the other ever more important.

REFERENCES

Anderson, J. R., Bothell, D., Byrne, M. D., Douglass, S., Lebiere, C., & Qin, Y. (2004). An integrated theory of the mind. *Psychological Review, 111*(4), 1036–1060.

Anderson, J. R., & Lebiere, C. (1998). *The atomic components of thought.* Mahwah, NJ: Lawrence Erlbaum Associates.

Anderson, J. R., Bothell, D., Lebiere, C., & Matessa, M. (1998). An integrated theory of list memory. *Journal of Memory and Language, 38*, 341–380.

Anderson, J. R., & Schooler, L. J. (1991). Reflections of the environment in memory. *Psychological Science, 2*, 396–408.

Archer, S., & Lavine, N. (2000). *Modeling architecture to support goal oriented human performance.* Interservice and Industry Training Systems Conference (I/ITSEC), Orlando, Florida.

Archer, S. G., & Adkins, R. (1999). *IMPRINT user's guide* (prepared for U.S. Army Research Laboratory, Human Research and Engineering Directorate).

Archer, S. G., & Allender, L. (2001). New capabilities in the Army's human performance modeling tool. *Proceedings of the Advanced Simulation Technologies Conference,* Seattle, WA.

Byrne, M. D., & Anderson, J. R. (2001). Serial modules in parallel: The psychological refractory period and perfect time-sharing. *Psychological Review, 108,* 847–869.

Ericsson, K. A., & Simon, H. A. (1993). *Protocol analysis: Verbal reports as data.* Cambridge, MA: The MIT Press.

Goodman, A., Hooey, B. L., Foyle, D. C., & Wilson, J. R. (2003). Characterizing visual performance during approach and landing with and without a synthetic vision display: A part-task study. In D. C. Foyle, A. Goodmam, & B. L. Hooey (Eds.), *Proceedings of the 2003 NASA Aviation Safety Program Conference on Human Performance Modeling of Approach and Landing with Augmented Displays* (NASA/CP-2003-212267). Moffett Field, CA: NASA Ames Research Center.

Hooey, B. L., & Foyle, D. C. (2006). Pilot navigation errors on the airport surface: Identifying contributing factors and mitigating solutions. *International Journal of Aviation Psychology, 16*(1), 51–76.

Laughery, K. R., Lebiere, C., & Archer, R. (2006). Modeling human performance in complex systems. In G. Salvendy (Ed.), *2005 Handbook of human factors and ergonomics.* New York: Wiley.

Lebiere, C. (1998). *The dynamics of cognition: An ACT-R model of cognitive arithmetic.* Ph.D. Dissertation. CMU Computer Science Dept Technical Report CMU-CS-98-186.

Lebiere, C. (1999). *Blending: An ACT-R mechanism for aggregate retrievals.* Presented at the *Sixth Annual ACT-R Workshop.* George Mason University, Fairfax, VA.

Lebiere, C. (2001). A theory-based model of cognitive workload and its applications. *Proceedings of the 2001 Interservice/Industry Training, Simulation and Education Conference (I/ITSEC).* Arlington, VA: NDIA.

Lebiere, C. (2005). Constrained functionality: Application of the ACT-R cognitive architecture to the AMBR modeling comparison. In K. A. Gluck & R. W. Pew (Eds.), *Modeling human behavior with integrated cognitive architectures: Comparison, evaluation, and validation.* Mahwah, NJ: Lawrence Erlbaum Associates.

Lovett, M. C., Reder, L. M., & Lebiere, C. (1999). Modeling working memory in a unified architecture: An ACT-R perspective. In A. Miyake & P. Shah (Eds.), *Models of working memory: Mechanisms of active maintenance and executive control.* New York: Cambridge University Press.

Newell, A., & Rosenbloom, P. S. (1981). Mechanisms of skill acquisition and the power law of practice. In J. R. Anderson (Ed.), *Cognitive skills and their acquisition* (pp. 1–56). Hillsdale, NJ: Lawrence Erlbaum Associates.

Rubin, D. C., & Wenzel, A. E. (1990). One hundred years of forgetting: A quantitative description of retention. *Psychological Review, 103,* 734–760.

Taatgen, N. A. (2001). A model of individual differences in learning air traffic control. *Proceedings of the fourth international conference on cognitive modeling* (pp. 211–216). Mahwah, NJ: Lawrence Erlbaum Associates.

7 Air MIDAS
A Closed–Loop Model Framework

*Kevin M. Corker, Koji Muraoka,
Savita Verma, Amit Jadhav, and Brian F. Gore*

CONTENTS

INTRODUCTION

In the early stages of system design, termed the conceptual design phase, there are typically multiple competing alternative designs to be evaluated, selected among, and merged or decided against. In this process, simulations of these alternatives are often used to embed the conceptual designs under consideration into their intended future operating environment. Computational representations and models of structures, of propulsion, of process control, and of aerodynamic systems are commonly used to explore variations in new conceptual designs for product or process enhancement. Modeling is a strategy for relatively quickly understanding the behavior of a system in cases where the complexity in its operational context makes analysis difficult or impossible. The usefulness of a model is determined by its ability to explain and bring a focus to observations, to simplify explanations, to predict, and to inspire discovery.

Models typically focus on some aspects of system behavior for the sake of comprehension, requiring a purposeful choice as to which behaviors require focus and which can be neglected based on the phenomena under study. Simulation is a process whereby models of system elements are brought into dynamic interaction in a common time or event reference frame. Simulation (the action of models in time) provides the systems engineer with an ability to set the system under study into action under specific conditions called collectively cases, or scenarios. With the models describing both the system elements and the scenario representation under computational control, the analysts can explore an array of behaviors or parameters of system performance to guide alternative selection and future development.

The systems under investigation described in this chapter are those associated with human operators interacting in an airspace through management and control of aircraft in civil transportation. As such, system elements of interest in the model are the airspace, the airports, the aircraft, subsystems on the aircraft and the human operators in the system who either pilot the aircraft or manage the airspace, and the movements of aircraft in that airspace. This list includes the physical components of a system for which there are an extensive array of traditional models. It also includes the human operators of those systems. It is here that the modeling and simulation state of the art is encountered.

The human operators' performance characteristics are determined by a complex interaction of cognitive processes and the physical world. The behavior of the physical components in these systems is predictable within a range of performance, and the sources of variation are known and can be accounted for in either the models or the simulation scenario. Humans, on the other hand, are not well modeled in their subsystem actions (perception, cognition, action) and are even less well understood in their interaction with each other and automated systems. In fact, humans are often maintained in highly automated systems specifically because of their ability to manage responses to the unexpected and to adapt to conditions that are changing in ways that had not or could not have been predicted by the systems' designers. Thus, human–system modeling is critical to the successful assessment of complex dynamic systems, and its application is the central theme of this chapter.

Traditional engineering models of human performance have considered the human operator as a transfer function and remnant in a continuous control mode of operation. They have concentrated on the interaction of one operator and a machine system with concern for system stability, accuracy of tracking performance, information processing of displays, and ability to handle disturbances. They are intended to provide guidance in design that determines whether the information provided, and the control system through which the operator performs his or her functions, allows successful performance with an acceptable level of effort (Baron & Corker, 1989). These models assume a closed-loop control in which the human operator observes the current state of the system and constructs a set of expectations based on his or her knowledge of the system. This internal model is modified by the most recent observation and, based on expectations, the operator assigns a set of control gains or weighting functions that maximize the accuracy of a command decision. In this loop, the operator is also characterized as introducing observation and motor noise (effector inaccuracies) and time delay smoothed by an operator bandwidth constraint.

However, as the human operator becomes increasingly served by automation in control of the complex dynamic systems of aerospace operation, a new representation of human–system interaction is needed to account for the cognitive and perceptual support provided to these operators. This development was led by Sheridan's work in supervisory control (Sheridan & Ferrell, 1975). The human operators' function in the distributed air/ground air-traffic management (ATM) system includes visual monitoring, perception, spatial reasoning, planning, decision making, communication, procedure selection, and execution. The level of detail to which each of these functions needs to be modeled depends upon the purpose of the prediction of the simulation model.

Human operators sharing control and information with automated aiding systems and with other operators in the control of complex dynamic systems will, by the nature of those systems, need to perform several tasks within the same time frame or within closely spaced time frames. Multiple operators performing multiple tasks are a challenge to the state of the art in human performance representation. What makes it absolutely essential to model multitasking and task management stems from the fact that there are limits to human bandwidth, information storage/retrieval capacity, and our ability to focus on attention resources that clearly affect performance.

The systems being reviewed in this chapter are focused on extending that range of performance in perception and decision making.

To support the analysis of such systems, we have undertaken representation of the "internal models and cognitive function" of the human operator in complex control systems. These systems are a hybrid of continuous control, discrete control, and critical decision making. These systems are characterized by critical coupling among control elements that have shared responsibility among humans and machines in a shifting and context-sensitive function (Billings, 1996).

Computational simulations of human–system performance complement human-in-the-loop (HITL) testing to uncover systemic flaws conducive to human error under a range of conditions. Fast-time simulations incorporating models of human performance support prediction of the sensitivities of human performance to changes in design, procedures, and training. When combined with HITL testing in nominal and off-nominal scenarios, fast-time simulations provide a complementary technique to develop systems and procedures that are tailored to the humans' tasks, capabilities, and limitations.

Using a computational human performance model (HPM), three distinct evaluations were undertaken with support from the National Aeronautics and Space Administration Human Performance Modeling (NASA HPM) project. The Air Man-machine Design Integrated Design and Analysis System (Air MIDAS) was used to evaluate airport surface (taxi) and flight operations, including the following:

- Prediction of performance errors in surface operations based on WM limitations and context-based response to scenario constraints;
- Prediction of information-seeking behavior in visual scanning with synthetic vision system (SVS) operations in approach and landing operations; and,
- Flight precision performance and accuracy in control of aircraft on approach with support from the simulated SVS.

In each of these modeling developments, the functionality that was provided for the HPM is documented and the output of the resultant model was compared to actual human performance in simulation. These simulations were supported by that part of the NASA Aviation Safety and Security Program (AvSSP) chartered with exploring the use of HPMs to examine the design of advanced aviation products supporting taxi operations and an enhanced visual system for approach and landing. The results of these efforts will be reviewed in this chapter.

We will begin by providing a description of the general HPM (Air MIDAS), its architecture, and its development to support the previously listed applications. We will then review the three applications and describe the required development and the results of the model runs. We will conclude by comparing the model's output with human performance in similar scenarios and suggest ongoing challenges with validation in the area of human performance modeling.

AIR MIDAS HUMAN PERFORMANCE MODEL

HUMAN PERFORMANCE MODELING: GENERAL ARCHITECTURE

Human performance models have been used to provide validated predictions of performance in complex operating environments such as highly advanced military systems (Atencio, 1998), nuclear power plant operations (Corker, 1995), and advanced concepts in aviation (Corker, Gore, Fleming, & Lane, 2000). The human performance modeling software tool Air MIDAS generated predictions of human operator performance using advanced display technologies that are aimed at improving the safety of aircraft operations.

The analyses can be exercised in two modes. First, in a single-run mode, the model's sources of variability are held constant (other than those being explicitly examined—for example, variation in an operational procedure). In this what-if exploratory mode, the effect of point changes in design can be evaluated. Second, the modeled system can be run for a large number of repetitions (Monte Carlo mode) with parameters of the modeled performance varying within prescribed ranges. The result of these runs is a distribution of system performance and an exploration of options. This Monte Carlo process is supported by the fact that the model of human performance provides ranges of durations for action and provides variation in what action is selected to be performed based on environmental conditions and resource constraints. The model provides predictions of operator behavior based on elementary perception, attention, working memory, procedural knowledge, and decision making. This modeling approach focuses on functional models of human performance that feed forward and feed back to other constituent models of the human system.

Air MIDAS

The Air MIDAS model (a joint U.S. Army, NASA Ames Research Center, and San Jose State University development effort) is a framework of models that predict human performance. The modeling framework is a software architecture that links functional models of human performance so that the output of one of those characteristic functions serves as the input to another in a closed loop between the simulation of the human operator and the environment in which he or she acts. The output of this integrated computational framework is a stream of behavior that has impact on the simulated world. This provides operational measures of the modeled world and a set of measures that characterize the perceptual, motor, and cognitive processes that generated that behavior.

Functions Represented in the Model

The Air MIDAS model is used in all of the studies reviewed in this chapter. Specific elements of the model are as follows:

- Perceptual process models were provided for visual and audition actions of flight crew and air traffic controllers.
- An updateable world representation (UWR) represents the operator's understanding of the world (i.e., information about environment, equipment, physical constraints, and procedures). The declarative information about the world is represented in a semantic net. The procedural information is held as decomposition, interruption, and completion procedures for goals, subgoals, and activities. Specific values in the world information serve to trigger activities in the simulated operator. It is worth noting here that the model does not embody

any representation of learning. The operator is assumed to be at some level of training with respect to the mission at hand. The level of training does not change across the simulation. The simulated operator's behavior does change in response to time stress and competency (as will be described).

- Active decision-making processes are represented as rules in propositional structure, as heuristics, or as software triggers (demons), which serve to trigger action in response to specific values in the environment.
- A scheduling mechanism imposes order on activities to be performed depending on their priority and the available resources to perform those activities. The scheduling mechanism also incorporates a switching mechanism that selects among control "modes" in which an operator can perform. These modes are a computational implementation of the COCOM model (Hollnagel, 1993; Verma & Corker, 2001) in which the model uses qualitatively different action sequences to perform required activity depending on the resource availability, time, and competence level of the operator.
- A set of queues manage current activities that are interrupted or waiting to be rescheduled.
- A representation of motor activity represents the process of performance of the selected action in the simulation.

Each of these functions is represented as an independent agent that communicates with each other in a message-passing protocol and keeps track of those interactions as a source of data about the functions of the model.

HOW THE SIMULATION PROCEEDS

The model is a discrete-event simulation in which events are defined as temporal increments (called "ticks") of a clock that sends a message to all agents to proceed with their functions at each event. The events/time base is a variable that can be set by the analyst using the model. Its lower bound is a 100-msec step size.

An activity is triggered in the model in two ways: by decomposition of goals to be performed, or by occurrence of a specific value in the environment for which there is a demon to respond and identify that value as significant and requiring action. When an activity is triggered in service of the mission goal or in response to environmental stimuli, the activity, prior to its having an effect in the simulation, is managed by the scheduler.

Activities are characterized by several defining parameters that include conditions under which the activity can be performed, its relative priority with respect to other activities, an estimate of its duration for scheduling, its interruption specifications, and the resource required to perform that activity. The scheduler is a blackboard process that evaluates these parameters and develops a time for the activities performance in the ongoing simulation. Following a multiple resource assumption, activity load is defined for visual, auditory, cognitive, and psychomotor dimensions (McCracken & Aldrich, 1984). Activities also require information for their performance. The information requirement is identified as knowledge either held in the operator's memory or available from the environment. If the information necessary for activity performance is available, and its priority is sufficient to warrant

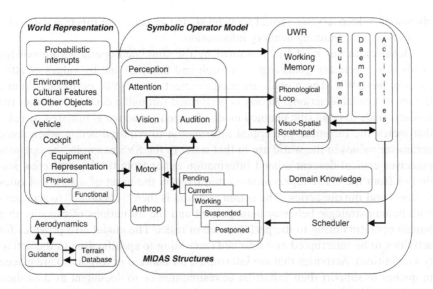

FIGURE 7.1 Air MIDAS constituent function agents and information flow. The agents that make up the Air MIDAS set of behaviors are represented as well as the linkage of information passed from one agent to another in the process.

performance, then the scheduler operates according to heuristics that can be selected by the analyst. In most cases, the heuristic is to perform activities concurrently when that is possible, based on knowledge and resource constraints.

Figure 7.1 illustrates the model's organization and flow of information among the model's components. The model is based on an agent architecture that is illustrated in the figure with the boundary of each of the hierarchically structured agents represented by an enclosure. Data are transformed in each agent and passed to other agents in a cyclic process. The "cost" of the transformation in terms of the operator resources required and time is calculated and archived for "workload" analysis.

For example, in aircraft operations, visual information, such as display information provided by the cockpit visual augmentation systems or the out-the-window (OTW) information, is perceived and attended to by the Air MIDAS simulated operator. The perceptual process follows a simulated scan pattern in which specific elements of the aircraft display suite or the external environment are queried for their state. The processes for perception and attention require both time and dedicated perceptual resources. The amount of information "perceived" and its accuracy are calculated as a function of the amount of time provided for its perception and the characteristics of the source of the information (either environmental or instrument).

This external information is then passed to the "updateable world representation" (UWR). What the operator knows about the world and his or her knowledge of required procedure makes up the "situation awareness" (SA). This information is held in a buffer in the UWR that represents working memory (WM). The WM is structured using the architecture suggested by Baddeley and Hitch (1974) with a phonological loop, a visuo-spatial scratchpad, and an executive function with rules for invoking and retaining memory information.

Behavior in our model is generated by the simulated operator's knowledge of his or her primary goals (e.g., approach and landing) and by values of attributes in the UWR. These goals are used to select action to be performed from a library of available actions. Action is performed through a decomposition of the mission goals to tasks (the planned mode of action) or action is initiated by rules that match the incoming perceptual information to required action when some parameter values in the world are in that state. In the SVS example, the operator perceives critical descent-related information either from the internal crew station or from the OTW representation that triggers the onset of a series of rules to select and initiate action. The action required is then "scheduled" to represent both human strategic behavior management and the limitations of capacity that human operators bring to the performance of tasks. The model also provides for activities to be interrupted and resumed (according to specification in the activity's definition). Activities that are interrupted (or aborted or forgotten) are placed in queues to support their initiation or resumption or to document as data their abortion and/or forgotten state. Once activities are initiated in the simulation, the transformation and information exchange defined in those activities takes place in the simulated world. The simulation world is then changed and the cycle of performance continues.

The forgoing provides an overview of the HPM used in these analyses. Particular details in the operation will be described as a function of the requirements for each of the three applications of this model to the analysis of advanced aviation systems.

SIMULATION 1: AIRPORT SURFACE (TAXI) OPERATIONS

We will first describe the scenario and the modeled components in the airport surface (taxi) operations simulation.

MODEL FUNCTIONS

In our surface (taxi) operations simulation we replicated NASA's full-mission HITL simulation of airport taxi operations at Chicago O'Hare Airport (KORD) (please see chapter 3 of this volume for a description of that experiment). In that simulation it was observed that the flight crews made navigation errors under low-visibility conditions while taxiing from the runway to the gate without the aid of cockpit navigation displays. Our model was intended to represent the processes by which those errors may have been committed. Our hypothesis for this representation was that the errors observed were a function of two elements of human performance. The first, we asserted, was associated with the memory of the details of the clearance, and the second was associated with a joint awareness between the ground controller and the flight crew with respect to the details of the clearance.

A simulation was created to generate predictions of human workload and time to complete various procedures in current-day operations. We encoded WM decay rates and capacity limits, as well as the required procedures to manage the rollout and taxi operations for the simulation. Included in these procedures were several heuristics to guide flight crew behavior in the case of missing or incomplete information, based on the work of Reason (1990).

The control of activity in Air MIDAS is sensitive to memory errors. Three types of memory error mechanisms were represented:

- Declarative memory errors included errors that occurred when simulated operators forgot to complete a procedure as a result of having too many procedures of the same type operating at a given time. The occurrence of this error was modeled by scheduling the simulation to cause multiple competing behaviors to occur concurrently and invoke the procedure scheduler (dropped tasks = memory loss).
- Memory load errors occurred as a result of information competing for WM space. When there were a number of items occupying WM, one item was shifted out of the limited capacity store by the subsequent information from other communication. This information was lost if it was not written down—for example, as with a lengthy taxi clearance.
- Updateable world representation discrepancy errors occurred when there was a worldview inconsistency between two operators. This error occurred when one simulated operator erroneously "thought" another operator had accurately received information.

The errors represented here were elicited from model behavior by increasing the number of items in WM and by increasing information transfers through active procedures. Each type of error emerged as a result of the scenario demands placed on the simulated operator.

AIRPORT SURFACE (TAXI) OPERATIONS SCENARIO

The scenario began when the ground controller sent a taxi clearance to the flight crew. The selected environment for simulation was a taxi route common to KORD operations (see chapter 3 for a review of the operating environment). We modeled three agents: the tower air traffic controller (ATC), the first officer (FO), and the captain. The procedures required of these three are provided in Figure 7.2.

HUMAN PERFORMANCE

This model was expected to generate predictions about human workload, errors that might result from that workload, and the time to complete various procedures. The environmental triggers, a key component in the emergent HPM, were coded in the Air MIDAS environment. Environment triggers (e.g., turns, signs, ATC calls) elicited behaviors. These served to identify risk factors that increase the probability of error. We manipulated the encoded WM decay rate, memory capacity, and information availability to the operators in the simulation. Error rates, performance times, and workload were output from the HPM.

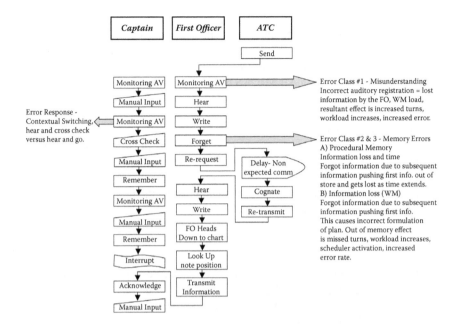

FIGURE 7.2 Overview of the scenario task flow guiding the development of human error within Air MIDAS.

DESIGN AND PROCEDURE

Having the classes of error discovered from the HITL simulation emerge from our model required the development of a rich external environment and a relatively complex set of procedures. We accomplished this by building a representation of pilot–pilot interaction, ATC–pilot interaction, and the airport surface environment. Realistic landing procedures were modeled, including environmental monitoring, changing radio frequency, contacting ATC and listening to the clearance, writing down the clearance, and intracockpit communication.

INDEPENDENT VARIABLES

In this experiment, we varied both the characteristics of the crew's memory and the demands made on their memory to determine if the modeled crew would produce behavior—specifically, error behavior—similar to that observed in the HITL experiment.

The manipulation of the crew's memory characteristics had two dimensions: the capacity of the WM store and the duration of the WM. We created a "low-capacity, short-duration" model and a "high-capacity, long-duration" model. The values assigned to these were intended to span the range of operator performance characteristic of this working environment (Baddeley, 1992; Smith & Corker, 1993). The low-capacity, short-term WM could contain five items and these were retained for 12 sec from time of entry into WM. The high-capacity, longer term model had a capacity of 10 items and these were retained for 24 sec.

The second independent variable manipulated was the point in time during taxi operations that the flight crew received the navigation instruction from ATC. The instructions were provided 24 sec, 50 sec, and 80 sec after exiting the runway. The memory manipulations were counterbalanced with the time that the crew received the navigation information. Each simulation scenario was run 10 times. For a more detailed description of this experiment, please refer to Gore (2002).

Results

The modeled operator made turn-direction errors at the two turn points in the taxi operations. These were similar to the errors made in the HITL experiment reported in chapter 3. We hypothesize that the reason for the errors in the first case was the perceptual and motor load in exiting the runway. In this case, the model sheds tasks (e.g., cross-checking turns), and the lack of cross-checking (in the probabilistic selection of action) resulted in a mistaken turn. In the second case, there was a memory information loss at a decision point to a taxiway. In this case, the operator model chooses a heuristic "take shortest route to the gate" to guide turn selection.

Visual demands were greatest at the first taxi intersection where the first crew error occurred. Cognitive demand was found to be associated with the second intersection, where the flight crew was relying more on their memory for accurate directions as opposed to using information perceptually available. Auditory demands also increased for the second intersection and that was due to increased communication between the flight crew because they needed to refresh their memory store as taxi-time increased from the time of receipt to the time of use. It is interesting to note that the flight crew forgot their navigation instructions when they relied solely on their memory.

Summary

The human performance modeling software tool, Air MIDAS, predicted performance effects of varying environmental conditions and effects associated with information availability. We feel that the behavioral predictions that emerge from Air MIDAS demonstrate that internal structures within Air MIDAS are sensitive to assumed psychological capacities and scenario demands. Procedural interruptions occur when operators are faced with procedures that compete for procedural memory resources. The availability of these resources is limited in time, and procedures become lost if time extends beyond an acceptable upper time boundary. The modeled procedural errors, manifested by incorrect turns, are consistent with the HITL surface operations simulation that found evidence of errors occurring because operators omitted or substituted parts of a required taxi clearance to get to the gate—a procedural memory error (see chapter 3; Hooey & Foyle, 2006).

SIMULATION 2: APPROACH AND LANDING USING A SYNTHETIC VISION SYSTEM

We undertook this analysis to predict the performance of pilots using an SVS. Many industry and government partners are developing technologies designed to aid the flight crew in the safe operation of the aircraft under conditions that in the past have been shown to contribute to increased hazards in aviation operations. Those technologies have a common purpose to aid the flight crew by providing information that either has not been available (e.g., improved traffic position information or rapid update of local meteorological conditions) or has been obscured and degraded (e.g., OTW visibility reduction in weather and at night). The advancements in computational techniques and sensor and communication technologies have resulted in an enviable design situation in which the amount and quality of information available must be carefully selected and designed to avoid overwhelming the flight crew.

As one element in this process, NASA and industry have been developing augmentative technologies as part of an SVS for commercial aircraft operations. In general, these systems are designed to generate a texture-mapped display of the Earth's surface and cultural features using a geographic database. Text and other symbology may be overlaid onto the terrain display to indicate any of the following: the aircraft itself, its velocity, a "follow-me" aircraft and its velocity, a "tunnel-in-the-sky" indication of the route, indications of other nearby aircraft, and flight control indicators (airspeed, attitude, elevation, etc.).

PROCEDURE DEVELOPMENT

The simulation represented an approach to the Santa Barbara Municipal Airport (SBA) in California under two visibility conditions (limited and clear) and two conditions of flight deck aiding (with and without the use of an SVS). Part-task simulation data from an HITL simulation that was completed by NASA Ames Research Center (see chapter 3) provided visual fixations for the PF. We produced general human performance models representing both PF and PNF and an approach air traffic controller. The procedures represented were associated with descent and approach.

PERCEPTUAL SYSTEM DEVELOPMENT

We included the affect of contrast legibility on visual search/reading time (Landy, 2002) in a reading rate model. With this model, we estimated visual acquisition time for text and flight deck display.

SCENARIO DEVELOPMENT

Scenarios were developed for approach and landing without SVS (baseline) and with the advanced SVS. These approaches were made in low-visibility mode under instrument meteorological conditions (IMC). The IMC approach used area navigation (RNAV), a global positioning system (GPS) precision style of approach. Area navigation is defined as a method of navigation that permits aircraft operation on any desired course within the coverage of station-referenced navigation signals or within the limits of a self-contained system capability, or a combination of these (NASA Web Tutor). In addition to the standard approaches, an approach condition that required a sidestep maneuver in low-visibility operations using the SVS system was used.

MODEL DEVELOPMENT

We developed a representation of the equipment and an augmented visual behavior model that incorporated visual scanning performed by pilots. The Air MIDAS perceptual functions developed included in-cockpit scanning both with and without the display augmentation of the SVS on the primary flight display (PFD) and on the navigation display (ND). In addition, an OTW visual capture model was implemented for detecting features associated with making decisions on approach involving aircraft stability and aircraft landing requirements. The visual scan patterns for the model (dwell fixations and dwell durations) were developed using the scan pattern data taken from the research data of Mumaw et al. (2000). An assumption was made that in cases in which the Mumaw et al. data included OTW fixations, those would be implemented as SVS scans in the SVS mode of operation. The fixations provided the aircraft state data used to populate the equipment representation in the model. An informal verification process was followed during model development with reference to the model calibration data from Mumaw et al.

A description of each of the visual sampling functions and its individual components is provided in Figure 7.3 and described next.

The major components of the visual processing model include a "tickable" database, which represents equipment that would be scanned by the PF and PNF. The flight crew agents, the PF and PNF, queried the database record. A number of instruments were incorporated in the simulation to provide the Air MIDAS

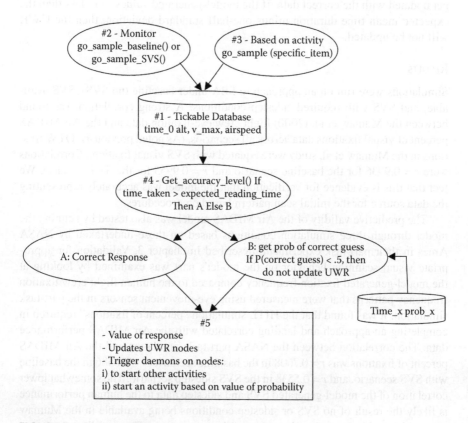

FIGURE 7.3 Flow of information for the vision model.

agent with required information. These data are extracted from the information provided by the PFD, mode control panel (MCP), ND, SVS, OTW scene, and other flight control information such as the flaps, the throttle, and the speed brake controls. The PFD provided the Air MIDAS operator with altitude, airspeed, attitude, heading, and the flight mode annunciator state. The MCP provided the vertical speed, the MCP altitude, and MCP altitude dial. The ND provided the aircraft true heading information. The OTW provided the simulated world landing information in terms of visibility of the runway. A more extensive examination of the SVS augmented display is provided in a later section. As represented in Figure 7.3, information from the database flows into the Air MIDAS agent's UWR through a scan pattern. The "go sample (specific target)" activity gets triggered whenever an activity requires information. For example, during the altitude change phase, the altimeter is scanned before every callout. There are two principal functions that need to be considered in information uptake: the expected dwell durations and the accuracy functions. Dwell durations used were from the Mumaw et al. (2000) data. Information uptake was based on an assumption that longer dwell durations will result in more accurate uptake of the information. If the model-generated dwell durations were more than the expected mean time duration (based on reading rate calculations) plus one-half standard deviation of dwell time, then the UWR will get updated with the correct data. If the model-generated values were less than the expected mean time duration minus one-half standard deviation, then the UWR will not be updated.

RESULTS

Simulations were run on an approach to SBA under baseline (no SVS), SVS available, and SVS with required sidestep conditions. A strong correlation was found between the Mumaw et al. (2000) flight crew eye-fixation data and the Air MIDAS percent of visual fixations data across all scenarios. (As noted previously, OTW fixations in the Mumaw et al. study were equated with SVS visual fixation.) Correlations were $r = 0.9936$ for the baseline scenario and $r = 0.9955$ for the SVS scenario. We feel that this is evidence for verification that the model was accurately representing the data source for the initial scan pattern based on procedures.

The predictive validity of the Air MIDAS model was also tested by running the model through three simulation conditions based on those undertaken by NASA Ames in their part-task experiment, described in chapter 3. Validation for appropriate visual scanning behavior on the model's part was examined by looking at the model-generated fixation frequency compared to the human flight crew fixation frequency patterns that were measured using eye-movement sensors in the part-task simulation. It was found that the HITL simulation "percent of fixations" required in completing an approach and landing correlated with the Air MIDAS performance data. The correlation between the NASA part-task simulation and the Air MIDAS percent of fixations was $r = 0.7608$ in the baseline scenario, $r = 0.6586$ in the baseline with SVS scenario, and $r = 0.5538$ in the SVS sidestep scenario. The somewhat lower correlation of the model-generated SVS and sidestep data to the human performance is likely the result of no SVS or sidestep conditions being available in the Mumaw data, which had been used as the source of the scan pattern. The model's scan pattern

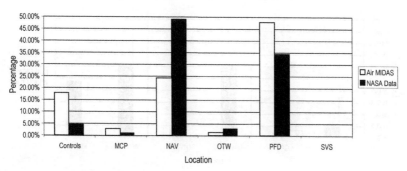

Comparison of Percent of Fixation Without SVS

FIGURE 7.4a Percentage of total fixations on specific instruments in a comparison between the model's performance and that of the human pilots (see chapter 3). Data shown are for a nominal approach into Santa Barbara Airport (SBA) under instrument meteorological conditions (IMC) without a synthetic vision system (SVS) (initial approach fix, IAF, to just before touchdown; Pilot 3, Scenario 4 in chapter 3).

in these conditions was predicted based on information requirements to achieve active goals.

In the baseline IMC scenario (without SVS) Air MIDAS predicted higher fixation rate on the controls and the PFD than did the human data produced by the HITL simulation. This suggests that the rules guiding human performance are different from those guiding the model's performance. The human pilot flies to a larger extent using the information on the ND, given the fixation pattern, than does the Air MIDAS pilot; the Air MIDAS pilot fixated on the PFD to a larger extent than did the human pilots. Figure 7.4a illustrates that mismatch.

In the IMC scenario with SVS, the Air MIDAS model also predicted lower dwells on the ND, the OTW scene, and the SVS displays than did the HITL simulation. This suggests that when flying with the SVS display, the human pilot looked at the SVS information to a greater extent than did the modeled pilot. We postulate that the human pilot received PFD information from overlays in the SVS and the model required the pilot to look at the PFD (not the SVS) for that information. Figure 7.4b provides an illustration of that difference.

The assumptions that guided the model's information-seeking behavior on approach were that the primary source of information for flight control was the standard cockpit instruments (e.g., PFD and ND displays). The SVS system was considered supplementary information to add to a flight crew's awareness, but not to be relied on for flight control. (This issue of the operational use of the SVS for flight command guidance was, in part, driven by concerns for certification of the terrain and cultural databases that are embedded in the SVS as flight critical software; such a designation brings with it a significant burden of safety and reliability checks.) This assumption was instantiated in the model's activity structure such that the goal for flight path maintenance, for instance, was satisfied by successful completion of subgoals that included seeking information from instruments, with the PFD and ND

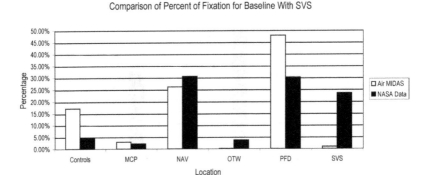

FIGURE 7.4b Percentage of total fixations on specific instruments in a comparison between the model's performance and that of the human pilots (see chapter 3). Data shown are for a nominal approach into Santa Barbara Airport (SBA) under instrument meteorological conditions (IMC) with a synthetic vision system (SVS) (initial approach fix, IAF, to just before touchdown; Pilot 3, Scenario 7 in chapter 3).

displays as the primary instruments interrogated for flight guidance. The SVS information was used to service a goal to maintain safe flight operations that included concern for other traffic and terrain obstacles as well as general ownship orientation in reference to the approach.

With these assumptions in place, the model underestimated the use of SVS by the human crews. We hypothesize that the human flight crews not only used the SVS for general orientation, traffic, and terrain awareness, but also used the overlaid heading and aircraft pitch and flight path guidance to effect their control. This difference in assumed operational use of SVS data can be seen in the fact that the model use of the PFD and ND displays does not change with the addition of the SVS display, as illustrated in Figure 7.4b compared to Figure 7.4a.

It should be noted that this simulation was undertaken in conditions in which the human performance model and the aircraft model were synchronized by a procedural script. That is, the HITL procedural performance established the procedural pacing for the simulation. This resulted in an essentially open-loop and scripted model performance. In the next simulation to be discussed, this procedure-based simulation was replaced with an active, dynamic aircraft model under the control of the Air MIDAS model. In this case, the actions of the model are taken in a closed loop around the aircraft performance. Schedule and activity completion are then dependent on the timing and dynamics of the human and aircraft system interaction. This closed-loop operation gives us a much better look at human–system performance demands and the consequences of not meeting those demands.

SUMMARY

The Air MIDAS simulation accurately produced the Mumaw et al. (2000) scan patterns and correlated well with the NASA part-task simulation. The model behavior is congruent with the human operators' performance across experimental conditions.

As noted, the sidestep visual fixation behaviors were derived from the model's goals in the maneuver. The modelers inserted the scan pattern based on the information needed for those goals to be completed and made estimates as to when that information was needed. Insofar as those estimates were inaccurate, the model's visual scan diverged from those of the human pilots.

SIMULATION 3: DETAILED SYNTHETIC VISION SYSTEM OPERATIONS WITH FULLY INTEGRATED AIRCRAFT DYNAMIC MODEL

The third simulation described here is an expansion of the model functionality to explore operational concepts and equipment. It uses the visual scan pattern developed in Simulation 2 and links the HPM to a dynamic aircraft model. This linkage provides an important capability to the simulation by enabling the aircraft to respond in a closed-loop fashion to the simulated pilot's commands. It places the model into a dynamically evolving procedural stream in which workload and scheduling become important components of the model's behavior. The information provided either in the standard operation or with the SVS system then becomes important to support the flight crew's goals in approach. Finally, the addition of the dynamic models allows designers and analysts to explore the response of the model to a variety of environmental conditions to identify the effects of those conditions on system performance.

MODEL FUNCTIONS

The purpose of this study was to generate predictions of human performance using the SVS under several conditions of approach and landing. These predictions are provided by a computational model of human–system performance.

To support these predictions in an accurate representation of the time-varying dynamics of approach to landing, a high-fidelity aircraft performance model, PC Plane model (Palmer, Abbott, & Williams, 1997), was integrated into the Air MIDAS knowledge-base and the aircraft model interacted with the SVS display generation models. The combined human and aircraft model and the flight's evolution in time served as a forcing function, or triggering mechanism, for emergent human performance in interaction with the display and control systems. The Air MIDAS visual scan pattern model, which has been developed and validated in the course of the Air MIDAS SVS project, was used to simulate the pilot's scan pattern to the instruments and the OTW scene. Approach, landing, and go-around, which are the most critical of normal flight phases, were selected for the simulation scenario, and human performance was predicted in terms of a number of dependent variables that include aircraft control performance (e.g., cross-track and vertical error), workload, decision making, and changes in the visual scan pattern change due to SVS usage.

The PC Plane flight simulation model framework was integrated into the existing Air MIDAS architecture as a part of the operator's world representation. PC Plane is a NASA-developed PC-based flight simulation software program providing representation of the flight management system and cockpit displays. PC Plane aircraft dynamics provide flight and system status to equipment components comprising

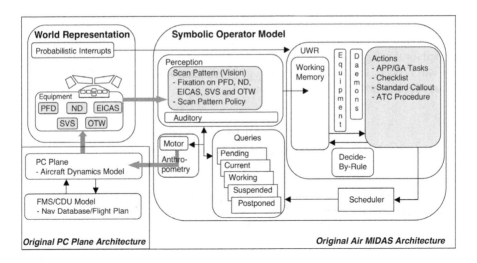

FIGURE 7.5 Overall Air MIDAS synthetic vision system (SVS) application architecture. Shaded areas are illustrative of newly developed detail in the world-equipment model, the pilot's scan pattern, and resulting actions associated with approach using the SVS system.

PFD, ND, engine indicating and crew alert system (EICAS), SVS, and an OTW view that was linked to aircraft dynamics. Visual scan pattern and flight crew cockpit tasks were implemented. The following sections describe the system architecture of the simulation environment as well as detailed implementation of PC Plane and the visual scan pattern model. Figure 7.5 illustrates the expanded detail of the Air MIDAS functional components to represent the dynamic interaction with the SVS.

System Architecture

Figure 7.6 depicts the system architecture of the Air MIDAS simulation environment to support analysis of SVS operations with a dynamic aircraft model. Three modules, including PC Plane aircraft dynamics, flight management system/control display unit (FMS/CDU), and Air MIDAS, were integrated into the simulation. A set of dynamic link library (DLL) functions that generate cockpit control input and time synchronization control to PC Plane through socket connection was prepared. DLL functions were invoked by the Air MIDAS module, which is written in LISP, through a JAVA network interface architecture. Time synchronization control function realized precise synchronization of Air MIDAS and PC Plane during simulation and enabled dynamic closed-loop simulation.

A Microsoft Office Access database architecture was used to share flight and aircraft system parameters between the symbolic operator model (SOM) and the UWR. The database comprises PFD, ND, EICAS, SVS, and OTW data sheets, and each sheet includes parameter values displayed on it. PC Plane updates all of the parameters in the database as time proceeds. The Air MIDAS visual scan pattern model reads data from a particular data sheet when the agent fixates on a corresponding display.

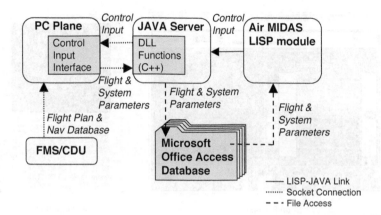

FIGURE 7.6 Software modules and system architecture. The shaded portions indicate new development for this simulation.

EQUIPMENT MODEL

PC Plane Architecture

The original PC Plane has the functions of an aircraft mathematical model, PFD, ND, and a processor for human control input such as flight control from mouse and game stick devices. Also, it has the function of communicating with external modules such as the FMS/CDU and the MCP. For the Air MIDAS UWR, the FMS/CDU module was used and PC Plane software was further enhanced by adding interface functions that process control input from the Air MIDAS pilot agent (instead of a human pilot). Also, a simulation-time control function was added to PC Plane. This enabled fast-time simulation synchronization with Air MIDAS instead of the real-time HITL simulation. The Boeing B757 aerodynamic and engine model of PC Plane was used for this study. The FMS/CDU module of Air MIDAS was connected to PC Plane through a socket connection to provide the navigation database and flight plan data. Figure 7.7 illustrates the flight displays and controls that the PF and PNF models used in flight operations.

Cockpit Displays

The cockpit display model was developed to simulate the PC Plane's flight and system status for the Air MIDAS pilot agent. We assumed displays shown in Figure 7.8 were provided on the aircraft. Air MIDAS does not have a vision function of depth perception or a transformation mechanism for visual image perception when looking at a plan-view display (such as the ND), to recognize the meaning of that information with respect to route of flight. Therefore, the display model was designed to provide the flight and system status by means of numerical values. These values are the sole source of position information in the Air MIDAS model.

The ND is scanned and the database provides position as a result of the scan. The Microsoft Office Access Database framework was used to share the parameter values between the Air MIDAS representation of the human operator (that is, the

(a) PC Plane   (b) FMS/CDU

FIGURE 7.7 PC Plane modules: Panel (a) provides a vertical situation display on the left, navigation/horizontal situation indictor on the right, with mode control switches at the top. Panel (b) is an illustration of the FMS/CDU interface that both members of the simulated flight crew would read and manipulate.

SOM) and PC Plane. The database comprises data sheets for PFD, ND, EICAS, SVS, and OTW. Each sheet includes flight parameter values that would be shown on an equivalent display. Each sheet not only included the values of the information, but also identified the source of the information (identified in the table as "area" source). Figure 7.9 summarizes the specification of the data sheets. All of the parameter values in the data sheets were continuously updated by PC Plane. However, the pilot's internal representation was only updated when the Air MIDAS vision model read part of them by "fixating" on an equivalent area of a display.

The SVS design used in this experiment differs from the SVS used in the NASA HITL simulation (chapter 3) and as described in Simulation 2 of this chapter. The SVS design was chosen to broaden the human performance and equipment models in the simulation series described here. For instance, this design includes a "tunnel-in-the-sky" display that other SVS instantiations did not.

AIR MIDAS SYMBOLIC OPERATOR MODEL

Flight and system information provided by the cockpit display models was passed into the symbolic operator model (SOM) through its visual perception models. To briefly review, the SOM scans for information to support its achievement of goals associated with the approach and landing plan. Information that supports activities is entered into the operator model memory. Information from the SVS tunnel in the sky, for instance, can provide navigation and guidance in the vertical and horizontal planes. That same information (though at a lower level of resolution) is also available from the ND and the vertical situation display. The information needed for the activity is available from two sources. In the case of the SVS tunnel-in-the-sky display, the same information is available from a single source. If the pilot model is seeking

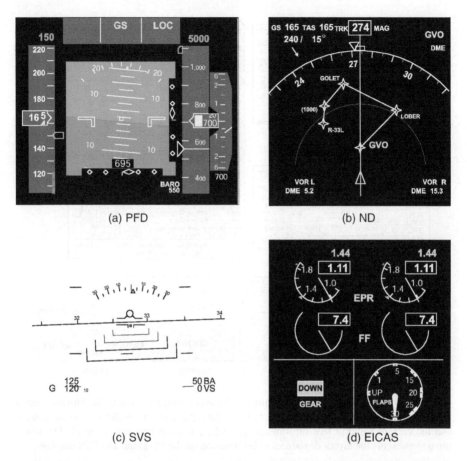

(a) PFD

(b) ND

(c) SVS

(d) EICAS

FIGURE 7.8 PC Plane displays images: (a) Primary fight display (PFD) provides aircraft control information, pitch, roll and yaw, speed, altitude, etc.; (b) navigation display (ND) provides navigation and world orientation information; (c) synthetic vision system (SVS) display including tunnel in the sky and artificial horizon; (d) engine indication and crew alerting system (EICAS), which can be configured to include many sources of system information, electrical, hydraulic, gear status, flaps, etc.

both horizontal and vertical information and is in a workload-constrained portion of the mission, the model will choose the time-optimal information seeking and select the SVS display. Once this information is perceived through the scan pattern, it is passed into the UWR and salient values of the data, defined by the analyst, are used to trigger cockpit activities.

For this study, WM storage nodes to accommodate PFD, ND, EICAS, SVS, and OTW were prepared. In WM, domain knowledge and rules were implemented to invoke actions regarding cockpit procedures such as (1) approach, landing, and go-around procedure; (2) standard callout; (3) checklist; (4) ATC communication; and (5) landing/go-around decision. In this research effort, perceptual processes associated

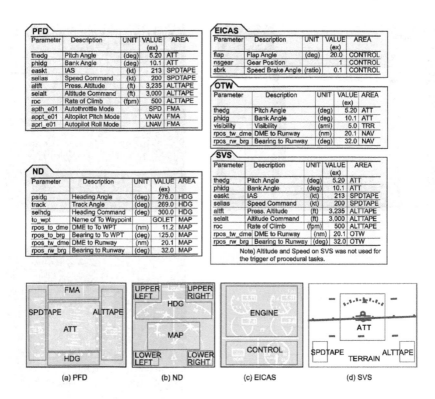

PFD

Parameter	Description	UNIT	VALUE (ex)	AREA
thedg	Pitch Angle	(deg)	5.20	ATT
phidg	Bank Angle	(deg)	10.1	ATT
easkt	IAS	(kt)	213	SPDTAPE
selias	Speed Command	(kt)	200	SPDTAPE
altft	Press. Altitude	(ft)	3,235	ALTTAPE
selalt	Altitude Command	(ft)	3,000	ALTTAPE
roc	Rate of Climb	(fpm)	500	ALTTAPE
apth_e01	Autothrottle Mode		SPD	FMA
appt_e01	Aitopilot Pitch Mode		VNAV	FMA
aprl_e01	Autopilot Roll Mode		LNAV	FMA

EICAS

Parameter	Description	UNIT	VALUE (ex)	AREA
flap	Flap Angle	(deg)	20.0	CONTROL
nsgear	Gear Position		1	CONTROL
sbrk	Speed Brake Angle	(ratio)	0.1	CONTROL

OTW

Parameter	Description	UNIT	VALUE (ex)	AREA
thedg	Pitch Angle	(deg)	5.20	ATT
phidg	Bank Angle	(deg)	10.1	ATT
visibility	Visibility	(smi)	5.0	TRR
rpos_tw_dme	DME to Runway	(nm)	20.1	NAV
rpos_rw_brg	Bearing to Runway	(deg)	32.0	NAV

SVS

Parameter	Description	UNIT	VALUE (ex)	AREA
thedg	Pitch Angle	(deg)	5.20	ATT
phidg	Bank Angle	(deg)	10.1	ATT
easkt	IAS	(kt)	213	SPDTAPE
selias	Speed Command	(kt)	200	SPDTAPE
altft	Press. Altitude	(ft)	3,235	ALTTAPE
selalt	Altitude Command	(ft)	3,000	ALTTAPE
roc	Rate of Climb	(fpm)	500	ALTTAPE
rpos_tw_dme	DME to Runway	(nm)	20.1	OTW
rpos_rw_brg	Bearing to Runway	(deg)	32.0	OTW

Note) Altitude and Speed on SVS was not used for the trigger of procedural tasks.

ND

Parameter	Description	UNIT	VALUE (ex)	AREA
psidg	Heading Angle	(deg)	276.0	HDG
track	Track Angle	(deg)	269.0	HDG
selhdg	Heading Command	(deg)	300.0	HDG
to_wpt	Name of To Waypoint		GOLET	MAP
rpos_to_dme	DME to To WPT	(nm)	11.2	MAP
rpos_to_brg	Bearing to To WPT	(deg)	125.0	MAP
rpos_tw_dme	DME to Runway	(nm)	20.1	MAP
rpos_rw_brg	Bearing to Runway	(deg)	32.0	MAP

(a) PFD   (b) ND   (c) EICAS   (d) SVS

FIGURE 7.9 Cockpit display area definition. The upper image illustrates the data associated with each display element including the types of data, the units in which these are presented, and the current value of that information and where that information is found. The lower image represents the layout of sources of information on the PC Plane and SVS displays.

with the SVS and/or OTW observations are critically important. The detailed structure is described next.

SCAN PATTERN MODEL

The Air MIDAS visual perception model simulates the pilot's information acquisition process from cockpit displays and OTW. This activity was assumed to comprise both periodical information sampling and directed information acquisition associated with demand from a particular cockpit task. During the flight, the pilot continuously monitors flight and system status based on the periodical information sampling. If a certain demand comes from a cockpit task—for example, the pilot's confirmation of his or her own action in setting a speed command—the pilot would intentionally focus on a specific type of information to support that confirmation. In this example, the pilot would look at the speed command indication on the PFD speed tape by interrupting the normal scan. Directed information acquisition of visual perception was implemented as a part of the Air MIDAS cockpit procedural tasks and periodical information sampling was implemented as a scan pattern model. This is described next.

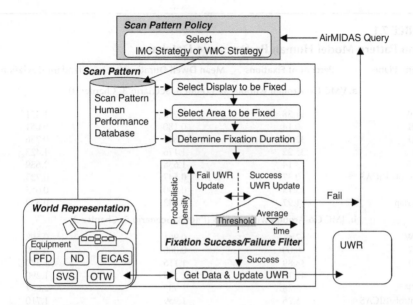

FIGURE 7.10 Visual perception model logic: The figure provides a flow diagram of the logic used to direct the flow of visual information into the updateable world representation (UWR).

A scan pattern model has been developed and validated in the course of ongoing Air MIDAS human performance modeling research efforts (Corker, Verma, Gore, Jadhav, & Guneratne, 2003). For this study, the scan pattern policy was a new addition and the scan pattern's specification of the display configuration and corresponding human performance database was modified to correspond to the simulated cockpit configuration. Figure 7.10 depicts the logic of the Air MIDAS scan pattern model. The scan pattern selects a display, selects the area (scene plane) to fixate its eyes, aggregates the values of displayed parameters, and then updates the UWR. Failure of data aggregation was also simulated, with a corresponding failure to update the UWR.

The scan pattern model shown in Table 7.1 was prepared based on the following assumptions:

- The pilot applies a different scan pattern according to the availability of OTW information. Two different scan pattern strategies, IMC strategy and visual meteorological condition (VMC) strategy, were prepared for with and without SVS configuration.
- Off-area of interest (Off-AOI) and overlapped fixation percentages were combined. Off-AOI scans signify inattention to the instruments. Overlapped AOI simply means that the operator is not foveating, but his or her peripheral vision can detect warnings and similar changes in the instruments.
- In order to account for differences in cockpit versions modeled in these simulations and in the NASA HITL experiment (chapter 3), which did not have an EICAS display, a generic reference—"fixation on control setting"—was used for the equivalent of an EICAS fixation.

TABLE 7.1

Scan Pattern Model Human Performance Database

Scene Plane	Percent of Fixations	Mean Dwell Duration (sec)	Standard Deviation
a. VMC Go-Around without SVS (Pilot 3, Scenario 3, Chapter 3)			
OTW	11.38	0.913	1.451
SVS	0.14	0.180	0.051
PFD	32.65	0.848	0.736
ND	35.22	1.176	1.423
MCP	3.14	1.658	2.639
Controls/EICAS	0.37	0.547	0.723
Off	15.13	0.430	0.653
Overlap	1.97	0.342	0.559
b. IMC Go-Around without SVS (Pilot 3, Scenario 5, Chapter 3)			
OTW	2.41	0.961	0.568
SVS	0.00	0.000	0.000
PFD	38.89	1.116	1.220
ND	47.12	1.169	1.290
MCP	3.28	1.935	2.634
Controls/EICAS	5.26	1.959	1.710
Off	2.91	0.277	0.250
Overlap	0.12	0.175	0.075
c. IMC Go-Around with SVS (Pilot 3, Scenario 9, Chapter 3)			
OTW	0.34	0.427	0.169
SVS	25.34	2.457	2.702
PFD	29.92	0.995	0.817
ND	32.21	1.137	0.945
MCP	4.19	3.176	3.241
Controls/EICAS	4.14	1.451	1.754
Off	3.80	0.403	0.653
Overlap	0.06	0.240	0.000

d. Fixation Rate Implemented for Each Display Area

PFD	Fix %	ND	Fix %	EICAS	Fix %	SVS	Fix %	OTW	Fix %
ATT	34	HDG	40	Engine	80	ATT	25	Terrain	33
SPDTAPE	27	MAP	40	Control	20	OTW	25	NAV	33
ALTTAPE	29	Up Left	5			SPDTAPE	25	ATT	33
FMA	6	Up Right	5			ALTTAPE	25		
HDG	4	Low Left	5						
		Low Right	5						

Notes: The scan pattern percentages (Table 7.1a–7.1c) were based on NASA human-in-the-loop (chapter 3) simulation data. The fixation rates in Table 7.1(d) were provided by subject matter expert pilots examining the procedures needed to perform an approach to landing. IMC = instrument meteorological conditions; MCP = mode control panel; ND = navigation display; OTW = out the window; PFD = primary flight display; SVS = synthetic vision system; VMC = visual meteorological conditions.

Once the display, its area, and duration to be fixated were determined, the fixation success/failure filter evaluated whether the fixation could successfully aggregate data or not. In Simulation 2, the filter was on reading rate. In this simulation (Simulation 3), all information seeking is subject to the filter process: We applied a threshold function such that it was assumed that all the data included in the area can be successfully perceived when the fixation duration was long enough or not perceived at all if the threshold value was not met. Based on this assumption, Air MIDAS updated parameters in the UWR when the fixation duration was longer than a specified threshold, but if shorter, it did not perform any UWR update. Since this assumption was associated with failure of perception and selection of a threshold would affect the results of the simulation, sensitivity analysis on threshold setting for information acquisition failure mode was performed. These threshold settings were examined through simulation runs determined to be set at a mean duration minus 1.0 standard deviation so that the rate of scan failure was about 10% or less. (Note that the Air MIDAS distribution of fixation times is not a normal distribution.) Examination of this threshold setting will be described in the Results and Analysis section.

A simple switching mechanism was employed to control whether and when the scan pattern should include an OTW element. In the case of the PF, in IMC conditions, OTW scans were initiated after PNF called "runway in sight." In the case of VMC, the OTW scan was included after descending to approach minimum. For the PNF, in VMC or IMC, the OTW scan was initiated after the final approach fix. After announcing "runway in sight", the PNF returns to an instrument scan and the PF takes up the OTW scan, as described. In the case of conditions that were not tested in the HITL experiment (chapter 3)—that is, the use of SVS in VMC conditions—we used the VMC baseline data and assumed that the pilots would prefer OTW data as opposed to SVS data in their terrain visual scan.

METHOD

A series of simulation runs (16 conditions and five runs per condition provided in Table 7.2) were performed to evaluate the impact of the SVS on crew performances, on scan pattern change associated with SVS implementation. Normal approach and go-around flight scenarios were simulated and two different decision altitudes (DAs) were prepared. The two different go-around triggers, including ATC command and lack of visibility of runway at DA, were prepared for the go-around scenarios.

SCENARIO

Flight Area

Figure 7.11 depicts the approach chart for the simulation assuming a GPS approach procedure to SBA, runway 33L. Initial position was located 5 nautical miles (nmi) northwest of the GAVIOTA initial approach fix, and flight using autopilot and autothrottle with vertical navigation (VNAV) and lateral navigation (LNAV) modes was assumed. Two DAs were assigned in order to examine the impact of SVS usage on DA selection for approach procedures. DA 650 ft is as high as the usual nonprecision approach decision altitude and DA 200 ft is as high as Category I instrument approach

TABLE 7.2
Simulation Cases

Case No.	Approach	SVS	DA (ft)	Weather Visibility Above (smi)/ Boundary Altitude (ft)/Visibility Below (smi)	Events	Description	Runs
1	Normal approach	Without	650	0.5/800/10.0		Baseline	5
2	Normal approach	With	650	0.5/800/10.0		Baseline	5
3	Normal approach	Without	200	0.5/350/10.0		DA @ 200	5
4	Normal approach	With	200	0.5/350/10.0		DA @ 200	5
5	Go-around	Without	650	0.5/800/10.0	ATC GA comm @ 750 ft	GA by ATC	5
6	Go-around	With	650	0.5/800/10.0	ATC GA comm @ 750 ft	GA by ATC	5
7	Go-around	Without	200	0.5/350/10.0	ATC GA comm @ 300 ft	GA by ATC	5
8	Go-around	With	200	0.5/350/10.0	ATC GA comm @ 300 ft	GA by ATC	5
9	Go-around	Without	650	0.2/650/0.2		GA by pilot	5
10	Go-around	With	650	0.2/650/0.2		GA by pilot	5
11	Go-around	Without	200	0.2/200/0.2		GA by pilot	5
12	Go-around	With	200	0.2/200/0.2		GA by pilot	5
13	Go-around	Without	650	0.5/800/10.0	ATC GA comm @ 900	ATC @ highWL	5
14	Go-around	With	650	0.5/800/10.0	ATC GA comm @ 900	ATC @ highWL	5
15	Go-around	Without	200	0.5/350/10.0	ATC GA comm @ 450	ATC @ highWL	5
16	Go-around	With	200	0.5/350/10.0	ATC GA comm @ 450	ATC @ highWL	5

Note: ATC = air traffic control; comm = communication; DA = decision altitude; GA = go-around; SVS = synthetic vision system; WL = workload; smi = statute miles.

decision height. A detailed missed approach pattern was not prepared as the scope of the go-around simulation was focused on the phase of making the go-around decision and initiating a go-around climb. The go-around simulation terminated when the aircraft achieved positive climb with gear up and flaps 5 configurations.

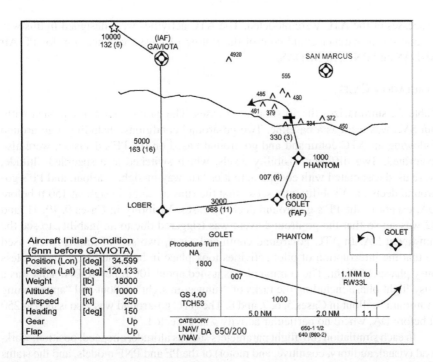

FIGURE 7.11 Flight area: Illustrates the approach to Santa Barbara Airport (SBA). The figure is a reproduction of a standard approach plate used on the flight deck.

The landing simulation terminated when the aircraft touched down. No other aircraft were assumed to be in the area.

Procedures

Cockpit activities were implemented as a part of Air MIDAS activities. These included (a) approach, landing, and go-around; (b) standard callout; (c) checklist; (d) scan policy and scan pattern; (e) ATC communication; and (f) go-around decision.

Among these activities, (a) through (c) are usually described in an aircraft operating manual; similar documents were used for Air MIDAS implementation. Scan pattern policy (d) was installed as a part of the scan pattern model and others were implemented as a part of Air MIDAS's domain knowledge and tasks. ATC communication (e) included approach clearance, landing clearance, or go-around command. The PNF was assumed to be in charge of the ATC communication tasks. Go-around decision (f) for this model was defined as a set of activities that decide whether to continue landing or perform the go-around based on the flight and system status such as visibility of the runway, tracking of nominal approach path, the stability of aircraft, etc. Go-around decision for Air MIDAS was scheduled so that it was taken immediately after passing DA.

Three human agent models, PF Air MIDAS, PNF Air MIDAS, and ATC Air MIDAS, were included in each simulation run. The ATC agent's set of activities was mostly communication and included providing the clearance message. No cognitive processes of the ATC were modeled. The ATC activities were designed in this way so that the researcher could control the timing of message generation for PF Air MIDAS and PNF Air MIDAS.

SIMULATION CASES

Table 7.2 summarizes the 16 simulation cases. The case of normal approach without SVS was used as a baseline. Two go-around conditions, including a go-around following an ATC command and go-around based on the PF's decision, were also examined. Two different visibility levels, which switched at a specified altitude, were used associated with cockpit activities "runway-in-sight" callout, and PF's go-around decision. Visibility was set so that the runway came in sight at 150 ft before DA, except in the PF's go-around decision cases. Visibility in Cases 9, 10, 11, and 12 was set so that the go-around event was triggered due to an inability to see the runway at DA. In ATC go-around command cases, two sets of timings were used so that the interruption of pilot activities took place in both the busier and the less busy phases of flight. The command was issued about 100 ft before DA. This was a busy flight phase including the tasks of "runway in sight" callout and "approaching minimum" callout in Cases 5, 6, 7, and 8. The ATC go-around was also issued at 250 ft before DA, where no particular activity was expected.

   In each simulation run, flight parameters such as altitude, airspeed, position, workload (visual, auditory, cognitive, and motor) of the PF and PNF models, and the status of visual scan including location and success or failure of the scan were recorded.

RESULTS AND ANALYSIS

Flight Profile and Task Sequences

In all runs, the landing or go-around mission was safely completed by the simulated Air MIDAS pilots. The following summarizes the procedures and task sequences required in the following scenario conditions: landing; go-around commanded by the ATC, and go-around decided by the pilot. Figure 7.12 illustrates the baseline procedures for these simulation tasks.

Normal Approach

Figure 7.13 (panels a and b) illustrates the flight path and task sequence of Case 2, Run 1 (see Table 7.2), one of the normal approach cases. Speed command setting, flap lever position, and gear position are the system parameters that were manipulated by the Air MIDAS pilots. After the flight started, airspeed was reduced by the VNAV-programmed airspeed command regardless of the setting of the speed knob. At 77 sec after descent initiation, the PF overrides the VNAV and commands an airspeed of 200 kts. When the airspeed was reduced to 218 kts, the PF model ordered flap deployment and the PNF model set the flap lever to 5°. At 134 sec into the descent, the PF model decreased the airspeed command to 160 kts and called for further flap extension.

   The timing of this action was before the aircraft had achieved the previous airspeed command since the starting condition of the task was specified by the completion of the flap deployment action and airspeed within certain margins (namely,

Scan Policy

PF: -Outside view should be included into his/her scan pattern after PNF calls out "Runway in Sight."
After passing Approaching Minimum:
- Outside View should be included into his/her scan pattern.
PNF: - After passing FAF, outside view should be included in his/her scan pattern. After runway or visual cue to identify the runway becomes insight and s/he callout it, he/she should perform instrument scan.
After passing Approaching Minimum:
- PNF should concentrate on instrument scan.

Standard Callout (Approach & Landing)

Flight Phase	PF	PNF
FAF	**** (GOLET)	****, **ft (GOLET, xx ft)
Field Elv.+1,000 ft (BARO)	(Roger)	(One Thousand)
Field Elev.+500ft (BARO)	Stabilized	(Five Hundred)
DA + 80 ft	Check	(Approaching Minimum)
DA	Landing/Go-Around	Minimum
Runway In Sight		Runway In Sight
MAP		Missed Approach Point
100ft RA		(One Hundred)

Landing Checklist

LANDING GEAR..................................DOWN
SPEEDBRAKE.......................................ARMED
FLAPS..XX

Approach & Landing

PF	PNF
Order "Flaps XXX" according to Flap Extension Schedule	Readback "Flaps XXX" Set Flap Lever XXX
Order "Gear Down" Order "Flaps 20"	Readback "Gear Down" Landing Gear DN Readback "Flaps 20" Set Flap Lever 20
Speedbrake Lever ARM Landing Flap "Flaps XX"	Readback "Flaps XX" Set Flap Lever
Set Missed Approach Alt on MCP Order "Landing Checklist" Callout "Checklist Complete"	Perform "Landing Checklist"
Monitor Approach Progress	

Go Around

PF	PNF
Callout "Go-Around" Push TO/GA Switch Order "Flaps 20"	Readback "Flaps 20" Set Flap lever 20
Confirm Go-Around Attitude and Increasing Thrust	
	Check appropriate GA-Thrust and correct Thrust Setting if necessary.
Positive Rate of Climb "Gear Up"	Check Positive Rate of Climb Readback "Gear Up" Gear Lever Up
⋮	⋮
Followings were omitted for simulation setting	

FIGURE 7.12 "Documented" cockpit procedures. These procedures were used to build the activity set that services goals in the approach and landing scenario.

between 210 and 190 kts). After passing 800 ft AGL (above ground level), which is the defining altitude for runway visibility in the baseline high-visibility scenarios, the PNF model called out "runway in sight" at 772 ft AGL and 784 sec. At that moment, the PNF model fixated OTW. At 796 sec, after passing DA (650 ft), the PF decided to land and called out "landing." The aircraft touched down on the runway at 845 sec.

Results for Normal Approach

Figure 7.13 (panels c and d) illustrates workload traces for the PF and PNF on approach. Both the PF and PNF models exhibited high workload before they completed the configuration of landing flaps, airspeed, and gear. The PF model also had high workload from DA to touchdown. The PF model experiences maximum visual workload of 7.0 when it performs speed reduction and orders flap extension. The maximum cognitive workload (7.0) was reached when the PF model made the landing/go-around decision. In the Air MIDAS model, when maximum workload is reached on any of its channel capacities, the scheduler mechanism in the model interrupts activities of lower priority when the capacity limit is reached. The scheduler also delays the initiation of new tasks when the capacity constraint is in effect. Finally, the scheduler returns to examine the queue of interrupted activities to see which should be resumed after the workload bottleneck is passed. The maximum auditory and motor workload was 5.0 in the simulated flight.

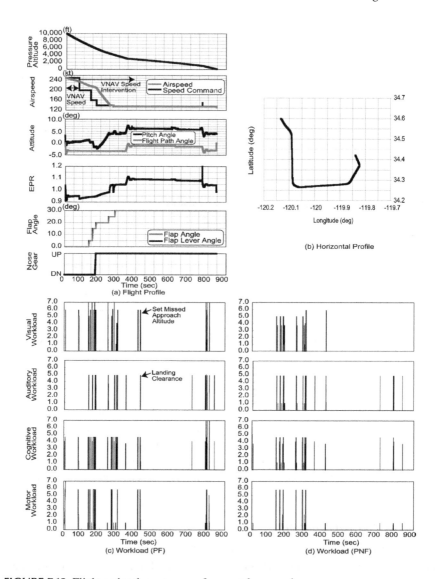

FIGURE 7.13 Flight and task sequences of a normal approach.

Go-Around According to ATC Command

Flight and task sequence of Case 8, Run 1 (see Table 7.2) is shown in Figure 7.14 (panels a and b) as an example of simulation results of a go-around due to ATC command scenario. At 812 sec after top of descent, the go-around command was issued by the ATC model. Both pilot models heard it and the PF model called out "go around" at 813 sec. Then the PF model pushed the go lever and set the pitch attitude to 10°. The PNF model set the flap lever to 5° at 819 sec, following the order from the PF model. After confirming positive climb, the PF model ordered gear up and the PNF set the gear lever up position at 822 sec.

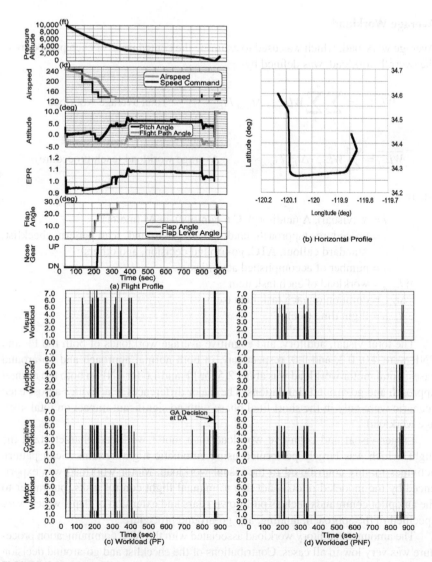

FIGURE 7.14 Flight and task sequence of Case 8 (Run 1): Go-around.

Results for Air Traffic Control Go-Around

In the case of the ATC-ordered go-around, the PF model had a higher level of visual, cognitive, and motor workload after hearing the go-around command (with auditory workload at 5.0 compared with the workload around DA in the normal approach flight). This was caused by the time criticality of the tasks required to perform the go-around. This suggests the impact of an externally ordered go-around as compared with the lower workload density after DA in normal approach flight.

Average Workload

Average workload, which was used to examine the contribution of each procedure to the overall workload, was defined by

$$\overline{WL_i}\,|_{total} = \sum_{j}^{j_{all}} \sum_{n}^{n_{all}} WL_{i,j,n} \cdot \Delta t_{i,j,n} / t_{total} \qquad \text{Total average workload}$$

$$\overline{WL_{i,j}} = \sum_{n}^{n_{all}} WL_{i,n} \cdot \Delta t_{i,n} / t_{total} \qquad \text{Average workload (each procedure)}$$

where

i = V (visual), A (auditory), C (cognitive), or M (motor)

j = procedures (approach, landing, and go-around, scan pattern, checklist, standard callout, ATC, go-around decision, and others)

n = number of accomplished activities

$WL_{i,j,n}$ = workload of each task item

$\Delta t_{i,j,n}$ = duration of each task item (sec)

t_{total} = flight time (sec)

The total visual, auditory, and cognitive average workloads of both the PF and PNF were 4.0, 0.2, and 1.0, respectively, in both normal approach and go-around cases. Total motor workload was about 2.5 for PF and 1.6 for PNF in both the normal approach and go-around simulations. The major differences in activities across each scenario were only in the short final phase with no significant impact on total average workload.

Larger visual and cognitive workload than motor workload characterized the flight, which was largely performed using automatic flight systems. Scan pattern activities mainly contributed to the visual workload. Motor workload was experienced by the modeled pilots—not due to manual flight control tasks, but rather to the autopilot commands such as pull-down gears and extension of flaps, which were specified in the approach, landing, and go-around procedure.

The amount of auditory workload associated with the ATC communication procedure was very low in all cases. Contributions of the checklist and go-around decision toward workload were also small. These procedures contained fewer task items than approach, landing, go-around, or callouts. It is important to note that lower average workload does not reduce the importance of the procedures. Scan pattern for instrument and OTW monitoring caused more than half of the visual workload and almost 30% of the cognitive workload for the PF and PNF. No major difference in scan pattern workload was observed between operations with SVS and without SVS.

SUMMARY

Prediction of human performance using the SVS in approach, landing, and go-around flight was performed by using Air MIDAS. In this simulation, we included aircraft dynamics that reacted to pilot control and drove flight displays in a closed-loop operation. Based on these dynamically changing values of flight parameters and aircraft situation and in order to respond to the dynamic requirements of closed-loop simulation in which flight crew decisions must be tailored to the evolving circumstances, several functional developments were made to the Air MIDAS model. The PC Plane aircraft simulation model was used for the world representation interacted by Air MIDAS pilot agents. A detailed scan pattern model was newly implemented and activities, including approach, landing, and go-around procedures; standard callout; checklist; ATC communication; and landing/go-around decision, were installed. Analyses of the 80 simulation runs are summarized as follows:

- The SVS supported approach, landing, and go-around phases regardless of DA and triggers of go-around, including the go-around initiated by either the PF at DA or by ATC's command. Use of the system will support approach and landing in conditions that would otherwise be unattainable without full instrument landing infrastructure and autoland capability. This is substantiated in that the procedures implemented for landing below minima resulted in effective landing as illustrated in Figures 7.12–7.14.
- We observed small delays of action initiation in flight control actions in the approach with the SVS relative to flights without SVS. This occurred because the chances of fixation on each display were decreased by adding SVS to the conventional display configuration and we included a "conventional instrument" cross-check for the SVS data. The cross-checking, while contributing to small increases in time, would increase the accuracy of the Air MIDAS internal situation model. No delays of task initiation were observed in the landing and go around phase, despite the time shifts in the approach phase.
- A scan pattern model (validated in Simulation 2) was subjected to a sensitivity analysis for the threshold setting for the information-acquisition model. It was concluded that a value of the mean minus 1.0 standard deviation resulted in a threshold of failure occurrence such that the error rate of scan perception was 10% or less.
- Analyses in this study did not include aspects of the SVS's information that could potentially enhance pilots' predictive SA. The information was used to guide current flight management behavior and did not support a model of future aircraft behavior.

CONCLUSION

The process of modeling human performance in air traffic and flight management systems is maturing to the point of providing a valuable and viable tool for both large-scale and small-scale analysis of operations, operational concepts, and specific aiding technologies. The process of performance modeling requires careful and detailed analysis of the air traffic management processes for each individual performance component in that system, and an equally detailed analysis of the interactions among those entities. In itself, this level of careful insight provides information important to the understanding of air traffic and flight management as it is actually performed (as opposed to how it is presumed to occur or how it is prescribed to occur).

The development of HPMs such as Air MIDAS, which is a framework of interacting agents representing the key human and system components, their behaviors, and their interactions, provides three benefits to the analysis of systems in early conceptual design stages:

- The description of humans and systems and their collaboration in the same mathematical, structural, and dynamic terms supports studies of functional allocation among human–automation systems and provides insight into the benefit that may be experienced with aiding systems.
- The analytic capability to define what information should be displayed to the human operator in the human–machine system as a consequence of his or her sensory/perceptual and cognitive characteristics in control can guide both the development and specification of intelligent systems for air traffic support.
- This approach supports a fundamental paradigm shift in which human–machine systems could be conceived as a joint entity coupled to perform a specific task or set of tasks, which may yield insight into what constitutes an effective and safe systems approach to the design for useful and usable systems.

The sequence of simulations described previously provides some insight into the processes undertaken to develop models of human–system integration that are of general utility in the analysis of systems early in conceptual design or initial implementation stages. In the first simulation, we attempted to develop functional elements in the human model that allowed error to emerge in ways that were consistent with actual human performance. The ability to generate errors is an important feature for predicting the performance of human operators in systems where they are critical components to system safety. In the second simulation, we developed a visual information-seeking model that could be compared and validated against information-seeking (eye-scan) behavior of actual human operators. This step is one of validation of a critical model component for the analysis of visual aiding systems (the SVS in this case).

In the third simulation, we linked that validated visual information-seeking model and the rest of the Air MIDAS human performance functions to a model of aircraft performance in a dynamic and closed-loop simulation. This closed-loop simulation allows us to predict the impact of information presentation on the performance of a time-critical set of tasks in approach and landing. Variations in the type of information available should have behavioral consequences for the performance of that task. In our simulation, landings could be performed with the SVS under conditions that a lack of information or guidance infrastructure would preclude without the SVS.

However, in using this HPM, we addressed a relatively specific question in the broader range of airspace analyses. A specification process is needed for what constitutes an appropriate level of detail in human performance and what behaviors or behavioral categories are of interest to the analyst. The range available is from micromodels of attention or cognitive performance to models of aggregate behavior in the operation of a section of the national airspace. This range provides a challenge for the appropriate selection of models to match analytic needs.

Detailed models can illuminate the impact of changes to procedure and equipment for individuals or teams of humans. The challenge is then how to aggregate the impact of those changes on much larger scales or operation with models of much more abstract representation.

One method to approach this issue is to develop protocols and methods for sensitivity analysis that examine the range over which individual performance enhancement can be expected to improve performance in all such cases. Another approach to this problem is to develop performance tables (based on more detailed model performance predictions) to be used in broader discrete event and stochastic simulation of larger airspaces. In this case, perceptual, cognitive, and motor performance for individual operators was called for and used to address the specific impact of changes to the visual information available to the flight crew as they interacted with ATC on approach. The level of detail in the model matched the analytic requirement.

Prior paradigms have used an incremental increase in testing complexity, fidelity, and cost from empirical results to prototype to full-mission HITL simulation, and then field testing to provide data for model development and validation. However, the rate of development of ATM systems, the tremendous economic pressure to implement and reap immediate benefits from technologies, and the significant complexity and cost of large-scale distributed air–ground tests suggest the development of other, more cost-effective methods of human factors research. Computational human performance modeling can be used early in the system definition and refined throughout the system life cycle. The models can become the repository of best practice as the system development unfolds. The benefit accrued from this approach (assuming well-developed and validated models) is that system and human characteristics can be quickly varied (e.g., based on an assumed technology change) and the impact of that change identified in the full-system context. Such models help focus the expensive and complex HITL simulation and field tests. In addition, performance at the edge of system safety can be explored in the computational human performance modeling paradigm. Such behavior is difficult to observe and study in any other way.

ACKNOWLEDGMENTS

The authors on the team had, as in any large team effort, specific areas of responsibility in the performance of these several years of research. The first author took the responsibility for project integration. Savvy Verma and Koji Muraoka were the primary architects of the functions needed to have the human performance model perform the required function in approach and landing scenarios. Brian Gore was instrumental in providing the performance context and link to the NASA simulations that served as underpinnings for the models performance, and Amit Jadhav was the primary coding architect and implementer.

The SJSU team wishes to thank Dr. Foyle and his team for the unflagging support of this effort in both administrative and technical support.

©PC Plane has been developed by NASA Langley Research Center and the NASA Ames Research Center.

REFERENCES

Atencio. A. (1998). *Short haul civil tiltrotor MIDAS* [online]. Available at http://caffeine.arc. nasa.gov/midas/original-midas/Tiltrotor_MIDAS.html.

Baddeley, A. D. (1992). Working memory. *Science, 255,* 556–559.

Baddeley, A. D., & Hitch, G. J. (1974). Working memory. In G. Bower (Ed.), *Advances in learning and motivation* (Vol. 8, pp. 47–90). New York: Academic Press.

Baron, S., & Corker, K. (1989). Engineering-based approaches to human performance modeling. In G. R. McMillan, D. Beevis, E. Salas M. H. Strub, R. Sutton, & L. Van Breda (Eds.), *Applications of human performance models to system design* (pp. 203–218). New York: Plenum.

Billings, C. E. (1996). *Aviation automation*: The search for a human-centered approach. Mahwah, NJ: Erlbaum.

Corker, K. M. (1995). *Models of human information requirements: When reasonable aiding systems disagree* (invited address). Topical meeting of the American Nuclear Society. Philadelphia, PA, June 25–29, 1995.

Corker, K. M., Gore, B. F., Fleming, K., & Lane, J. (2000). Free flight and the context of control: Experiments and modeling to determine the impact of distributed air–ground air traffic management on safety and procedures. *Proceedings of the 3rd USA–Europe Air Traffic Management R & D Seminar,* Naples, Italy: USA–Europe Air Traffic Management.

Corker, K. M., Verma, S. Gore, B. F., Jadhav, A., & Guneratne, E. (2003). *SJSU/NASA coordination of Air MIDAS safety development human error modeling: NASA Aviation Safety Program Phase 1: Integration of Air MIDAS human visual model requirement and Phase 2: Validation of human performance model for assessment of safety risk reduction through the implementation of SVS technologies* (HAIL Laboratory Report 2307-2005). San Jose, CA: SJSU, Human Automation Integration Laboratory.

Foyle, D. C., Andre, A. D., McCann, R. S., Wenzel, E., Begault, D., & Battiste, V. (1996). Taxiway navigation and situation awareness (T-NASA) system: Problem, design philosophy and description of an integrated display suite for low-visibility airport surface operations. *SAE Transactions: Journal of Aerospace, 105*(1), 1411–1418.

Gore, B. F. (2002). An emergent behavior model of complex human–system performance: An aviation surface related application. *VDI Bericht 1675, 1*(1), 313–328, Düsseldorf, Germany: VDI Verl.

Hamilton, D., & Bierbaum, C. (1990). Task analysis/workload (TAWL)—A methodology for predicting operator workload. *Proceedings of the Human Factors and Ergonomics Society 34th Annual Meeting* (pp. 1117–1121). Santa Monica, CA: HFES.

Hollnagel, E (1993). *The context of control.* Pacific Grove, CA: Duxbury Press.

Hooey, B. L., & Foyle, D. C. (2006). Pilot navigation errors on the airport surface: Identifying contributing factors and mitigating solutions. *International Journal of Aviation Psychology, 16,* 51–76.

Hooey, B. L., Foyle, D. C., & Andre, A. D. (2000). Integration of cockpit displays for surface operations: The final stage of a human-centered design approach. In *Proceedings of the AIAA/SAE World Aviation Congress.* Warrendale, VA: SAE International.

Landy, M. S. (2002). *Vision and attention for Air Midas* (NASA Final Report NCC2-5472). Moffett Field, CA: New York University, Dept. of Psychology and Center for Neural Science.

McCracken, J. H., & Aldrich, T. B. (1984). *Analysis of selected LHX mission functions: Implications for operator workload and system automation goals* (Technical Note ASI 479-024-84(b)). Fort Rucker, AL: Anacapa Sciences, Inc.

Mumaw, R., Sarter, N., Wickens, C., Kimball, S., Nikolic, M., Marsh, R., et al. (2000). *Analysis of pilot monitoring and performance on highly automated flight decks* (NASA

Final Project Report: NAS2-99074). Moffett Field, CA: NASA Ames Research Center.

Muraoka, K., Verms, S., Jadhav, A., Corker, K. M., & Gore, B. F. (2004). *Human performance modeling of synthetic vision system use* (Final Report No. 2903-100-04). San Jose, CA: SJSU Human Automation Integration Laboratory.

NASA Web Tutor: http://www.allstar.fiu.edu/aero/RNAV.htm.

Palmer, M. T., Abbott, T. S., & Williams, D. H. (1997). Development of workstation-based flight management simulation capabilities within NASA Langley's Flight Dynamics and Control Division. In R. S. Jensen & L. A. Rakovan (Eds.), *Proceedings of the Eighth International Symposium on Aviation Psychology*, 1363–1368. Columbus: The Ohio State University.

Reason, J. T. (1990). *Human error.* Cambridge, UK: Cambridge University Press.

Sheridan, T. B. & Ferrell, W. R. (1975). Man-machine systems. Cambridge, MA: MIT Press.

Smith, B., & Corker, K. (1993). An architecture and model for cognitive engineering simulation analysis: Application to advanced aviation automation. *AIAA Computing in Aerospace 9 Conference*, October 21, 1993, San Diego, CA.

Verma, S., & Corker, K. (2001). Introduction of context in human performance model as applied to dynamic resectorization. (2001). *Proceedings of the Ohio State University, Proceedings of the 11th International Symposium on Aviation Psychology*, Columbus, OH, March 5–8.

8 D-OMAR
An Architecture for Modeling Multitask Behaviors

Stephen E. Deutsch and Richard W. Pew

CONTENTS

INTRODUCTION

Research in human performance modeling leading to the distributed operator model architecture (D-OMAR)* began at BBN Technologies almost 20 years ago. Developments in experimental psychology, cognitive science, and cognitive neuroscience have been important in elaborating and lending support to the underlying theory on which the models have been based. As one of the research groups in the National Aeronautics and Space Administration Human Performance Modeling (NASA HPM) project, we provided an architecture in which we developed and demonstrated an evolving series of aircrew human performance models (HPMs). As the HPMs for the aircrew were developed and the results of the human subject trials (discussed in chapter 3) replicated, the theory that establishes the architecture for the models contributed to our improved understanding of the underlying sources of the observed behaviors. In particular, the models were used as a tool to help better understand how errors seen in the human subject trials came about.

With these insights in hand, it was then possible to make further use of the models to examine equipment design issues and procedure structure with the goals of reducing the incidence of aircrew error and providing support to recover from errors that do occur. The contributions to the overall findings of the research, particularly in the area of understanding human error, were made possible through a combination of the modeling effort and the development of the underlying theory that led to the models of human behavior.

The NASA HPM project had two phases. Each was planned as a derivative of human subject trials that provided the modeling teams with a rich source of empirical data that supported model development and refinement (see chapter 3).

The focus of the first phase was on surface (taxi) operations subsequent to landing at Chicago's O'Hare International Airport (KORD; see chapter 3). In the BBN research effort, we developed models for the aircrews and the air traffic controllers, with procedures to support approach, landing, and taxi operations. The modeling was accomplished using the D-OMAR human performance modeling environment to represent the behaviors of the aircrews and air traffic controllers. D-OMAR was also used to model the aircraft and their flight decks, the air traffic control (ATC) workplaces, and the essential features of the airport (KORD) and the local airspace. Our approach to the research task was to produce models that represented the robust performance typical of commercial aircrews and then probe the models to seek to *replicate* and *explain* a subset of the classes of error that were observed in the surface operations human subject trials.

In the second phase of the research project, we began by emulating the synthetic vision system (SVS) human subject trials (chapter 3) that examined pilot performance in the use of baseline and SVS-equipped flight deck configurations. Approach and landing procedures from the taxi scenarios were refined to employ the newly modeled SVS-equipped flight deck configuration. The essential features of the Santa Barbara Municipal Airport (SBA) and the local airspace were added to the model. The initial goal was to provide models whose performance was comparable to that

* The D-OMAR software and documentation is available as OpenSource at http://omar.bbn.com.

of the human subjects in executing the 10 baseline and SVS-equipped scenarios described in chapter 3.

The analysis conducted in the course of emulating the human subject trials provided the basis for the remainder of this phase of the research effort. The analysis of the trial data suggested that providing pilots with the SVS as a second attitude instrument (i.e., an integrated display showing horizon, pitch, altitude, altitude rate, and speed) induced a change in their instrument scan pattern. While the tasks they executed remained the same, pilots adjusted the balance between attending to the attitude instruments and the navigation instrument more toward the attitude instruments. In seeking to mitigate this effect, we "designed" an enhanced SVS that combined the traditional primary flight display (PFD) and the SVS functionalities in a single attitude instrument. We then used the modeling environment to explore the aircrews' use of the new instrument in the SVS scenarios.

The goal of the new SVS design was to attempt to restore the baseline scan pattern—a more even distribution in attitude-instrument versus navigation-instrument scan time while preserving the advantages of the SVS. We introduce the term "aviation scan" here, representing the active visual scan of information required to "aviate" the aircraft. This aviation scan information includes pitch and attitude information, as well as altitude, altitude rate, and speed information. This follows the convention (e.g., Billings, 1997) that distinguishes among information required to aviate, navigate, and communicate.

A second focus for the analysis was on pilot response when the SVS was found to be out of alignment with the out-the-window (OTW) view. The procedural rule in place for the human subject trials required that the pilot execute a go-around. That two of the three pilots did not immediately follow the rule and elect a go-around was not a complete surprise—pilots will often pursue a landing in adverse conditions. We thought it important to further investigate possible pilot responses on finding the SVS to be misaligned. In modeling the SVS misalignment scenario, we examined two pilot responses: following the rule by immediately initiating a go-around and continuing the approach and landing.

In the next section we introduce the D-OMAR modeling environment and provide a detailed look at the HPMs developed for the aircrews and air traffic controllers. The section provides a discussion of the integrated perceptual–cognitive–motor architecture underlying the models and discusses how elements of the architecture support multitasking behaviors representative of the work of commercial aircrews. We then discuss our approach to understanding human error and, more specifically, the ways that errors can intervene in otherwise successful human performance.

With the needed background material in place, we describe the surface operation errors that were modeled and the understanding of the sources of these errors as developed through the modeling effort. We then turn our attention to the SVS scenarios. The section traces the refinement of the HPMs, the aircrew and air traffic controller procedures that they execute, and the strategy for building the emulation of the SVS human subject trials. The section traces the design and evaluation of an enhanced SVS whose goal was to restore the baseline pilot scan pattern altered by the introduction of the SVS as a second attitude instrument. We then discuss the motivation for, and the modeling of, alternative pilot responses to the detection of the SVS misalignment as seen in the human subject trials.

The final section provides a discussion of conclusions derived from the model research effort concerning the ways in which HPMs can contribute to improved aviation safety.

MODELING HUMAN PERFORMANCE IN D-OMAR

The development of HPMs in D-OMAR (Deutsch, 1998; Deutsch & Adams, 1995; Deutsch & Pew, 2002, 2004) has been a cross-disciplinary endeavor based on research in experimental psychology, cognitive science, cognitive neuroscience, and recent research in the theory of consciousness. There is a theoretical framework underlying the architecture that has played an important role in gaining an understanding of observed aircrew behaviors. A broad range of individual human functional capabilities are represented whose complex interactions generate the models' behaviors. To provide insight into the sources of the model behaviors exhibited in the surface operation and SVS scenarios, we introduce a few of the central features of the models and the architecture that provides the framework for their development. We discuss the architecture for the models, how multiple task behaviors are developed within the architecture, how the distributed model of working memory operates in the process-dominated architecture, and the role of vision in conducting flight deck operations.

COGNITIVE ARCHITECTURE FOR THE D-OMAR MODELS

When we use the "cognitive architecture" shorthand, we are speaking of an architecture that integrates perceptual, cognitive, and motor functions. The architecture, at once, facilitates and constrains the construction of a model. It provides the operational framework for the computational elements of the model and imposes limitations on selected model capabilities. The architecture reflects the main tenets of a *theory* for how perceptual, cognitive, and motor capabilities interoperate to produce human-like behaviors.

Most human performance modeling environments are framed by a particular architecture; D-OMAR is different in this respect. Our sense is that there is still much to learn about cognitive architectures and that there is much to be gained by leaving room to explore alternative approaches to their construction. Rather than a specific cognitive architecture, D-OMAR is a flexible modeling environment in which to explore alternate architectural constructs. It comprises a discrete-event simulator and a suite of representation languages that may be used selectively to instantiate an architecture and construct HPMs.

The D-OMAR representation languages provide the foundation for establishing a cognitive architecture and instantiating HPMs. They are also the basis for building the simulation environment entities with which the HPMs interact. The Simple Frame language (SFL) is used to define the objects and agents in the simulation environment. It is a direct descendent of KL-ONE (Brachman & Schmolze, 1985), also developed at BBN. The basic elements of SFL are *concepts* and *roles*. Concepts are analogous to the classes of object-oriented programming language. Roles are the basis for defining the slots of a concept.

The difficult task in model development is producing truly human-like expert behaviors. The behaviors are defined in the Simulation Core (SCORE) language, a language for expressing the goals and procedures of an agent. A simulation agent is defined as an SFL concept. It is built on a SCORE *basic-agent* object that provides the agent with the capability to execute SCORE language procedures within the simulator. All the active players in the simulation environment are agents. They may be aircrew models or active airspace entities (e.g., the aircraft or radar systems). A rule language, the Flavors Expert (FLEX; Shapiro, 1984), is included to complete the D-OMAR suite of languages, but was not used in the current architecture.

MODELING HUMAN-LIKE MULTITASK BEHAVIORS

The basic-person and domain-specific skills that form the foundation for our HPMs must play together in a coherent manner that faithfully produces human-like behaviors. They must exhibit multitask behaviors that are coherent and reasonably bounded. The bounds on what can be accomplished concurrently take several forms. A typical behavior may be to set aside a flight deck conversation in order to respond to an ATC communication, while, at another level, two competing tasks may each require the use of the pilot's eyes to guide a manual operation. In the first instance, it is a matter of *protocol* that is established. In the second instance, it is contention for a physical *resource* that must be resolved. Using a simple priority-based conflict resolution process, a broad range of thoughtful and skill-based task processes are interleaved without the need for a central executive responsible for scheduling future actions.

In designing an architecture for our models in D-OMAR, we have defined a computational framework in which to assemble functional capabilities that operate in parallel, subject to appropriate constraints. The individual functional capabilities work together to address specific task demands and the conflicting demands of multiple tasks are resolved in a human-like manner. The desired behaviors have a combination of proactive and reactive components: The aircrews have an agenda that they are pursuing, but must also respond to events as they occur.

The multitask behaviors exhibited by D-OMAR models are derived from a dynamically changing network of procedures that addresses aircrew goals and responds to impinging events. From a bottom-up perspective, there is an assembly of individual perceptual, cognitive, and motor capabilities that are recruited as procedures to address current goals and subgoals. Neumann's (1987) functional view of attention, and the localization of mental operations in the brain, as put forward by Posner, Petersen, Fox, and Raichle (1988) are important contributions supporting this capabilities-based approach to modeling human behaviors. Taken together, they point to the functional components in task execution as taking place at particular local brain centers, with the coordinated operation of several such centers required to accomplish any given task.

The form of the coordination among computational elements is of particular importance in the architecture. Specific functional capabilities are modeled as procedures that represent the work of the various centers acting concurrently in support of the completion of a task. The architecture must provide for coordination among the procedures necessary to accomplish a task. It must also provide for the movement of information among the procedures. A *publish–subscribe* protocol is used to address both of these requirements. The unit of information transfer is a *signal* that has a name specifying the signal *type* and a message consisting of attribute values.

As a procedure generates attribute values, they are assembled as a message in an appropriately type-named signal that is then *published*. Procedures requiring informational elements of a particular signal type *subscribe* to the signal based on signal type. As a consequence of subscribing to a signal type, the subscribing procedure is alerted to publication of the signal and the message content may then be processed by the receiving procedure. More than one procedure may subscribe to a signal type; each will be notified when the signal type is published. The publish–subscribe protocol thereby creates dynamic links in the network of computational elements—the procedures that represent the model's collective functional capabilities.

Goals and their subgoals and plans are used to establish the network of procedures. Together, they represent the things that a person knows how to do: *basic-person skills* (e.g., coordinated hand–eye actions used to set a flight deck selector) and *domain-specific skills* (e.g., the aircrew adjusting flap settings consistent with current airspeed during the approach). Active goals represent the pilot's proactive agenda for managing aircrew tasks. These top-level goals activate a series of subgoals and procedures. The goals and subgoals represent the objectives of the actions to be taken; the procedures are the implementation of the actions to achieve the goals. Procedures may include decision points at which to adjust to variations in the immediate situation. Concurrent execution of goals and procedures is initiated using the SCORE language forms *race, join,* and *satisfy*.

A pilot's overall agenda is thus implemented by the dynamic network of procedures established by the goal–procedure hierarchy and linked by the publish–subscribe protocol. Only a subset of the procedures is active at any one time; most are in a wait state. Many of the procedures in a wait state are designed to respond and channel responses to impinging events. Others are established to monitor for the fulfillment of expectations or take remedial action when expectations are not met.

Within this framework, the demand for the allocation of attentional resources comes from two, sometimes competing, sources: top-down, operator-driven activation stemming from the currently active goals and procedures and bottom-up, stimulus-driven (typically auditory, visual, or haptic) procedure activation initiated by signals representing sensory inputs. Attention allocation is a special case of the general requirement to resolve conflicts between procedures that compete to execute.

Two elements are needed to set up the framework for conflict resolution. First, the procedures are classified as potentially being in conflict with one another. Second, a priority, perhaps dynamically changing, is associated with each procedure. By virtue of being active, a procedure has sufficient priority to block a newly initiated procedure of the same or lower priority. In this case, the new procedure will be initiated when the active procedure terminates. If the new procedure has higher priority than the active procedure, the active procedure may be suspended or terminated (as in a winner-take-all competition) and the new procedure activated. When the new procedure completes, a suspended procedure will resume. The blocked procedure may be defined to resume operation at the point at which it was interrupted or at an earlier point in its execution.

The conflicts, as described, are notable for being automatically resolved; the decision process represented is a well-practiced process and thoughtful intervention is not needed. A thoughtful cognitive act of deciding on the next action is modeled as just that: a thoughtful decision-making procedure that determines the action to follow.

THE DISTRIBUTED MODEL OF DECLARATIVE MEMORY

The architecture that we have created in D-OMAR is very much an architecture grounded in *process* (Edelman, 1987, 1989) with declarative memory relegated to a supporting role. Indeed, Glenberg (1997) has identified memory for process—the remembrance of what we know how to do—as memory's principal function. Hence, expertise (what the model knows how to do) is a major component of long-term memory. Much of this expertise is skill based. Bartlett (1958), Logan (1988a), and Bargh and Chartrand (1999) have extended the traditional view in thinking about human skills. In addition to perceptual and motor skills, they suggest that there are *cognitive skills* that play a significant role in all human endeavors. The expertise of the aircrews and, in particular, their cognitive skills are thus represented as process—goals and procedures. Within this framework, long-term declarative memory (e.g., knowledge) exists as slots of the goal or procedure for which it is needed. There is little domain knowledge that is context free. By encapsulating a knowledge element within the goal or procedure, we take advantage of the fact that the goal or procedure forms the context that grounds the knowledge. With this background in place, we describe how the declarative elements of *working memory* function in the process-based architecture.

On the flight deck, each aircrew member will typically have several ongoing tasks at various stages of completion. In pursuing the completion of each task, there will usually be several goals and procedures concurrently active. Using an example based on the SVS scenarios, as the aircraft proceeds along the flight leg to the approach fix PHANTOM (see the Santa Barbara approach plate in chapter 3) in instrument meteorological condition (IMC) with an 800-ft cloud ceiling, at least two of the captain's tasks are concerned with aircraft altitude. Let us say that the altimeter is currently reading 1,021 ft above field elevation (AFE). The more immediate task is concerned with monitoring the descent to the target altitude of 1,000 ft for the flight leg to PHANTOM. A second task also concerned with tracking the current altitude has as its goal making the decision to land—a decision that will be made as the aircraft approaches the decision altitude (DA) of 650 ft. There is other work in process, but these two tasks are sufficient to suggest how the altimeter reading—a working memory element—is processed in two distinct procedure streams in the model.

For the captain and first officer (FO) models, the altitude reading is available from the altimeter. The models' *aviation scan* provides at least the aircraft's heading, altitude, and speed in numeric form. In the vernacular of the model architecture, upon completion of the aviation scan the numeric altitude value of "1021" is *published*—that is, the altitude is part of an aviation scan message that is the product of the aviation scan procedure, where "aviation scan" is the message type. The published message is then available to all subscribers to the message type.

In our example, at least two of the captain's procedures are *subscribed* to aviation scan messages: The first is concerned with monitoring the aircraft's descent to the target altitude of 1,000 ft for the current flight leg to PHANTOM; the second is preparing to make the decision to land as the aircraft subsequently approaches the 650-ft DA. The aviation scan procedure publishes the aviation messages, which, in effect, "wake" the procedures that are subscribed to the message type. The publishing of the products of the proactive visual process triggers cognitive processes requiring that information to facilitate their next actions.

From the captain's perspective, the value "1021" is processed in a unique manner in each procedure. For the task of monitoring the descent to the target altitude, "1021" immediately becomes "I need to attend carefully to the aircraft's altitude over the next few seconds to assure that we level off at 1,000 feet." For the decision-to-land task, "1021" becomes "I've got a little time, but I should be breaking out of the cloud cover shortly and I then need to acquire the airport and runway to support the decision to land." In each case, the published altitude value of "1021" is immediately, significantly, and uniquely *reinterpreted* by the subscribing procedures. The value "1021" is just an intermediate value in concurrent processes of rapid transformations.

Working memory requires a home in an architecture, but it cannot be properly captured as a single box in an architecture diagram. In this D-OMAR architecture, we posit that working memory is *distributed* across brain centers. The model's publish–subscribe protocol is designed to mirror the movement of information through the brain's perceptual centers and on to cognitive centers for further interpretation, action selection, and action execution. Moreover, we assert that these working memory items do not have a separate existence as database items, but rather are each encapsulated by local processes—the procedures that operate on them at each stage of their migration, interpretation, and transformation.

VISION AS A COMPONENT WITHIN MULTITASK BEHAVIORS

Demands on the pilot's visual system are varied and often complex. A broad range of these visual capabilities at the object level is included in the models. Some of a pilot's actions are purely visual. There are basic processes that take in information from the major flight deck instruments and the view out the windscreen. The viewing of a flight deck instrument can provide a generic aircraft attitude or navigation status check, or the monitoring of an instrument for a specific target value. The view out the windscreen can be to assure that there are no traffic conflicts (i.e., seeing *nothing* can be the desired outcome), acquiring a specific object (e.g., sighting the runway on the approach to support the decision to land), or tracking the aircraft's alignment with the runway during the final approach.

For some actions, the visual component plays a supporting role. The execution of flight deck operations requires coordinated hand–eye actions to set switches, adjust selector settings, and reposition control levers. A different form of hand–eye coordination is needed in using the tiller to guide aircraft taxi operations.

Reading and, more broadly, the interpretation of graphical display information are additional visual tasks. The approach plate and airport diagram are used as sources of information to support approach, landing, and taxi operations. Some of the information from these sources is purely textual; some involves the interpretation of annotated graphical information. Elements of the PFD, navigation display (ND), and SVS require graphical interpretation.

Vision plays a central role in the modeling of a pilot's multitask capabilities. The pilot's procedure for each of these vision-supported activities has a priority associated with it. Within this framework, the vision system is a resource and the tasks that require its capabilities compete based on their assigned priority. As modeled, the scans of the windscreen, the PFD, the ND, and the SVS (when it is present) are modeled as separate procedures, all with the same priority. The product of these procedures is an *emergent* background scan pattern. It has not been necessary to adjust the scan priorities dynamically.

Scanning out the windscreen for the runway and setting a selector or control lever are examples of *purpose-driven* procedures: They are the product of an intention to take immediate action. They are invoked at a priority higher than the background scan procedures. The elected action takes place immediately, interrupting the background scan pattern. Once the purpose-driven action has completed, the background scan pattern resumes.

Actions with a visual component that extend over time operate with a slight variation. Visually tracking the runway to support the landing operates at the purpose-driven action priority, but to accommodate the extended time duration, it allows intervals at which elements of the background visual scan can intervene. The interval between these purpose-driven scans varies depending on the particular circumstances. For example, as the aircraft closes on a target altitude, the interval between altimeter readings decreases until the scan is terminated as the target altitude is established. After the purpose-driven scan is terminated, the background scan pattern serves to monitor for the maintenance of the current altitude.

The mix of a background scan pattern interleaved with purpose-driven scans is much like that identified by Bellenkes, Wickens, and Kramer (1997). They refer to what we have established as the background scan pattern as "minding the store." What we have called a purpose-driven scan, they have referred to as attending to the "action"—the changing parameters for the particular maneuver. Bellenkes et al. found that skilled pilots maintained their "minding the store" scans as they attended to the "action." It is these skilled behaviors that emerge from our model.

HUMAN ERROR MODELING CONCEPTS

HPMs have much to offer in seeking to reduce the frequency and consequences of human error in commercial aviation. The research that has supported the development of the models and their architectures has provided new insights into the sources of human error. In use, the models provide a low-cost method to augment human studies. They can extend the range of operating conditions that may be explored as they are used to examine new procedures and equipment.

Our approach has been to model the interactions between skilled human operators and the complex equipment in their workplaces. We then build on the initial models to address teamwork—the interactions among multiple operators and their systems. At each level, we have sought to identify and understand the sources or causes of human error. With the causes identified, the next step has been to explore procedure and/or equipment changes to prevent the error or mitigate its effect. The approach may be used as well to seek performance improvements not specifically associated with error.

We start the process by mapping the human performance space—building the means for the expression of sound behaviors in our models and then developing scenario events that lead to humanly plausible errors based on what is known about the strengths and weaknesses of human operator capabilities and performance. Rather than attempting to model human error per se, we have endeavored to build robust models of human capabilities and then probed the seams of the resultant model behaviors for susceptibility to error.

ERROR MECHANISMS

Our approach to error mechanisms relies on three independent, but clearly interrelated perspectives on human performance:

- An error taxonomy according to Reason (1990);
- An analysis of human information processing fallibility derived from Adams, Tenney, and Pew (1991); and,
- The decision-making analysis of Rasmussen (1976).

We have adopted the error taxonomy of Reason (1990), which is a simple taxonomy (Figure 8.1) beginning with unintended and intended actions. Unintended actions are broken out into slips and lapses, and intended actions into mistakes and violations. Reason's taxonomy serves to classify error types broadly. We then sought to understand the fundamental sources of these types of errors.

In our monograph, there is an extensive discussion of the real-world constraints that impact the effectiveness of human information processing. The following quote captures the flavor of the problems:

> We know that memory is limited. We also know that list maintenance is effortful and fallible, more so, if the list must be ordered, still more if the membership of the list must be dynamically reordered and modified during retention. These considerations suggest that proper maintenance of the queue of pending tasks would require considerable cognitive effort, even if it consisted of nothing more than a simple list of things to do. But unlike any simple list, the memory for the tasks on the queue must somehow include or point to larger complexes of knowledge and experience that underlie each. After all, it is only through access to these fuller representations that operators can update and keep track of changes in task demands, and performance parameters with unfolding events. Further it is only with reference to these fuller representations that the operator can set and adjust the logical and temporal constraints that dictate when, why and how each task can and should be carried out. (Adams et al., 1991, pp. 62–63)

The monograph helps to identify the fundamental human information processing stages most likely to be error inducing: attention, discrimination, memory, situation awareness (SA), and planning. Attention management leads to errors resulting from a failure to attend to relevant information or from attention distraction away from the

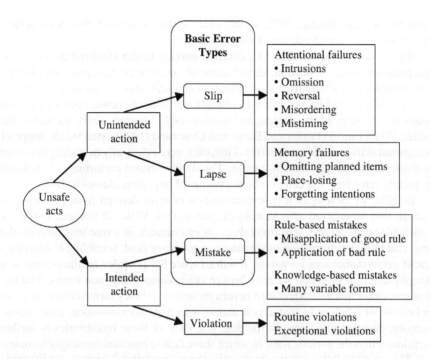

FIGURE 8.1 Error taxonomy derived from Reason (1990).

relevant information. Attention and memory interact when a distraction away from information in immediate focus leads to forgetting of a procedural step. Attention and SA interact when the aircrew or controller fails to attend to the relevant information to interpret a situation appropriately. Long-term memory and planning interact when the aircrew lacks relevant knowledge or related data and fails to undertake or delays the planning or information seeking required to obtain it.

Errors rarely occur unless the crews are either too busy or not busy enough. Accidents rarely occur as a result of a single human error or system failure; rather, it is the concatenation of too many things to do combined with fallible information processing that leads to incidents and accidents. Excessive workload can result from high traffic, system failures, weather conditions, or confusing airport layouts, to name just a few sources. These factors can lead to excessive workload by generating communication requirements that take time, require additional processing capacity, and are themselves subject to fallible interpretation, or by creating complex, disproportionate demands on attention.

Operators working as teams add another layer of complexity. Acting as team players exacerbates the multitasking demands and adds additional dimensions to the error space. Teams may work more or less well, in part determined by their communication patterns; good teams make effective use of implicit as well as explicit communication (Serfaty, Entin, & Johnston, 1998). Shared goals and understanding of the situation further support effective teamwork (Orasanu, 1994; Salas, Dickinson,

Converse, & Tannenbaum, 1992; Serfaty et al., 1998). Failures of these team performance processes are an additional source of error.

Rasmussen (1976) has laid out a decision-making ladder idealized as activation, observation and data collection, identification of system state, interpretation of situation, evaluation of alternatives, task definition and selection of goal state, procedure selection, and procedure execution. His insight was that, of course, operators seldom follow all the steps of the process and he identified important *shortcuts* within the ladder. When Logan (1988a) and Barge and Chartrand (1999) extended the scope of recognized skills to include cognitive skills, they were essentially extending the range of shortcuts that people actually employ. The shortcut-based performance is fast and frequently error free, but can also be the source of important classes of error.

In 1981, we developed a conceptualization of error derived from Rasmussen's analysis that we referred to as Murphy diagrams (Pew, Miller, & Feehrer, 1981). For a particular task context, a Murphy diagram enumerates, in a tree structure, all the ways that human information processing can go wrong (and, according to Murphy's law, if performance can go wrong, it will). Figure 8.2 provides an illustration of a Murphy diagram in which we have broken out taxiway navigation errors. The aircrew can either make a correct turn or turn incorrectly. If they turn incorrectly, it can be because of a failure in spatial orientation, attention, discrimination, expectation/ memory, or communication. In the diagram, each of these possibilities is further partitioned into the various ways in which these failure mechanisms might be manifest. For example following the "expectation/memory failure" branch, we list three ways in which a memory failure could result in the wrong turn. First, the FO could have failed to write down the controller's instructions and remembers them incorrectly. Second, he or she could have written them down but made a transcription error.

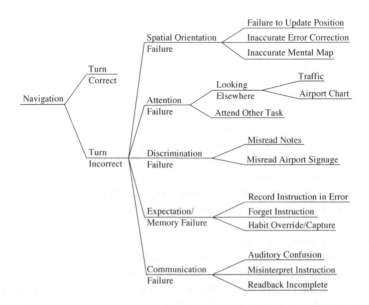

FIGURE 8.2 Murphy diagram for taxiway navigation errors.

Third, he or she could be moderately familiar with the airport and let expectations of where to turn to reach the gate take precedence over the controller's instruction. The other branches can be followed in a similar manner.

We did not code these tree structures into our error model; rather, we have coded the mechanisms that drive model behaviors with the expectation that the errors that do occur will lie within these tree structures. Had an error fallen outside the current tree structure, it would then have been necessary to revise the tree. Thus, we create procedures in our model wherein the FO is obligated to record in writing the taxi directives when they are received from the controller and to read them back at appropriate times. The model provides the opportunity for competing goals to play out in alternate ways, some leading to errors. As the scenarios are explored, the errors committed by the models are expected to be consistent with the Murphy diagrams.

THE RELATIONSHIP BETWEEN HUMAN ERROR AND SUCCESSFUL PERFORMANCE

It is our belief that the same features of human performance that lead to robust, intelligent behavior can also induce error. When viewed from this perspective, human error can be discovered in thoughtful, detailed models of robust, successful human performance!

Several key factors are essential to this model-based exploration of error in the execution of commercial airspace procedures:

- Theoretically grounded architectures and HPMs;
- Models capable of the breadth of performance required by airspace procedures;
- Models whose performance errors reflect human error;
- An approach to locating the sequences of errors that are at the root of unsatisfactory outcomes; and,
- An approach to finding potential remedies that successfully alters error sequence outcomes.

Several key human performance capabilities have important implications when attempting to achieve a better understanding of the sources of human error and when seeking to prevent errors or develop error-mitigation strategies:

- *Perception/discrimination*: Our models represent perception at the level of object recognition, identification, and interpretation. Errors can occur when objects fail to be recognized because discrimination among alternative objects fails or when a sign or symbol is misinterpreted. Receipt of verbal communication is also a perceptual act that can be received and interpreted correctly or incorrectly.
- *Situation awareness (SA)*: Situation assessment represents the collection of all the sensory, perceptual, and cognitive activities up to the point of making a decision related to selecting and executing an action. Situation assessment leads to SA. The process provides many opportunities for error.
- *Intention formation*: Model goals are activated by contextual events or as the result of following through procedurally with a sequence of goals called for by a defined task. In high-tempo activities, multiple goals compete for

execution. Their priorities may be changing dynamically and are continually being reassessed. The goals drive the intent of the crewmember. Errors can arise when events that trigger goal contention cause incorrect intended actions or bypassing or deferring actions.

- *Attention*: Goals driving thoughtful procedures that are highest in priority receive attention; hence, events initiating new goals or changes in the priorities of current goals can lead to changes in the focus of attention. The resulting goal or procedure interruptions, or the activation of new competing procedures, can lead to a variety of attentional errors.

- *Memory/forgetting/interference*: Forgetting can be explored by a process of heuristically guided search—postulating specific instances of forgetting, inserting them into the scenario, and examining the effect on subsequent events.

- *Expectancies*: Behavior is strongly influenced by an individual's previous experiences. Successful performance depends on being able to anticipate what is most likely to happen next and to have formulated a plan or an intention for responding. In many cases, the proactive disposition of the operator improves the operator's ability to perform routinized tasks under high-workload conditions involving multiple tasks competing for processing resources. However, this can also be a source of error, referred to as habit capture (Norman, 1988) or "strong, but wrong" (Reason, 1990), wherein less salient cues indicating a preferable alternate course of action are missed.

- *Communication*: Communication has a central role in the modeling of multitask behaviors. It is an important factor that can lead to several failure types. As a source of interruption, it may distract a crewmember, leading to an error—crew activities can be interrupted and not correctly resumed, leading to failures such as missed checklist items.

- *Execution*: Crewmembers follow through with action by executing motor movements. In the aircraft, these may be movements of levers or actuating switches or selectors on the flight deck control panels. Errors of execution arise as the result of "slips," instances where all the decision making and intentions are correct but the wrong button is pushed or the wrong selector setting is made. At times, a highly automated movement sequence may be completed when the context calls for a less familiar one. Task execution has a window of opportunity: a time when it becomes pertinent and a time when it is too late. Executing a task outside its opportunity window represents an error.

- *Influences of context*: Highly practiced intellectual and physical skilled performance is context dependent—human errors are not derived from equally likely statistically random events. Each context constrains behavior to a particular "family" of task choices and therefore may induce errors specifically related to the immediate context.

- *Workload*: When too many tasks compete for attention within a time window, crewmembers make choices or change their strategies for managing the tasks. Excessive workload calls for strategic workload management (Adams et al., 1991). Whether error occurs depends, in part, on whether the task design is forgiving. The ability of tasks to be cued for later action when there are too many to do provides a forgiving context. When tasks cannot be cued, the crewmember may choose to prioritize and skip less important tasks. Alternatively, he or she may choose to complete each task, but to execute it less thoroughly or completely than desired. Each of these alternatives represents a failure to complete the action as planned, but not all such failures will have safety consequences. Too many tasks for the time available can also lead to "priority churn": undue time devoted to assessing and reordering priorities.

• *Effects of stress*: Stressors, which can include fatigue, excessive workload, high task criticality, fear of retribution for error, personal, non-job-related stress, or uncomfortable environmental conditions, have the effect of amplifying potential error tendencies. Fatigue can lead to mental blinks—brief periods during which the cognitive apparatus simply closes down and information is not fully processed or acted upon. Most stressors can lead to a narrowing of focus and the risk of overlooking potentially critical environmental events.

MODELING HUMAN ERROR IN AIRPORT SURFACE (TAXI) OPERATIONS

The surface operations human subject experiments (chapter 3) provided the framework for the first phase of the HPM research project. Our goal was to model the behaviors of the human subjects—professional aircrews—and use the models and the theory on which they are based to better understand the sources of human error as seen in the experiments. Our approach was to model the robust behaviors of the aircrews and then probe the models for potential sources of error. As we use models to gain better insight into the sources for human error we will be better positioned to revise procedures, equipment, or training to reduce the incidence of error. Model-based testing can then be used to evaluate design changes.

We first outline the development of the aircrew procedures for the approach, landing, and taxi operations. We then look at how the tasks being executed by the captain and FO provided them with different perspectives on their common objectives and how they share information to compensate for each other's shortfalls.

To understand how errors can intrude on otherwise robust behaviors, we describe the emergence of an "intention to act" by tracing the resolution of the contention among multiple "intentions to act." We then examine two particular classes of error from the experiment trials. For each, we describe the modeling of the occurrence of the error and how the theory leads to an explanation of the source of the error.

MODELING ROBUST APPROACH, LANDING, AND TAXI PROCEDURES

We used the D-OMAR simulation framework for implementing the approach, landing, and taxi scenarios. HPMs were developed for the captain, FO, and the approach, tower, and ground controllers. The modeled Boeing-757 flight deck includes the instruments and controls necessary for the crew to execute the approach and landing procedures. The principal displays include the PFD, airspeed and altitude instruments, the ND, and lights providing landing gear status. Modeled flight controls include the mode control panel (MCP); switches for the autopilot; and levers for the throttles, flaps, landing gear, and speed brakes. The development of the D-OMAR model for the airport (KORD) and local airspace was based on NASA-provided materials that included airport diagrams, approach plates, and runway and taxiway signage information (see chapter 3).

The nominal task sequence (as presented in chapter 3) provided a sequential listing of the approach, landing, and taxi tasks performed by aircrews during the surface operations experiments. It describes the scenarios as beginning about 12 nautical miles (nmi) out in the approach sequence. The captain was the pilot flying (PF); the FO, as pilot not flying (PNF), was handling communications with ATC. The aircrew was engaged in intracrew conversation and party-line communication with the approach controller. They were conducting a basic instrument landing system (ILS) approach.

The modeled scenarios were initiated approximately 12 nmi before touchdown because this was the point in the approach sequence at which the first information on the subsequent taxi operations was obtained and processed by the aircrew. The approach controller provided the landing clearance and preferred runway exit. These were acknowledged by the FO, who then made a note of the runway exit. Even as the aircrew was primarily concerned with safely executing the approach and landing procedures, this early taxi-related information was processed to a level that it might well influence upcoming taxi operations. The aircrew knew the planned runway exit and the concourse and gate to which they were headed—the basic elements needed to suggest a likely taxi routing to their destination.

The modeled captain and FO continued a standard approach and landing. As the aircrew transitioned to taxi operations, the FO reviewed his or her notes and the airport diagram and notified the captain of the aircraft's location with respect to the designated runway exit. The captain slowed the aircraft to taxi speed, monitored the runway signage for the exit, and prepared to take the exit from the runway. As the aircraft approached the runway exit, the FO notified the tower controller that the aircraft was clearing the runway, the aircrew switched their radio frequencies to that of the ground controller, and the FO notified the ground controller that the aircraft had cleared the runway. The ground controller responded with the sequence of taxiway segments leading to their assigned concourse. The captain monitored the communication while the FO took notes on the taxi routing.

At this point, we encountered the first instance of a requirement for a *coping strategy*. Many of the high-speed exits at KORD have a very short run to the first intersection. When combined with the lengthy taxiway routing instructions often required at KORD, this can leave little time for preparation for the first turn. We encountered this first when modeling a landing on runway 9R using high-speed exit M6 with an immediate left turn onto taxiway M (see the KORD airport diagram in chapter 3). The FO was head-down, writing out the taxi directives and, as initially modeled, was late in providing the captain with information on the upcoming turn. The captain was also listening to the taxi routing and could either go with what he or she heard or slow the aircraft and get confirmation on the turn from the FO. The coping strategy that we modeled had the captain go ahead with the turn as heard and notify the FO of the turn as it was executed.

The aircrew then followed a basic pattern for subsequent taxi operations. As each turn was completed, the FO reviewed his or her notes for the next turn, checked the airport diagram for turn location, and notified the captain of the upcoming turn, providing additional details as required by the geometry of the taxiway layout. As the captain controlled the progress of the aircraft along the taxiways, the captain and FO (when not head-down) scanned the OTW view for traffic. The captain monitored and called out the airport signage and announced the turn as it was executed. The pattern then repeated until the final taxiway before the concourse, at which point the scenario terminated.

As expected, the modeled nominal process proved very robust. With the aircrew procedures in place, the challenge was to probe the model to find event sequences that produced errors consistent with those in the baseline surface operations scenarios.

LOCAL AND GLOBAL SITUATION AWARENESS DURING TAXI OPERATIONS

As the captain and FO meet their responsibilities during taxi operations, the inherent nature of the tasks that they perform provides them each with a different sense of their immediate location and their location with respect to their assigned taxi routing. They each achieve and maintain different levels of local and global SA. When working well as a team, they will strive to fill each other's gaps in awareness. In building the aircrew models, we felt that it was essential to reflect this level of teamwork.

In modeling the captain's tasks, he or she was modeled as predominantly head-up during taxi operations. The captain announced each turn as it was executed to keep the FO informed of their immediate location. Meanwhile, the FO used the airport diagram and kept written notes on the runway exit and taxiway routing. The use of these aids enabled the FO to provide the captain with a more global view of their taxiway routing. As modeled, the FO noted upcoming turns that had short lead times and intermediate taxiway crossings for turns with longer lead times. One effect of providing this level of detail in good crew performance was to make the taxiway procedures just that much more robust and error that much less likely.

INTENTION TO ACT

A classical view of the taxiway process might be that in approaching a turn the captain has a planning problem whose resolution is then followed by plan execution. Do we in fact need to make a turn at the upcoming intersection and, if so, which way? There might be a schema in place for executing the next turn with slots to be filled in for the name of the next taxiway and the direction to turn. In this view of the process, error might come about by incorrectly filling the slot for the next taxiway name or the direction of the turn to make. We would like to argue in favor of an alternate view in which there are typically several *intentions to act* concurrently in process.

Multiple Intentions to Act

The intentions may be established at different points in time. One or more of them may lead to a correct turn or to making an error at the intersection. A winner-take-all process leads to the execution of one of the intentions to act and the correctness of the outcome is the product of the winning intention. We label the process "intention to act" to suggest that the process is not necessarily the product of a conscious, thoughtful decision process. There can be the immediacy of an automatic, atomic process rather than a sequential process of planning and acting.

In the nominal case in which the FO has provided the correct prompt for the turn, the turn is, most likely, simply executed in response to the prompt and most likely matches the captain's pre-existing intention. In lieu of the FO's prompt, the captain will act on a pre-existing intention or, alternatively, pause and query the FO on the next turn. In the taxi environment it is reasonable to expect that the captain has an intention to act in place and ready to be acted on.

We now build the case for the idea that in performing relatively simple tasks like correctly executing the next taxiway turn, there may be several competing intentions to act. Most may be automatic processes that require little or no conscious deliberative thought. They emerge from different ongoing processes competing to determine the action taken. Occasionally, the winning intention will determine an action that is in error. During the course of this study, we have attempted to identify the sources for these errors and provide reasoned explanations on why the errors emerge.

For most of us there is a broad range of everyday activities that we perform quickly and effortlessly; they are employed automatically and involve little thought or conscious awareness (James, 1890; Logan, 1988b). Logan characterizes this automaticity—the execution of these activities—as fast, effortless, autonomous, stereotypic, and unavailable to conscious awareness. They are autonomous in the sense that the acquisition of these skills comes about independently of any deliberate intention to learn them.

Logan (1988b) developed the "instance theory of automaticity" for how automatization emerges. He identifies a role for *each* learned instance acquired in the automatization process. For any given problem, there may be several remembered solutions. When presented with a problem, there is a concurrent attempt to access the remembered instances of previous solutions *and* an explicit problem-solving computation. Due to the remembrance of each solution instance, there may be several correct retrievals in process. The first successful process determines the time required to solve the problem. The memory access is a fast process; the deliberative process is comparatively slow. If the memory access is successful in retrieving a solution, there will be a rapid response to the posed problem. If the memory access is not successful, the response based on the algorithm will be slower. Through experience, more and more solutions are acquired and, at some point, the deliberative process is simply not a contender.

Logan (1988a) further argues that the memory traces that support automaticity are at work in much of what we do. He suggests that we "look more broadly for automatic processes. They need not be restricted to procedural knowledge or perceptual-motor skill but may permeate the most intellectual activities in the application environment." Here we are suggesting that the captain's procedures for addressing the next turn in the taxiway sequence may sometimes be characterized as automatic and that while these will often lead to correct behaviors, they may sometimes lead to errors such as those seen in the surface operations experiments.

Intentions to Act as a Source of Error

Our review of the surface operations experiment data pointed to two important factors that deserved particular attention in our modeling effort. Chapter 3 identified the importance of the location of the destination gate and its relation to the taxi route. Seven errors occurred in 534 instances of turns *toward* the concourse, whereas five errors occurred in only 48 instances of required turns *away from* the shortest route to the concourse. At any given intersection, the aircrews had a bias to turn toward their destination concourse. When the correct turn was one away from the concourse, there was a greater tendency toward making an error. The second factor was the basic observation that time pressure can lead to error. The time pressure of a second turn closely following a first turn was an important factor in each of the errors that we generated in the modeling effort.

Four sources of intention to act were identified and modeled. Each was initiated at a different point in the taxi sequence. The first source was *episodic memory*—a constituent of autobiographical memory and a source for habit-based actions. Similar situations have been encountered in the past and we typically have a ready source of responses that have worked well. These responses have proven successful and are generally able to carry us through many of the activities of the day. When they fail, Reason (1990) refers to these habituated processes as "strong-but-wrong." In our particular case, the aircrews have a history of previous landings at KORD and the taxi routings that they have executed to reach their destination gates.

A second source of intention to act is *context-based expectation*, driven by partial knowledge. Explicit partial information provided in the current situation prompts a particular intention. In the taxi scenarios, the captain knows the destination concourse and, based on this knowledge, may reasonably have an expectation that the next turn will take them on the shortest route to that destination. These particular situation-specific information points are sufficient to set up an intention for the next turn.

The third source of intention to act is the *remembrance* of the taxi sequence provided by the ground controller. The content of the remembrance may be correct or incorrect.

The fourth and most immediate intention to act in the taxiway situation is the explicit prompt by the FO based on written notes on the taxi directives. In the nominal case, the FO's prompt will match the captain's intention and will lead to error-free performance.

We modeled the contention between these intentions as a winner-take-all process mediated by priority and explored the impact of varying the priorities of the contending intentions.

MODELING AND UNDERSTANDING TWO
CLASSES OF TAXI NAVIGATION ERROR

The error analysis focused on two error sequences that each had a pair of turns in rapid succession. For the first case, there were two instances of the same error as crews took high-speed exit M7 from runway 9R (see airport layout in chapter 3). At the first intersection after the high-speed exit, each captain turned left toward the destination concourse rather than right as directed by the ground controller. In the second case, there were two scenarios that shared a similar turn sequence: after turning onto taxiway F in the first instance and M2 in the second instance, there was a quick right turn onto taxiway B. In each of the scenarios, one of the captains turned left rather than right. The errors were noteworthy because the captains each turned *away* from their intended concourse rather than toward the concourse as directed.

ERROR DRIVEN BY EXPECTATION BASED ON PARTIAL KNOWLEDGE

Our hypothesis is that the incorrect turn following the high-speed exit (M7) was driven by the captain's expectation that the shortest route to the concourse was the route to be taken. The captain incorrectly turned to the left just after the high-speed runway exit. The captain's earlier discussion with the approach controller and the FO on the proposed runway exit gave rise to this intention to act. At that point, the captain knew the runway exit (M7) and the location of the concourse, and might reasonably have expected to turn left from the high-speed exit at taxiway M, taking the aircraft toward the concourse. It became one intention contending to be executed at the first turn after exiting the active runway.

As the scenario played out, the FO completed the task of taking notes on the taxiway sequence and then prompted the captain on the first turn following the runway exit. The FO's prompt triggered a new, contending intention to act on the captain's part. While the new intention may or may not have been consistent with preexisting intentions to act, in the nominal case, the FO's prompt and the captain's intention would match, leading to a correct turn. Given the correct prompt by the FO, it was possible, but unlikely, that the captain would incorrectly execute the turn.

To open a window for an error to occur, we constructed a situation that occupied the FO, thus preventing him or her from prompting the captain on the upcoming turn. The very short run to the first turn after the high-speed exit played an important role by limiting time available to accomplish necessary tasks. The FO was busy taking notes on the taxiway routing just provided by the tower controller, yet he or she had time to provide the prompt for the upcoming turn. A "mistake," the failure to *preset* the radio frequency for the transfer to the ground controller, imposed additional workload on the FO, incurring a delay sufficient to prevent him or her from providing the on-time prompt.

As the scenario played out, the FO was completing his or her notes on the taxi routing and had not provided the captain with the explicit prompt on the upcoming turn. Based on the coping strategy described earlier, the captain might have a correct intention to act based on having attended to the ground controller's taxi directive *and* an incorrect intention based on the expectation of receiving a shortest route to the concourse gate. Much of the time the coping strategy, based on the remembrance of the taxi directives, would be expected to win the winner-take-all competition and lead to a correct turn. Some of the time, the expectation-based intention to act—the turn toward the concourse—might be acted upon, leading to a taxiway error. Hence, a reasonable, grounded source for an error consistent with the surface operations experiments has been identified and modeled.

ERROR DRIVEN BY HABIT

The second scenario examined the surprising cases in which an aircrew incorrectly turned *away* from the shortest route to their destination concourse. The intention to take the shortest route to the gate would have led to the correct behavior, yet it was not the acted-upon intention. In the two instances of this error at similar intersections, the captains turned away from the concourse.

To account for these errors, we speculated that a crew whose company's concourses were on the opposite side of the airport from those required by the scenario might incorrectly turn toward their company's concourse, thereby exhibiting an error based on habit. Requiring an aircrew to proceed to a gate opposite in direction from their company gates might be considered an artifact of the particular scenario, but in a commercial air travel environment that has seen many mergers and new companies, it is not uncommon for aircrews to find themselves working for companies with gate locations that are new to them.

Once again, we manipulated the situation in a realistic manner such that the FO was not able to provide a timely prompt to the captain on the upcoming turn. Conflicting taxiway traffic was present on the FO's side of the aircraft. The FO informed the captain of the presence of the traffic and continued to monitor the other aircraft. Consequently, the FO was delayed in going head-down to review his or her notes on the upcoming turn and relating it to the airport diagram. Following this initial delay, the FO's prompt on the upcoming turn was immediately interrupted by a message from the ground controller directing the other aircraft to hold short of the upcoming intersection, allowing the first aircraft to proceed with the turn. Very slight changes in timing of the interruption would have opened the window for a timely prompt that might well have led to a correct turn.

In the absence of the prompt, there were still multiple intentions to act. As modeled, there were intentions to act based on the remembrance of the ground controller's taxi directive and on habit based in episodic memory. A turn based on a correct remembrance would have led to a correct turn. When the captain's habit-based intention to act won the winner-take-all competition, an error was committed. An informal post hoc analysis of the human subject trial error (B. Hooey, personal communication) provided support for the speculation that the model represented: Experiment subjects had made incorrect turns toward their companies' gates.

In the two cases modeled, the theory underlying the models provided insights into the mechanisms at work that led to the errors as seen in the human subject trials. As such, they suggest that another timely mode of prompting might well fill in to prevent error when the workload prevents the FO from providing the required prompt.

MODELING THE NASA SYNTHETIC VISION SYSTEM TRIALS

We began the second phase of the human performance modeling research project by using the D-OMAR simulation framework to build the modeling components needed to emulate the NASA part-task simulation environment for the SVS experiments (chapter 3). We then developed the HPMs for the aircrews that successfully executed the complete set of NASA part-task scenarios. Findings developed during this initial modeling effort provided the basis for the additional SVS studies outlined in this section.

Modeling the Synthetic Vision System Scenarios

The SVS experiment trials included 10 scenarios that selectively examined three variables: display configuration, visibility conditions, and approach events (chapter 3). Our approach to developing the scenarios was first to construct the basic aircrew and ATC procedures to support the area navigation (RNAV) approach (Keller & Leiden, 2002a; see also chapter 3) using the baseline flight deck under visual meteorological condition (VMC). With these basic scenario elements in place, we then added cloud cover to the model to represent instrument meteorological condition (IMC). For the modeled IMC scenarios (excluding the go-around condition), the *actual* cloud ceiling was at 800 ft; hence, the breakout from the clouds occurred 150 ft above the 650-ft DA specified by the approach plate.

With the nominal VMC and IMC scenarios operational, we were ready to address the "late-runway-reassignment" condition (described in chapter 3). We developed an extended scenario with two closely spaced aircraft on the approach to SBA runway 33L. Our goal was to create a realistic situation that forced the late-reassignment option. As the modeled *lead* aircraft landed, it blew a tire and temporarily held on the active runway. The aircrew notified ATC of the problem and ATC then addressed the situation by offering the *trailing* aircraft the opportunity to sidestep to the parallel runway 33R.

For the VMC "go-around" condition, the tactic employed for setting up the condition was a variation on the late-reassignment scenario. A notice to airmen (NOTAM) was issued within the simulation to notify the controller that runway 33R, the runway parallel to 33L, was closed for repairs. With the lead aircraft temporarily blocking runway 33L due to a blown tire, the controller had no option other than requesting that the trailing aircraft execute a go-around.

To model the IMC "go-around" condition, the reported 800-ft cloud cover was modeled as actually extending all the way to ground level. Under these conditions, neither the airport nor the runway could be visually acquired OTW as the aircraft descended through the established DA, and the captain was forced to elect a go-around.

Lastly, we added the SVS display (chapter 3) to the flight deck. As in the human subject trials, a single SVS was provided for use only by the captain. Based on the cognitive task analysis (Keller & Leiden, 2002b; also see chapter 3), the aircrew procedures were extended to include the use of the SVS. With the SVS-equipped flight deck and aircrew procedures in place, we were then able to run the four IMC scenarios using the SVS-equipped flight deck that were described in chapter 3.

The modeled approach and landing procedures followed the standard progression through the airspace transiting from approach controller to tower controller to ground controller and terminating as the aircraft completes the landing and taxis to its assigned concourse. In the go-around scenarios, the aircrews followed the missed approach plan established on the approach plate for runway 33L (see chapter 3) with the scenarios terminating as the specified left turn was initiated during the climb out.

As each modeled scenario begins, the captain is the PF and the FO is the PNF. The approach controller clears the aircraft for the approach to SBA runway 33L and provides information on VMC or IMC depending on the scenario. Using information from the approach controller and the approach plate, the captain continues the approach by reviewing the runway and weather information with the FO. The captain briefs the go-around procedures based on the information read from the approach plate.

The aircrew then focuses on navigation as the aircraft proceeds along the flight path under the control of the flight management computer (FMC). As they approach each fix, the captain calls for a new MCP altitude setting for the next leg based on information read from the approach plate. The aircrew then monitors the aircraft's heading and altitude changes as the aircraft transitions to the new flight segment. They continue to monitor the heading and altitude until the target values are fully established. As the approach progresses, the captain calls for a series of flap settings consistent with their airspeed and position along the approach path.

The approach controller provides the radio frequency for the tower, the aircrew resets the radio frequency, and the FO contacts the tower. The tower controller immediately grants the aircrew the clearance to land. The captain asks for the final flap setting, that the landing gear be lowered, and that the speed brakes be armed. At this point, the captain and FO then complete the execution of the prelanding checklist.

As the aircraft continues its descent, the FO monitors the aircraft's altitude and makes the required altitude callouts. The captain is responsible for the OTW sighting of the runway and making the decision to land. Under VMC, the captain can readily acquire the airport and the runway OTW well before decision altitude. For the SVS scenario, the captain uses the SVS to acquire the runway before the aircraft breaks out of the cloud cover, but must visually acquire the runway OTW to support the decision to land. In IMC, the captain anticipates the breakout from the cloud cover at 800 ft and has adequate time to acquire the runway before the 650-ft DA. Depending on the outcome of the decision to land, the captain will either take "manual" control of the aircraft to complete the landing or elect the FMC-programmed go-around maneuver.

Assessing Model Behaviors for Validation

The focus of our validation was on assuring that the models for the captain and FO collaborate effectively, respond appropriately to ATC directives, and robustly execute the specific approach and landing procedures as skilled human aircrews do. D-OMAR simulation tools provide explicit measures of these behaviors that were essential in the *assessment* and *validation* of model performance. Aircrew behaviors are frequently multitask behaviors—the successful or, occasionally, less successful integration of the demands of several ongoing tasks. Each of these tasks may be made up of several steps that require the coordinated actions of several human functional capabilities (e.g., maintaining an intracrew conversation and the coordination of hand–eye actions to set the altitude selector). As the detailed model behaviors were developed, the assessment process was used to assure that the varied situations were addressed by reasonable, human-like aircrew model performance.

D-OMAR graphical display tools provide multiple levels of visibility into model behaviors. An event-recording subsystem captured the data that supported the evaluation process. As key events were recorded, they were displayed in the simulation control panel's trace pane. A plan view allowed an observer to monitor the progress of the aircraft along its flight path. A Gantt chart display provided detailed information on goals and procedures as executed by the captain and FO. It provided insight into how each crewmember's goals and procedures played out in time to generate their actions. An event timeline provided detailed insight into the behaviors of the publish–subscribe protocol used to coordinate procedure execution.

Each of the assessment tools has opened a new window of visibility into model performance; with each new window, our experience has been that, while we see new aspects of successful model performance, we also find new views into model shortcomings. Validation will be as good as the capabilities of the tools used to assess behaviors; hence, considerable effort has been devoted to these tools. The validation will be meaningful to the extent that the assessment covers the range of situations appropriate for the intended use of the models.

THE SYNTHETIC VISION SYSTEM IMPACT ON PILOT SCAN PATTERNS

The findings developed in reviewing the part-task simulation data and in modeling the scenarios laid the foundation for the further research effort. The first finding pursued was the recognition that the addition of the SVS as a second attitude instrument altered time-tested pilot instrument scan patterns. When the SVS was added to the flight deck, the captain, as modeled, alternated between the SVS and PFD as the target for the basic aviation scan. One impact of the new scan pattern was that less time was devoted to the ND. This finding with respect to the model triggered a further review of the human subject data. The same effect was then found in human subject data; a finding identified in the modeling process suggested further analysis of the human subject data. The effect does not appear to be a product of the SVS itself, but rather derivative from the symbology added to the SVS display by which it becomes a target for an *aviation* scan. Our goal in following up on this finding was to seek a means to restore the more efficient scan pattern as used by the pilots for the baseline flight deck configuration, while preserving the advantages provided by the SVS.

WORKLOAD IMPLICATIONS OF THE SYNTHETIC VISION SYSTEM
AS A SECOND ATTITUDE DISPLAY

The modeled aircrews and human subjects tended to spend less time attending to the ND when the SVS was available even as they were principally monitoring their progress along the flight path. One explanation might be that the aircrews had sufficient time to accomplish their navigation task and were simply using the ND to the extent that it was needed. Nevertheless, it was possible that there was an underlying problem.

On the SVS-equipped flight deck, the captain, in scanning the two attitude instruments, was drawn into spending more time attending to the redundant pieces of information (i.e., heading, altitude, altitude rate, and speed) than had been allocated when using only the PFD without the SVS—the *habituated* pattern can become a redundant two-instrument scan. In situations where time pressure is high, having two instruments from which to obtain required information can impose additional workload on the pilot. Does the pilot stick with the time consuming habituated scan or try to save time by switching to a single-instrument, but less practiced scan? Even electing to consider the options has a cost. Moreover, the extra cognitive effort required to switch to a single instrument scan may negate its potential inherent advantage. This is an imposition on the aircrew that should be avoided if possible.

AN ENHANCED SYNTHETIC VISION SYSTEM AS THE SINGLE PRIMARY FLIGHT DISPLAY

The addition of the SVS to the flight deck with its capability to provide a "clear day" view under all conditions has the potential to lead to greater operating safety; the potential impact on pilot scan patterns is one for which mitigation should be sought. One solution is to consider a single attitude display combining PFD and SVS functionality—an enhanced SVS. With the enhanced SVS as the single attitude instrument, the baseline scan pattern might then be restored.

Modeling provided a cost-effective means to explore pilot performance using the enhanced SVS. The necessary steps were to define the information requirements to be met by the enhanced SVS instrument and refine modeled pilot behaviors to use the new single PFD instrument. With these changes in place, it was then possible to rerun the SVS scenarios using the new flight deck configuration and examine the impact on pilot scan patterns.

During much of the approach, the pilots are concerned primarily with heading, speed, and altitude along each flight leg. In the model, each crewmember performs a background aviation scan task to assure that these critical values are being maintained. As the aircraft transitions from one flight leg to the next, the aircrew is concerned with monitoring the change to the new heading and altitude. Each crewmember executes procedures to monitor the aircraft's transition actively until the new flight leg is fully established.

Completing the design for an operational enhanced SVS would be a significant undertaking. Fortunately, at the information-seeking level, "designing" an enhanced SVS was a simpler task. If we were to remove the standard PFD from the flight deck, we had to assure that each of its information items would continue to be available. The SVS immediately provides the three values of interest for the basic aviation scan: heading, altitude, and airspeed. As described in chapter 3, the SVS also provides a vertical speed indicator, a roll indicator, a pitch ladder, localizer dots, and a flight-path predictor. The mode annunciators were the only missing information sources. For the purposes of running the modeled SVS scenarios using the enhanced SVS, it was sufficient to add the mode annunciators to the SVS. With this simple change accomplished, the essential PFD functionality was established in the enhanced SVS and the PFD could be removed from the flight deck configuration.

THE ENHANCED SYNTHETIC VISION SYSTEM SCENARIO TRIALS

Our goal in pursuing the design of the enhanced SVS was to provide an SVS instrument that would restore the balance in the pilot PFD–ND scan pattern as seen with the baseline flight deck. To obtain the scan data through which to assess the success of the design, we repeated the four NASA part-task SVS-equipped scenarios using the enhanced SVS. The modeled aircrews readily executed each of the scenarios as required.

Table 8.1 provides representative eye-tracking data for a flight segment between the final approach fix (FAF) and DA. Goodman, Hooey, Foyle, and Wilson (2003;

TABLE 8.1

Pilot and Model Eye-Tracking Data: IMC Nominal Approach—FAF to DA

	1	2	3	4	5	6	7	8	9
	Pilot 1	Pilot 2	Pilot 3	Model	Pilot 1	Pilot 2	Pilot 3	Model	Model
Condition	IMC	IMC	IMC	IMC	SVS	SVS	SVS	SVS	Enh SVS
Off	2.22	4.08	2.84		1.80	5.07	1.53		
OTW	0.00	11.57	2.03	10.39	0.52	4.37	2.57	8.55	10.79
SVS					19.22	33.53	49.04	28.37	37.60
PFD	39.98	46.72	42.54	39.60	41.21	35.96	24.76	29.53	
ND	33.45	32.40	44.64	39.62	26.90	19.53	18.16	28.66	39.57
Other	24.35	5.22	7.94	10.39	10.36	1.55	3.95	4.89	12.04

Notes: DA = decision altitude; Enh SVS = enhanced synthetic vision system; FAF = final approach fix; IMC = instrument meteorological conditions; SVS = synthetic vision system; OTW = out the window; PFD = primary flight display; ND = navigation display.

see also chapter 3) provided the human subject data for the baseline and SVS trials. The model data for the baseline, SVS, and enhanced-SVS trials were derived from the D-OMAR model trials. The rows of the table present the percentage of dwell time that each subject, human or model, devoted to the OTW view, to each of the principal flight deck instruments, and to "other" identified areas in their field of view. For the human subjects, the row labeled "off" accounts for the percentage of dwell time for which the eye cursor was centered on an undefined area or for which the data were invalid (e.g., subject blinks). Columns 1 through 3 provide data from the baseline-IMC approach for the three human subjects; column 4 provides data for the D-OMAR model. Columns 5 through 7 provide data from the SVS-equipped IMC approach for the three human subjects; column 8 provides data from the D-OMAR model. Column 9 provides model data for the IMC scenario using the enhanced-SVS display.

As can be seen in the table, usage of the SVS varied considerably among subjects. Yet, there was consistency in one aspect of subject response to the addition of the SVS to the flight deck: The percentage allocation of dwell time devoted to the ND decreased while the summed allocation to the two attitude displays, the PFD and the SVS, increased. For the trial using the enhanced SVS, the percentage dwell time allocation closely approximates the percentage dwell times for the baseline IMC trial using the single attitude display. The model trials demonstrate that when the enhanced SVS is used as the single attitude display, the balance in time devoted to the attitude display (in this case, the SVS) and the ND reverted to the usage seen in the baseline configuration of the PFD and ND, respectively.

Evaluating alternate flight deck instrument configurations and their implications for aircrew performance can be very costly. In this instance, modeling enabled us to design and evaluate the proposed enhanced SVS quickly. The modeling outcome suggests that the enhanced SVS will avoid the potential workload implications of the PFD–SVS configuration.

Alternate Responses to Synthetic Vision System Misalignment

A second finding was derived directly from the human subject trial data and was reflected in the subsequent model development. Our related work on aircrew decision making (Deutsch, 2005) pointed to a strong bias on the part of aircrews to complete an in-process approach and landing in spite of cues suggesting that a go-around may be the better course of action. Orasanu, Martin, and Davison (1998) have referred to this response as a plan continuation error. For the NASA part-task trials, the instructions to the subjects were quite explicit: If they detected a terrain mismatch between the SVS view of the runway and OTW view, they were to execute a go-around.

With this as background, we elected to examine the flight paths of the aircraft in the terrain-mismatch scenarios following the point at which the human subjects recognized the SVS-mismatch condition. For each scenario trial, NASA provided room-camera and eye-camera view VCR tapes that included an audio track. They also provided part-task simulation data on aircraft position and attitude. In reviewing the eye-camera tapes, we found that shortly after the FO notified Pilot 3 that the aircraft was approaching DA, the pilot looked out the windscreen, recognized that they were "not aligned visually" and immediately decided: "We're going around." The pilot acted quickly in following the specified response to the SVS misalignment situation.

When the FO notified Pilot 1 that the aircraft was approaching DA, he responded, "Airport in sight, runway in sight," followed shortly by, "SVS does not line up with the picture out the window," and then "I'm going to take it down a little bit lower, but it's putting me in a bad position." The subject subsequently proceeded with the approach and it was only when the FO announced "200 feet" that the pilot responded, "Okay, I'm off center, going around."

When the FO notified Pilot 2 that the aircraft was approaching DA, the pilot responded with "runway in sight" and subsequently announced *contrary* to the misalignment rule, "Landing." It was only when the FO notified the captain that "[it] looks offset to me" that the pilot responded, "It is" and initiated the go-around. The intent of Pilots 1 and 2 was to proceed with the landing. Had the simulated aircraft been more realistic in terms of the operation of the controls and the ability to monitor control settings (monitoring throttle settings, in particular, was difficult), Pilots 1 and 2 might well have succeeded in aligning their aircraft with the runway and completed the landing.

As seen in the pilot responses, on finding that the SVS is misaligned, the decision to pursue the landing or elect a go-around is not straightforward. Two of the three pilots elected to set aside the "go-around rule" and pursue the landing. Several factors may have played into their decisions. The situation was not unlike the sidestep scenario in which they were explicitly asked if they would sidestep to the parallel runway. The offset and approach timing were similar and the human pilots (and the models) were readily able to execute the necessary sidestep maneuver. In addition, they may have been reluctant to climb back into the cloud cover when dependent on an instrument that was then known to be misaligned.

While we were not surprised by the finding that two of the three pilots pursued the approach and landing well past when the decision to go-around should have been made, we did initially develop pilot procedures that faithfully followed the misalignment rule. We subsequently changed the decision-making process to reflect the human pilots' behaviors more accurately. Cues considered in making the decision included the detected offset from the runway, distance to the runway threshold, and the "priority" that the modeled pilot assigned to pursuing the landing. With the evaluation of these cues in place, the model sometimes elected to go around and sometimes successfully completed the landing. Developing procedures for the mitigation of SVS misalignment problems deserves more attention; it is a point at which there is very little time for thinking through solutions to problems and well-practiced responses need to be readily available.

CONCLUSIONS

We close this chapter by highlighting three conclusions established in the course of our human performance modeling research effort.

1. *D-OMAR served very well as a human performance modeling framework for addressing aviation safety research challenges.* D-OMAR provided excellent support for developing the approach, landing, and taxi scenarios. Aircrew and ATC models were developed that executed the ILS landing and taxi operations for the taxi phase of the study at KORD. The models were readily extended to include an RNAV approach and the use of the SVS at SBA.

Approach, landing, and taxi operations are flight phases in which a disproportionate number of incidents and accidents occur (NTSB, 1994). The aircrew models that have been developed to execute these operations have enabled sources of human error, as well as system and procedural vulnerabilities, to be identified and better understood. These new capabilities in the model framework and in the aircrew models are supportive of the goal of the project: the improvement of the HPMs so that they can be used more effectively as we seek to further improve aviation safety (see chapter 1).

2. *HPMs can be used as an exploratory tool to better understand the sources of human error leading to incidents and accidents.* The surface operation experiments provided excellent empirical data on aircrew taxiway errors. The open question was then: Can we better understand the underlying sources of the errors that were committed? Our approach to answering this question was to build models of robust aircrew performance and then probe the models for vulnerabilities to error. The process, relying on the models and the theory that supported model development, yielded important insights into the sources of human error.

3. *The model-based "design" and evaluation of the enhanced SVS demonstrated the potential that HPMs have for forecasting potential system vulnerabilities and exploring mitigation strategies.* Our long-term goal is to continue to improve our understanding of pilot behaviors as represented in HPMs and to use the models and that knowledge to explore the means to identify accident precursors and establish procedural and system changes to prevent them or mitigate their impact. The response to the finding related to pilot scan patterns when the SVS was added to the flight deck is representative of how human performance modeling has been productively employed. In extending the model's procedures to use the SVS, we uncovered an unanticipated effect that led us to revisit the human subject data. The subsequent analysis of the human subject data further supported the hypothesis that introducing

the SVS to the flight deck led to more time being devoted to the attitude displays (the PFD and the SVS) and less time devoted to the ND.

The question addressed became: How might the advantages of the SVS be secured, while at the same time not creating a flight-deck environment that altered the balance in pilot scan patterns related to aircraft aviation and navigational tasks? Using the D-OMAR modeling environment, we were readily able to design and evaluate an enhanced SVS that combined the basic capabilities of the traditional PFD and the SVS. Exercising the model produced the anticipated shift back to the more efficient scanning behavior observed in the baseline condition while preserving the advantages of SVS capabilities.

These results are important not only for the substantive predictions they generated, but also as a concrete example of the ways in which HPMs can contribute to design decisions early in the design process. Using the modeling framework, we were able to "design" an enhanced SVS and demonstrate the validity of the hypothesis that the enhanced SVS would restore the baseline pilot scanning behavior. The cost in time and effort in exercising the model to demonstrate the validity of the hypothesis was a small fraction of the cost of building a revised human-in-the-loop simulation, running a new set of subjects, and analyzing the resulting eye-movement data.

REFERENCES

Adams, M. J., Tenney, Y. J., & Pew, R. W. (1991). *Strategic workload and the cognitive management of advanced multi-task systems.* Wright Patterson AFB, OH: Crew Systems Ergonomics Information Analysis Center.

Bargh, J. A., & Chartrand, T. L. (1999). The unbearable automaticity of being. *American Psychologist, 54,* 462–479.

Bartlett, F. C. (1958). *Thinking. An experimental and social study.* London: George Allen & Unwin.

Bellenkes, A. H., Wickens, C. D., & Kramer, A. F. (1997). Visual scanning and pilot expertise: The role of attentional flexibility and mental model development. *Aviation, Space, and Environmental Medicine, 68,* 569–579.

Billings, C. E. (1997). *Aviation automation: The search for a human-centered approach.* Mahwah, NJ: Lawrence Erlbaum Associates.

Brachman, R. J., & Schmolze, J. G. (1985). An overview of the KL-ONE knowledge representation system. *Cognitive Science, 9,* 171–216.

Deutsch, S. E. (2005). Reconceptualizing expertise: Explaining an expert's error. *Proceedings of the 7th Naturalistic Decision Making Conference.* Amsterdam.

Deutsch, S. E. (1998). Interdisciplinary foundations for multiple-task human performance modeling in OMAR. *Proceedings of the Twentieth Annual Meeting of the Cognitive Science Society.* Madison, WI.

Deutsch, S. E., & Adams, M. J. (1995). The operator model architecture and its psychological framework. *Proceedings of the 6th IFAC Symposium on Man–Machine Systems.* Cambridge, MA.

Deutsch, S. E., & Pew, R. W. (2002). Modeling human error in a real-world teamwork environment. *Proceedings of the Twentieth-fourth Annual Meeting of the Cognitive Science Society* (pp. 274–279). Fairfax, VA.

Deutsch, S. E., & Pew, R. W. (2004). Examining new flightdeck technology using human performance modeling. *Proceedings of the 48th Meeting of the Human Factors and Ergonomic Society Meeting.* Santa Monica, CA: HFES.

Edelman, G. M. (1987). *Neural Darwinism: The theory of neuronal group selection*. New York: Basic Books.

Edelman, G. M. (1989). *The remembered present: A biological theory of consciousness*. New York: Basic Books.

Glenberg, A. M. (1997). What memory is for. *Behavioral and Brain Sciences, 20,* 1–55.

Goodman, A., Hooey, B. L., Foyle, D. C., & Wilson, J. R. (2003). Characterizing visual performance during approach and landing with and without a synthetic vision display: A part-task study. In D. C. Foyle, A. Goodman, & B. L. Hooey (Eds.), *Proceedings of the 2003 Conference on Human Performance Modeling of Approach and Landing with Augmented Displays* (NASA/CP-2003-212267) (pp. 71–89). Moffett Field, CA: NASA.

James, W. (1890). *Principles of psychology*. New York: Holt.

Keller, J., & Leiden, K. (2002a). Information to support the human performance modeling of a B757 flight crew during approach and landing: RNAV (Tech. Rep.). Boulder, CO: Micro Analysis & Design.

Keller, J., & Leiden, K. (2002b). Information to support the human performance modeling of a B757 flight crew during approach and landing: SVS addendum (Tech. Rep.). Boulder, CO: Micro Analysis & Design.

Logan, G. D. (1988a). Automaticity, resources, and memory: Theoretical controversies and practical implications. *Human Factors, 30,* 583–598.

Logan, G. D. (1988b). Toward an instance theory of automatization. *Psychological Review, 95,* 492–527.

Neumann, O. (1987). Beyond capacity: A functional view of attention. In H. Heuer & A. F. Sanders (Eds.), *Perspectives on perception and action*. London: Lawrence Erlbaum Associates.

Norman, D. A. (1988). *The psychology of everyday things*. New York: Basic Books.

NTSB (National Transportation Safety Board). (1994). *Aircraft accident report: A review of flightcrew-involved, major accidents of U.S. carriers, 1978 through 1990* (NTSB/SS-94/01). Washington, DC: U.S. Government Printing Office.

Orasanu, J., Martin, L., & Davison, J. (1998). Errors in aviation decision making: Bad decisions or bad luck? *Proceedings of the Fourth Conference on Naturalistic Decision Making*. Warrenton, VA.

Orasanu, J. M. (1994). Shared problem models and flight crew performance. In N. McDonald, N. Johnston, & R. Fuller (Eds.), *Aviation psychology in practice*. Aldershot, UK: Ashgate.

Pew, R. W., Miller, D. E., & Feehrer, C. E. (1981). *Evaluation of proposed control room improvements through analysis of critical operator decisions*. Palo Alto, CA: Electric Power Research Institute.

Posner, M. I., Petersen, S. E., Fox, P. T., & Raichle, M. E. (1988). Localization of cognitive operations in the human brain. *Science, 240,* 1627–1631.

Rasmussen, J. (1976). Outlines of a hybrid model of the process plant operator. In T. B. Sheridan & G. Johannsen (Eds.), *Monitoring behavior and supervisory control*. New York: Plenum.

Reason, J. (1990). *Human error*. Cambridge, England: Cambridge University Press.

Salas, E., Dickinson, T. L., Converse, S. A., & Tannenbaum, S. I. (1992). Toward an understanding of team performance and training. In R. W. Swezey & E. Salas (Eds.), *Teams: Their training and performance*. Norwood, NJ: Ablex Publishing Company.

Serfaty, D., Entin, E. E., & Johnston, J. H. (1998). Team coordination training. In J. A. Cannon-Bowers & E. Salas (Eds.), *Making decisions under stress: Implications for individual and team training* (pp. 221–245). Washington, DC: American Psychological Association.

Shapiro, R. (1984). *FLEX: A tool for rule-based programming* (BBN Rep. No. 5843). Cambridge, MA: BBN Technologies.

9 Attention-Situation Awareness (A-SA) Model of Pilot Error

Christopher D. Wickens, Jason S. McCarley,
Amy L. Alexander, Lisa C. Thomas,
Michael Ambinder, and Sam Zheng

CONTENTS

INTRODUCTION

Pilot errors result from multiple causes. According to taxonomies developed by Norman (1981) and Reason (1990), *slips* are easily detectable errors in which an unintended action is executed, often as the result of poor design. Decision "errors" result from a variety of breakdowns, biases, or tendencies in human information processing. In aviation, such decision errors are more likely to produce fatalities than errors of other classes (Wiegmann & Shappell, 1997). However as has been noted clearly by Woods, Johannesen, Cook, and Sarter (1994), considerable caution must be exercised in labeling decisions that produce unfortunate outcomes as "errors" because of the great biases of hindsight.

The focus of the present modeling effort is on *errors of situation awareness* (SA), with particular emphasis given to those errors related to attention allocation. There are several reasons for this focus:

- An information processing analysis suggests that situation assessment (diagnosis) is a critical component of decision making (Wickens & Hollands, 2000, pp. 293–336).
- Endsley's (1995) three-stage model of SA places considerable emphasis on the importance of attention and perception in supporting effective situation assessment. Indeed a review found that errors of perception and attention (Stage 1 SA) were responsible for a majority of SA-related aircraft accidents (Jones & Endsley, 1996). Errors of Stage 1 SA may include attentional tunneling, which is often responsible for controlled-flight-into-terrain (CFIT) accidents, and the failures to monitor automation, which can leave an operator surprised at changes of system state (Sarter & Woods, 2000).
- Given the criticality of attention in supporting SA (Sarter & Woods, 1995), some of the most effective computational models of human performance have been developed in the area of attention allocation (Carbonnell, Ward, & Senders, 1968; Klatzky, 2000; Moray, 1986).
- Recent analysis has revealed the critical role of attention in the monitoring and supervision of flight deck automation (Parasuraman, Sheridan, & Wickens, 2000; Sarter & Woods, 1995; Sklar & Sarter, 1999; Wickens, 2000; Wickens, Gempler, & Morphew, 2000).
- Eye-movement monitoring can serve as a direct index of attention allocation (Bellenkes, Wickens, & Kramer, 1997; Wickens, Goh, Helleberg, Horrey, & Talleur, 2003; Wickens, Helleberg, & Xu, 2002). We are able to build upon this in formulating the current model.

However, in spite of the prevalence of computational models of attention and visual scanning in the literature and the vast quantity of recent research on SA, there appear to be no prior efforts to integrate these into computational models of SA. This is our intention in the current effort. Thus, the rationale for our focus is as follows:

1. Faulty pilot judgments are a critical element for understanding flight safety.
2. Many such faulty judgments result from a breakdown in SA and assessment—that is, a failure to attend to and integrate appropriate sources of information in order to update the assessment of the situation.
3. Attention and belief updating are amenable to computational modeling.

In the following pages, we first describe a two-component computational model of attention and situation awareness (A-SA), then discuss applications to two different domains: (1) taxiing at Chicago's O'Hare International Airport (KORD), where we infer the allocation of attention and use this to draw inferences regarding pilots' awareness of where they are on the airport surface; and (2) use of a synthetic vision system (SVS) in approach and landing, where we directly assess visual scanning, apply the A-SA model to model the allocation of visual attention (Stage 1 SA), and show how the model fit directly predicts performance-based assessments of SA. The SVS domain is presented in two separate applications: one used to develop the model and described only briefly, and the second used to demonstrate statistical validity of the model and described in more detail.

The reader will note some evolution and change in the model across the two domains, as it needed to be altered slightly based upon the aviation data stream of events available for modeling. Nevertheless, the common components of the model across both domains should be quite evident.

MODEL ARCHITECTURE

Foundation of the Model

The underlying theoretical structure of the A-SA model is contained in two modules, one governing the allocation of *attention* to events and channels in the environment, and the second reaching an inference, situation assessment, understanding, or *belief* regarding the current and future state of the aircraft within that environment. The first module corresponds roughly to Endsley's (1995) Stage 1 SA (perception), while the second corresponds to Stages 2 (understanding) and 3 (prediction). In dynamic systems, there is a fuzzy boundary between Stage 2 and Stage 3 because the understanding of the present usually has direct implications for the future, and both are equally relevant for the task.

The elements of the attention module are contained in the *SEEV* model of attention allocation developed by Wickens, Helleberg, Goh, Xu, and Horrey (2001); Wickens et al. (2003); and Horrey, Wickens, and Consalus (2006), which estimates the probability of attending, $P(A)$, to an area in visual space, as a linear weighted combination of four components:

$$P(A) = sS - ef\ EF + (ex\ EX \times vV) \qquad (9.1)$$

Here, coefficients in upper case describe the properties of a display or environment, while those in lower case describe the weight assigned to those properties in the control of an operator's attention. These elements indicate that the allocation of attention in dynamic environments is (1) driven by bottom-up attention capture of *salient* (S) events (e.g., a flashing warning on the instrument panel will capture the pilot's attention), and (2) inhibited by the *effort* (EF) required to move attention (e.g., a pilot will be less likely to scan to an instrument on the overhead panel, head down, or to the side, where head rotation is required, than to an instrument located directly ahead on a head-up display, HUD).

The SEEV model also predicts that attention is driven by the (3) *expectancy* (EX) of seeing (4) *valuable* (V) events at certain locations in the environment. For example, a high level of expectancy for a given channel (location) results because the channel in question has a high information bandwidth; that is, events occur frequently on the channel, such as the rapid oscillation of the attitude indicator on a light aircraft when flying through turbulence. A high weighting for expectancy means that the property of channel bandwidth exerts a strong influence in driving the pilot's attention. A high level of value means that information supports a very important task, such as maintaining the stability and lift of the airplane. The name of the SEEV model is derived from the first letter of each of the four terms listed previously (salience, effort, expectancy, value).

Equation 9.1 represents a *static analytic* model of attention allocation. Figure 9.1 contains a more detailed representation of the *dynamic* form of the full attention and situation awareness (A-SA) components of the process or simulation model. As shown in the figure, the model combines two distinct but interacting modules, one of which seeks to represent the dynamic SEEV-driven attentional processes by which the pilot collects information from the environment; the other, based on the output of selective attention, seeks to represent the cognitive processes by which attended information is integrated to establish and update the pilot's level of SA.

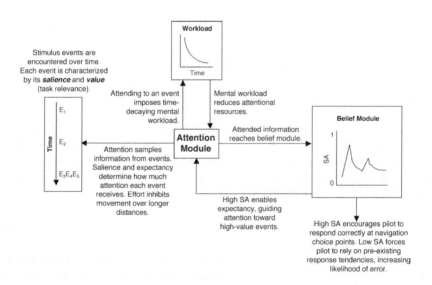

FIGURE 9.1 A-SA model for taxi error prediction.

On the left side of Figure 9.1, events (E) of varying salience occur over time across a set of locations or *channels* (areas of interest: AOI), with channels differing in bandwidth (frequency of events). The bandwidth of a channel determines the *expectancy* component of SEEV. Attention is allocated to events as they occur and is distributed in parallel to multiple events that occur simultaneously. Because the total amount of attentional resources available at any moment is limited, however, the need to attend to multiple channels in parallel reduces the amount of attention that any particular channel can receive. Thus, when simultaneous events occur, the amount of attention allocated to any single event is (a) proportional to the *value* of the event in supporting SA [V(SA)] added to the salience (S) of the event, and (b) inversely proportional to the number of other channels and events competing for attention (Bundesen, 1990; Levison, Elkind, & Ward, 1971). Note that the impact or weight of events in the attention module is also diminished or inhibited by cognitive workload as represented in the upper box, which is assessed by a decaying function of the number of events that have occurred within the past 10 sec. When attention moves, an effort component can inhibit the movement of attention across longer distances, relative to shorter ones.

Attended events provide *evidence* to the situation assessment or belief component of the model, illustrated by the box on the right in Figure 9.1, which incorporates Hogarth and Einhorn's (1992) anchoring and adjustment model of belief updating. Each attended event is modeled to support, refute, or be neutral to the operator's belief about his or her current state of affairs (e.g., belief of where the pilot is on the airport surface, belief in the presence of a traffic hazard in the forward view, or belief in the need for a flight control correction). Such a belief will decay or decline with the passage of time—slowly if little else is occurring and the pilot is fully concentrating on maintaining SA (time constant = 60 sec), and rapidly if there is high workload and distraction (time constant = 5 sec). More details of this distinction are provided later.

As shown in the figure, the output of the belief module is a value of SA that ranges between 0 and 1, where 1 indicates perfect SA and 0 denotes a complete lack of SA. This value is used both to guide future attentional scanning (the SA expectancy arrow feeding back to the attention module) and to determine the likelihood that the pilot will behave correctly when forced to select between potential actions at behavioral choice points. In our taxiway application, these choice points occur when the pilot must select the proper course at a taxiway intersection, where errors are frequent.

Finally, as shown at the bottom of the figure, when there is a total loss of SA, pilot behavior will be totally governed by pre-existing or "default" response tendencies (e.g., "when in doubt, go straight" or "when in doubt, guess at random") and will therefore be subject to error. The behavioral tendencies assumed by the model will necessarily be domain specific and, as such, are not integral to the model's architecture, but will be determined by the analyst for a particular application.

The distinction between the static equation (or analytic model) and the dynamic simulation depicted in Figure 9.1 is important. The static SEEV model as implemented only yields estimates of the long-term average distribution of attention across different AOIs. We refer to this measure as the probability of attending [P(A)] , or the percent dwell time (PDT) (Wickens et al., 2001, 2003). It is computationally simple, but does not produce estimates of the second-by-second scan path, between different AOIs. It is these transitions that are captured by the dynamic simulation model. The latter model as inhibited by effort can be run with multiple iterations, and produces estimates of the long-term averages of transition frequencies between AOIs (first-order Markov, estimating the movement of attention) as well as of the overall PDT within each AOI (zero-order Markov).

We describe three applications here, all using the dynamic model; the third also uses a static analytic version of the SEEV model. The first application is applied to explaining the errors made in aircraft taxiing in a simulation of low-visibility surface operations at KORD, provided by NASA (chapter 3). The second, described only briefly, is applied to a simulation of landing with an SVS display, where we model the attention module. The third is applied to a simulation of SVS landing trials, where we model the attention module and examine its implications for SA-based performance, and we offer a full validation.

APPLICATION 1: AIRCRAFT TAXI ERRORS

The application of the model to predicting taxiway navigation errors in a KORD simulation study is described in more detail in Wickens and McCarley (2001).

ATTENTION MODULE

In this application, pilots are assumed to encounter "events" as they progress along the taxiway, as shown on the left side of Figure 9.1. As noted, each event can be characterized by its salience or conspicuity, which is maximum for auditory events (Spence & Driver, 2000) and diminishes for visual events as these are available further in peripheral vision. An event is also characterized by its value in supporting awareness of where the pilot is (or which way he should turn) along the taxiway. Each parameter is coded by the analyst within a range of 0–1.0, using expert judgment.

Attention within the model is considered to be a graded resource that can be allotted simultaneously in different quantities to different events. The proportion of attention allotted to an event is determined by the event's attentional weight relative to the summed attentional weights of all concurrent events (Bundesen, 1990; Luce, 1959). SA is assumed to be self-reinforcing, such that a high level of SA serves to guide scanning toward stimuli or events that are themselves conducive to high SA. This tendency captures the influence of top-down factors such as expectancy and the confirmation bias on evidence seeking (Wickens & Hollands, 2000).

BELIEF MODULE

Situation awareness is updated any time a piece of evidence—a task-relevant item or group of items—is collected. Each piece of evidence is characterized by a value V from 0 to 1, indicating the degree to which the item is relevant for good SA. The change in belief effected by the items encountered at a given time step is determined by a weighted mean of their evidence values, with the value V for each item weighted by the amount of attention allotted to the item. This weighted mean will be referred to as the net evidentiary value (NV) for the time step. For example, a taxiway sign has a high NV because it explicitly signals the identity of an intersection point. After the NV has been determined for a given time step, SA is updated via an anchoring and adjustment process like that described by Hogarth and Einhorn (1992). Thus, the current value of SA is adjusted upward or downward in accordance with the NV of the newly encountered evidence. As noted by Hogarth and Einhorn, belief updating via such an anchoring-and-adjustment process captures the effects of order of information presentation (i.e., anchoring and recency effects) on information integration.

After SA has been updated in response to the evidence encountered within a given 1-sec interval, the model proceeds to the next interval using the newly calculated value of SA. If additional evidence is encountered, SA is again updated per the processes described previously. Across intervals where no evidence is encountered, SA is assumed to decay. The decline of SA in the absence of new evidence is intended to reflect the fact that preservation of SA is a resource-demanding process requiring rehearsal and that, even in low-workload situations, such rehearsal is not likely to be continuous. We estimate the time constant of the decay function to be relatively long (60 sec) and approximating the characteristics of *long-term working memory* (Ericsson & Kintch, 1995), a construct thought to underlie SA (Wickens & Hollands, 2000). Under distraction, this time constant is the traditional constant of working memory, approximately 5 sec (Card, Moran, & Newell, 1986; Peterson & Peterson, 1959).

Finally, we assume that the level of SA that exists when the pilot encounters a taxiway intersection determines the likelihood that choice behavior (i.e., the choice of route to follow) will be based on processed event information. Otherwise, choices will be based on analyst-specified pre-existing response tendencies such as "when in doubt turn toward the terminal." Behavior that is not based on processed event information may be erroneous. The probability of an erroneous response thus increases when SA is poor.

TEST APPLICATION

In the application of the data to predicting taxiway errors (Wickens & McCarley, 2001), we coded video and audio tapes of a typical pilot during the NASA human-in-the-loop (HITL) simulation run (chapter 3), noting the time at which certain "events" occurred, and coding the salience and net evidentiary value of each, as indicated by the prototype sequence in Table 9.1. This allowed us to compute a value of SA at each 1-sec interval, increasing with relevant events and degrading with the passage of time—slowly when nothing else was occurring and rapidly in the presence of irrelevant distracting events (see the "irrelevant discussion" at time = 45 sec of Table 9.1).

SA was assumed to reflect the probability that the pilot, when reaching a choice point (e.g., a two- or three-way taxi intersection in Table 9.1), would recognize the correct response option and behave appropriately as a result. In cases where imperfect SA did not lead the pilot to choose the correct response option, the pilot was assumed to select randomly from among the options available. The probability of a correct choice was thus equal to the probability of the pilot choosing correctly because he or she recognized the correct option (good SA), plus the probability that the pilot failed

TABLE 9.1

Example of Scenario Timeline for Taxiway Application

Time (sec)	Object/Event	Salience	Evidentiary Value
00	Rehearsal of instructions	1.0	1.0
20	Irrelevant discussion	1.0	0.0
30	2-way intersection	0.5	0.5
32	Sign	0.5	0.5
35	Sign	0.5	0.5
40	Branch	1.0	0.5
45	Irrelevant discussion	1.0	0.0
48	3-way intersection	1.0	0.5
48	3-way intersection	1.0	0.5
50	Visually salient traffic	1.0	0.0
52	Visually non-salient traffic	0.5	0.0
58	Point at which pilot should exit	1.0	1.0

to recognize the correct option as such (poor SA) but happened to guess correctly among the *N* options.

In this application, we did not actually use the model to predict performance (i.e., to validate against errors made). Rather, we ran the model through various iterations to show a sensitivity analysis—that is, to illustrate how SA and choice accuracy would wax and wane across time and to demonstrate how these changes would be influenced by changes in model parameters such as varying the impact of salience or SA decay rate.

Figure 9.2 shows the output of the A-SA model, driven by the sequence of events in Table 9.1 and played through multiple iterations in a Monte Carlo simulation. In the top panel, the two lower functions demonstrate the waxing and waning of SA, with two different decaying time constants for long-term working memory in the nondistracted circumstances, set at 60 and 20 sec. The latter function shows lower overall SA. The impact of distraction, with its 5-sec decay constant, is evident in the steeper drops that appear periodically throughout the sequence. Note in particular the sharp drop at 20 sec, corresponding to the irrelevant conversation seen in the second line of Table 9.1.

The four vertical bars along the time axis of Figure 9.2 represent the time of encountering four intersections or choice points. Finally, the uppermost of the three functions represents the model run outputs when the pilot is supported by the T-NASA taxiway aid (Hooey & Foyle, 2006), a system that contains a head-up display (HUD) with a three-dimensional augmented reality taxiway image, and a head-down, three-dimensional exocentric map of the airport surface showing a continuously updated representation of aircraft position. The lower panel of Figure 9.2 shows the level of choice accuracy (calculated as 1.0 − error rate) predicted at the four choice points for each of the three simulation runs (and thereby corresponding to the three functions in the top panel).

While these functions were not formally validated against performance data in a quantitative manner, they do indicate the loss of SA associated with the passage of

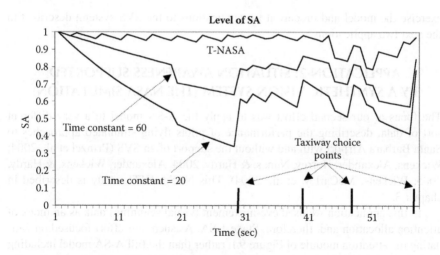

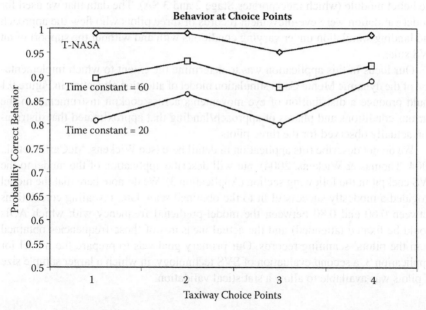

FIGURE 9.2 Model-predicted situation awareness (SA) value (top graph) and choice point accuracy (bottom graph) as a function of time and of different model-based assumptions described in the text.

time and with distraction, the recovery of SA following relevant events, the sensitivity of SA to different assumed decay rates, and the benefit of SA improvement with a display aid (T-NASA). The latter was explicitly designed to support SA and has been shown in empirical data to greatly reduce the number of taxiway errors (Hooey & Foyle, 2006). Most importantly, this first application gave us an opportunity to exercise the model and prepare it for applications to the SVS system, described in the next two applications.

APPLICATION 2: SITUATION AWARENESS SUPPORTED BY A SYNTHETIC VISION SYSTEM (THE NASA SIMULATION)

The focus of our second effort was to apply the A-SA model to a very different sort of data, describing the performance of pilots flying simulated approaches to Santa Barbara Airport with and without the support of an SVS (Prinzel et al., 2004; Wickens, Alexander, Horrey, Nunes, & Hardy, 2004; Alexander, Wickens, & Hardy, 2005; Wickens, McCarley, et al., 2004). This NASA HITL study is described in chapter 3.

In this application we used eye-movement (visual scanning) data as an index of attention allocation and, therefore, Stage 1 SA. As such, our effort focused on validating the attention module of Figure 9.1, rather than the full A-SA model including the belief module (which incorporates Stage 2 and 3 SA). The data that we used for model validation were eye-movement records of three pilots who flew the approach and landing simulation under varying conditions with and without the support of an SVS suite.

Our focus in this application was to determine the extent to which implementation of the dynamic Monte Carlo simulation model of attention allocation in Figure 9.1 could produce a distribution of eye movements across cockpit instruments, in the various conditions and phases of approach/landing that approximated that distribution actually observed for the three pilots.

We do not describe this application in detail here (see Wickens, McCarley, et al., 2004; Thomas & Wickens, 2004), but will describe application of the model to the SVS cockpit in the following section (Application 3). We do note here that the model provided a modestly successful fit to the obtained scan data, revealing correlations between 0.60 and 0.80 between the model-predicted frequency with which AOIs would be fixated (attended) and the actual measures of those frequencies obtained from the pilots' scanning records. Our primary goal was to prepare the model for Application 3, a second evaluation of SVS technology, in which a larger sample size of pilots was available to allow a statistical validation.

APPLICATION 3: SYNTHETIC VISION SYSTEM MODELING REVISITED (THE ILLINOIS SIMULATION)

In the third application, we applied the A-SA model to the visual scanning data of eight pilots flying an SVS simulated landing at the University of Illinois, described in detail in Wickens, Alexander, Horrey, et al. (2004); Wickens, Alexander, Thomas, et al. (2004); and Wickens, McCarley, et al. (2004). The purpose of the application was to expand upon the modeling lessons learned from Application 2 to employ for modeling the attention data on an SVS simulation at the University of Illinois. Our sample size was sufficiently large to allow some statistical evaluation of the results, and the simulation also contained an explicit objective assessment of SA (traffic awareness, as measured by probe questions) that could serve as a performance validation measure.

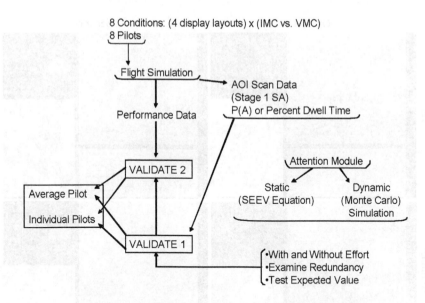

FIGURE 9.3 Conceptual flow of validation of the A-SA model for Application 3.

The overall approach of Application 3 is outlined in Figure 9.3. Participants flew eight different missions (approach and landing scenarios) that were created by crossing four different display formats, with either instrument meteorological conditions (IMC; clouds, terrain not visible) or visual meteorological conditions (VMC; clear, terrain visible). The goal was to vary the amount of display SA support and the distribution of information across AOIs in a way that would challenge the model to predict differences in attention allocation.

As shown in the figure, the 64 cells (8 × 8) of data yielded by the flight simulation produced flight performance data and visual scanning data across the AOIs; these scan data could be used to estimate Stage 1 SA ($P(A)$, which we assess by PDT). The attention module was implemented in software to predict both the static distribution of attention (the SEEV equation shown in Equation 9.1) and the dynamic allocation of attention from the Monte Carlo simulation. The outputs of these two were compared, and the simulation was also rerun under a variety of conditions (described later) to examine the effects of effort, display redundancy, and the expectancy–value relationship on scanning.

As shown to the left of the figure, two validation efforts followed. One was designed to validate the scan models by comparing model predictions of $P(A)$ against actual PDT distributions of individual pilots and of the average pilot, assessed in the experiment. The other was to validate the extent to which differences in scanning, between pilots, could predict performance-based measures of SA. Details of this modeling process are described, following a description of the experimental simulation (see also Wickens, Alexander, Thomas, et al., 2004; Wickens, McCarley, et al., 2004).

Overlay   Separate

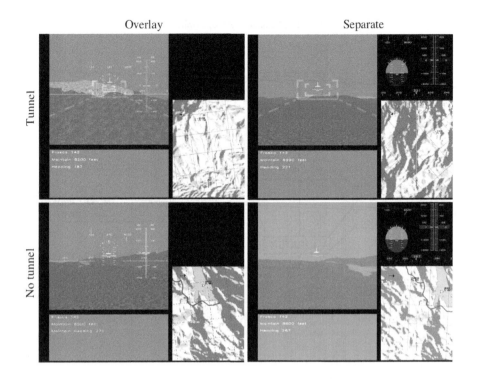

FIGURE 9.4 Four displays used in Application 3, defined by the characteristics of overlay (left column) versus separate (right column), and tunnel (top row) versus no tunnel (bottom row).

EXPERIMENTAL SIMULATION DETAILS

In a flight simulator simulating a Piper Archer III single-engine aircraft with a 120° view of the outside world, the eight pilots whose visual scanning data were measured flew curved step-down approaches to a runway in the terrain-challenged environment around Yosemite National Park. Across a set of conditions, pilots flew with each of the displays shown in Figure 9.4, which represents the four display suites in a 2 × 2 array. All four versions of the display suites included an SVS depiction of the terrain in the upper left. The SVS also rendered the traffic visible within its field of view. The two versions of the display suite on the upper row of Figure 9.4 contained a three-dimensional predictor symbol, a tunnel or "highway in the sky" (HITS) to guide flight path, overlaid on the SVS terrain. The two versions in the bottom row had no tunnel, but provided guidance through a vertical situation display (VSD) and compass and a map path located in the navigational display (ND; lower right of each panel).

The commanded trajectories were provided in a "data link box" at the lower left of each of the four panels. The two display suites shown in the left column had the instrument panel overlaid upon the SVS display. Those on the right had the panel separated and positioned in the upper right corner of the suite. Each display suite was flown under VMC (with the outside world visible) and under IMC (with the outside world obscured). Each pilot received one practice flight, followed by the eight simulation flights of the eight conditions, in random counterbalanced order. Each flight lasted approximately 8 min.

We were interested in the extent to which each of the two features (tunnel and overlay) defining the four display suites supported tasks, including lateral and vertical flight path tracking, detection of traffic presented in the SVS-hosted cockpit display of traffic information (CDTI), and detection of two critical "off-nominal" events (Foyle & Hooey, 2003; Wickens, 2005) in the upper left panel in each display suite. The first off-nominal event was a runway offset in which the SVS guidance brought the pilot toward a landing displaced from the true runway. This could only be detected if the pilot looked outside and realized that the SVS display was directing to a landing beside the true runway. The second off-nominal event was a "rogue airplane" trial (Wickens et al., 2002) in which the flight path directed the aircraft to fly directly into the flight path of another aircraft that was only depicted in the outside world as if its transponder were inactive. All of these provided performance measures to be predicted by the model. Visual scanning was measured for all eight pilots who completed the simulation with valid scanning data.

PERFORMANCE AND SCANNING RESULTS

Full details of the performance results are provided in Wickens, Alexander, Thomas, et al. (2004) and Wickens, McCarley, et al. (2004). Importantly, we found that the two tunnel display suites supported much better flight path tracking performance and somewhat better traffic detection performance than did the two display suites using the distributed guidance from VSD on the instrument panel and from the navigational display. We found not only that the overlay failed to improve tracking performance, but also that the clutter it created substantially inhibited traffic detection. Regarding the off-nominal events, in our design eight pilots (including some whose scanning was not measured) encountered the rogue aircraft while flying with the tunnel. Of these, four detected it and four failed to detect it (as indicated by the absence of any deviation maneuver in their flight path). Scanning data were available from four of the eight pilots, two of whom detected the rogue aircraft and two of whom did not (Thomas & Wickens, 2004). Of the 12 pilots who encountered the runway offset with the SVS display, 5 failed to notice it (see Wickens, 2005).

Figure 9.5 presents a measure of visual scanning behavior—the PDT values of the gaze of the eight pilots with valid scan data—as a function of the four display conditions and the two visibility conditions (IMC with outside world obscured, VMC with outside world visible) across the five most important AOIs: the four panels of each display suite shown in Figure 9.4, and the outside world. The variance in PDT across conditions and across AOIs is immediately evident, as is the interaction between these two variables. These significant effects are described in detail in Wickens, Alexander, Horrey, et al. (2004) and Wickens, Alexander, Thomas, et al. (2004). Most importantly, we were able to use these scanning data for our validation of the A-SA scan model, which we describe in this application as follows.

DYNAMIC SIMULATION MODELING

To implement the dynamic simulation model, the following steps were taken, closely paralleling the steps taken in Application 1, but departing from them in two important

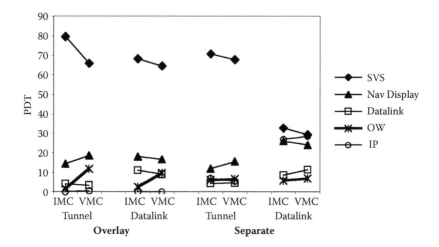

FIGURE 9.5 Percentage PDTs for the five AOIs shown in the legend (right) as a function of the eight display conditions.

respects. First, because we did not have an event time line and hence could not code the salience of specific events, the salience component of the SEEV model was deleted. Second, because events occurred along well-defined channels (the four for each display suite shown in Figure 9.4, plus the outside world), it was possible to calculate a predicted attention attractiveness of each channel, as determined by its expectancy and value. More specifically, we treated the components of the SEEV model shown in Equation 9.1 as follows:

S: As noted, we did not code *salience.*

EF: The *effort* needed to shift attention between two AOIs was coded by the distance between them, with a larger attention shift being presumed to require greater effort. These distances were assigned ordinal integer values.

EX: *Expectancy* was coded by the bandwidth of events within each AOI. In cases where a given AOI hosted two sources of information, we added the bandwidths of the two. For example, the SVS panel in all conditions contained an event stream representing changes in attitude of the airplane (reflected in oscillation of the horizon) as well as appearance of traffic airplanes.

V: *Value* was coded as follows: Each AOI was assigned a value proportional to the *importance* of the task that it served (aviate = 3, navigate = 2, maintain hazard awareness = 1) and weighted by the *relevance* of the AOI in question to the task. These relevance measures were assigned by the modeler.

Thus, the expected value of the amount of attention allotted to a given AOI is predicted by Equation 9.2:

$$P(A) = BW_{AOI} \times \sum (relevance_{t-AOI} \times value_t) \qquad (9.2)$$

where the summation is across all tasks served by an AOI. The subscript t indicates "task" and the subscript $t–AOI$ indicates an AOI-task pair.

We note that Equation 9.2 does not contain an "effort" term. There are two reasons for this. First, effort is not and should not be a component of expected value. Second, effort characterizes the distance between a pair of AOIs, rather than the attractiveness of a single AOI. Its inclusion in modeling will be described later.

Thus, the coefficients of the dynamic simulation model determine the attentional attractiveness of each AOI. On each iteration of the simulation, attention moves from one AOI to another or remains fixed on the current AOI. The probability that a given AOI will be attended is determined by the attentional attractiveness of that AOI relative to the summed attractiveness of all AOIs, scaled by the effort needed to move it. These factors determine a probability of attention movement to a given AOI. When the model is iterated multiple times, these probabilities yield a distribution of transitions (first-order Markovs) as well as a distribution of fixations within each AOI (zero-order Markov). The latter corresponds to PDT.

Table 9.2 presents the coefficients that were used to run the model for each condition. Across the top of the table are labeled the eight experimental conditions formed by the 2 (tunnel present vs. absent) × 2 (displays overlaid vs. separated) × 2 (weather IMC vs. VMC) design. The first set of rows (bandwidth) represents the actual bandwidth in hertz of the different AOIs. One can notice, for example, that the outside world (OW) has 0 bandwidth (BW) during IMC, since there is nothing at all to be seen there. At the very bottom of the table, the three tasks are shown in terms of their priorities or values (v): Aviating (AV) has highest priority (1), navigating (NAV; keeping on the flight path) is given second priority (2), and hazard awareness (HAZ, noting appearance and change in traffic and terrain) is given lowest priority (3). This is because it is assumed by the pilot that the flight path is hazard free, particularly from terrain. Hence, adhering to the flight path should generally guarantee safety.

Moving back upward, in the middle rows of the table are shown the *relevance* coefficients whereby the task served by each AOI is depicted; in particular, the relevance of the AOI information for a given task is shown. For the most part, these relevance coefficients are either 1 or 0. However, they are assigned a value of 0.5 when a given task can be served by two AOIs. Thus, for example, in all VMC conditions (odd-numbered columns), the outside world (OW) is partially relevant to aviating (AV) since the true horizon is visible and can be used for controlling attitude.

Table 9.3 presents the output of the model. Within the top section are presented the actual values of the computations across the eight conditions (using the same condition-coding initials, in the same order as in Table 9.2) for each of the five AOIs. Within the second section, each of these 40 values is normalized within a column so that they are expressed as proportions. This then generates a set of 40 predicted values of $P(A)$. The third section of the table presents the actual values of $P(A)$ observed from the human pilots' eye-movement data means (Figure 9.5).

The predicted values of mean $P(A)$ across the multiple AOIs were correlated against the obtained values to provide two important measures of model validation. First, all 40 data points were correlated, as represented by the scatter plot in Figure 9.6. Here, each point represents a unique combination of an AOI and a condition. As can be seen, there is a strong degree of linearity in the relation between predicted

TABLE 9.2
Parameter Matrix for Application 3

			Experimental Conditions							
			Tunnel Overlay	Tunnel Overlay	Tunnel Separated	Tunnel Separated	Datalink Overlay	Datalink Overlay	Datalink Separated	Datalink Separated
			VMC	IMC	VMC	IMC	VMC	IMC	VMC	IMC
Parameter	AOI	Task	TOV	TOI	TSV	TSI	DOV	DOI	DSV	DSI
Bandwidth (BW)	SVS		0.62	0.62	0.62	0.62	0.62	0.62	0.62	0.62
	IP		0.81	0.81	0.81	0.81	0.81	0.81	0.81	0.81
	ND		0.18	0.18	0.18	0.18	0.18	0.18	0.18	0.18
	DL		0.05	0.05	0.05	0.05	0.05	0.05	0.05	0.05
	OW		0.5	0	0.5	0	0.5	0	0.5	0
Relevance (R)	SVS	AV	1	1	1	1	1	1	1	1
	SVS	NAV	1	1	1	1	1	1	0	0
	SVS	HAZ	1	1	1	1	1	1	1	1
	IP	AV	0	0	0	0.5	0	0	0.5	0.5
	IP	NAV	0	0	0.5	0	0	0	1	1
	IP	HAZ	0	0	0	0	0	0	0	0
AOIs	ND	AV	0	0	0	0	0	0	0	0
	ND	NAV	0.5	0.5	0.5	0.5	1	1	1	1
	ND	HAZ	0.5	0.5	0.5	0.5	0.5	0.5	0.5	0.5
	DL	AV	0	0	0	0	0	0	0	0
	DL	NAV	0	0	0	0	1	1	1	1
	DL	HAZ	0	0	0	0	0	0	0	0
	OW	AV	0.5	0	0.5	0	0.5	0	0.5	0
	OW	NAV	0.5	0	0.5	0	0.5	0	0.5	0
	OW	HAZ	0.5	0	0.5	0	0.5	0	0.5	0

Notes: Experimental Conditions: D = Datalink; I = Instrument Meteorological Conditions (terrain invisible); O = Overlay; S = Separated; T = Tunnel; V = Visual Meteorological Conditions (terrain visible)

AOIs: DL = Datalink Display (command target information); IP = Instrument Panel; ND = Navigation Display; OW = Outside world; SVS = Synthetic Vision System

Tasks: AV = Aviate (3); NAV = Navigate (2); HAZ = Maintain Hazard Awareness (1)

TABLE 9.3

Attentional Weights (Top Section), Predicted Values of P(A) (Middle Section), and Experimentally Observed Values of P(A) (Bottom Section) for each AOI Across the Experimental Conditions Tested in Application 3

Attentional Weights		Experimental Condition							
		TOV	TOI	TSV	TSI	DOV	DOI	DSV	DSI
	SVS	3.72	3.72	3.72	3.72	3.72	3.72	2.48	2.48
	IP	0.00	0.00	0.81	0.81	0.00	0.00	2.835	2.835
	ND	0.27	0.27	0.27	0.27	0.45	0.45	0.45	0.45
	DL	0.00	0.00	0.00	0.00	0.10	0.10	0.10	0.10
	OW	1.50	0.00	1.50	0.00	1.50	0.00	1.50	0.00
	sum	5.49	3.99	6.3	4.8	5.77	4.27	7.37	5.87
Predicted P(A)		TOV	TOI	TSV	TSI	DOV	DOI	DSV	DSI
	SVS	0.68	0.93	0.59	0.78	0.64	0.87	0.34	0.42
	IP	0.00	0.00	0.13	0.17	0.00	0.00	0.38	0.48
	ND	0.05	0.07	0.04	0.06	0.08	0.11	0.06	0.08
	DL	0.00	0.00	0.00	0.00	0.02	0.02	0.01	0.02
	OW	0.27	0.00	0.24	0.00	0.26	0.00	0.20	0.00
	sum	1.00	1.00	1.00	1.00	1.00	1.00	1.00	1.00
Observed P(A)		TOV	TOI	TSV	TSI	DOV	DOI	DSV	DSI
	SVS	0.66	0.80	0.68	0.71	0.65	0.68	0.29	0.33
	IP	0.00	0.00	0.05	0.07	0.00	0.00	0.28	0.27
	ND	0.18	0.14	0.15	0.12	0.17	0.18	0.24	0.26
	DL	0.03	0.04	0.04	0.04	0.09	0.11	0.11	0.09
	OW	0.12	0.02	0.07	0.06	0.10	0.03	0.07	0.06
	sum	1.00	1.00	1.00	1.00	1.00	1.00	1.00	1.00

Notes: Experimental Conditions: T = Tunnel, D = Datalink, O = Overlay, S = Separated, V = VMC (terrain visible) I = IMC (terrain invisible)

AOIs: SVS = Synthetic Vision System, IP = Instrument Panel, ND = Navigation Display, DL = Datalink Display (command target information), and OW = Outside world

and obtained values, suggesting validation of the model. The correlation is $r = 0.93$, indicating that the model accounts for $r^2 = 86\%$ of the variance in the data. The one characteristic most noteworthy in the scatter plot is the cluster of six data points near the bottom, around the predicted score of 0.2. This cluster represents scans to the ND, which falls well below the regression line and hence was scanned less than predicted.

Second, correlations were computed within each condition, each now based upon $N = 5$ (the number of AOIs). These separate correlations are presented in Table 9.4, showing the correlation and the r-squared values. We note the uniformly positive correlation values that are, with only one exception, greater than 0.80. It is also noteworthy that the two lowest fits (those with substantially less than 80% of the variance accounted for) are the two separated display conditions without the tunnel guidance (DSV in VMC and DSI in IMC). These have in common that the sources of

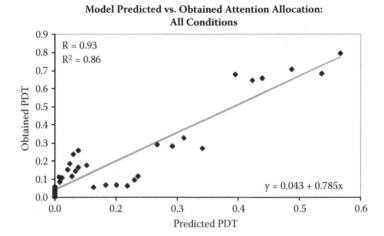

FIGURE 9.6 Scatter plot of model-predicted versus obtained PDT for Application 3. Each data point in the plot represents the unique AOI by condition combination, with data averaged across all eight pilots.

information are most distributed (two right panels of Figure 9.4), and the VMC condition (DSV, the lowest fit of all) has the greatest distribution because it includes the outside world as an AOI. We may infer that the large number of AOIs to scan in these two conditions avails a greater opportunity for individual differences in scanning strategy, hence lowering the consistency of results across pilots (lower reliability of scan data) and therefore lowering the validation correlations with model predictions.

MODEL ADJUSTMENTS

The previous discussion indicates that our model was reasonably valid, at least as assessed by the traditional criterion of performance variance accounted for (generally over 80%, and over 90% for the overall model fit). Importantly, the coefficients for the model were established a priori and were not adjusted to maximize model fit. With this as a baseline model, we then went on to implement four different adjustments to the model, described next.

TABLE 9.4
Correlations Between the Model Predicted and the Obtained Data, Overall, and for the Eight Individual Conditions. Each Correlation is Based Upon Five Data Points, Corresponding to the Five AOIs

	Overall	TOV	TOI	TSV	TSI	DOV	DOI	DSV	DSI
Correlation (r)	0.93	0.93	1.00	0.90	0.98	0.93	0.99	0.57	0.80
r-squared	0.86	0.87	0.99	0.81	0.96	0.86	0.98	0.33	0.64

Notes: Experimental conditions: T = tunnel; D = data link; O = overlay; S = separated; V = visual meteorological conditions (terrain visible); I = instrument meteorological conditions (terrain invisible)

The Role of Effort

In Application 2, described previously, we examined the model fit with and without the effort parameter (see Wickens, Alexander, Thomas, et al., 2004; Wickens, McCarley, et al., 2004) imposed to penalize scanning transitions across longer distances, and found that for the three pilots examined in that study, removing the effort penalty did not degrade the model fit. Hence, we repeated that exercise for the model fitting in Application 3 and augmented it by including different variations in how effort might be coded. We proceeded to run seven different versions of the dynamic model, adjusting the effort penalty imposed for scanning: a "no-effort" version (1) contrasted with three versions (2–4) in which the effort to transition between any two AOIs imposed a penalty (of three different levels); a version (5) in which added penalties were imposed for longer distances (the version described earlier); a version (6) in which a particular heavy penalty was associated with transitions between the outside world and the four AOIs on the instrument panel; and a version (7) that combined penalties (5) and (6).

This comparison of model fits revealed that the simple no-effort model provided the best fit to the data of any of the alternatives, replicating the general results of Application 2. In the context under study, *effort does not appear to inhibit optimal scanning*. Interestingly, the one model version that came closest to matching the fit of the no-effort model was version (6), which imposed a penalty solely on transitioning between the instrument panel and the outside. It is of note that this transition is the only one that also imposes visual accommodation differences, as well as scanning distance.

The general finding that the model fit is not improved by incorporating the effort parameter is important in two respects. First, it indicates that our pilots, all well-skilled flight instructors, scanned their work domain in an optimal fashion—a conclusion in harmony with that reached by Moray (1986) in his summary of scanning models: namely, that those studies using well-trained experts tended to reveal more optimal models of scanning than those using more "naïve" participants. Stated another way, all information sources in the cockpit are of sufficient importance that the pilots scan them regardless of the effort required to do so. Second, it indicates that a relatively simple and parsimonious three-parameter (expectancy and value mediated by expectancy) model is adequate to capture Stage 1 SA. We consider the implication of this fact as follows.

Combining Expectancy and Value

Equations 9.1 and 9.2, used in the current validation, express a multiplicative relation between expectancy and value. As such, it is consistent with the notions, emanating from decision theory, of "expected value" defined by the product of the two terms. An alternative possibility is to make expectancy and value additive, such that their combined influence is represented by the formula $E + V$ (rather than $E \times V$). This computational difference, consistent with the additive expression of the original SEEV model (Wickens et al., 2001), can be justified from a psychological viewpoint since it retains the influence of either term even if the other has a value of zero. In other words, the additive model provides a mechanism whereby an AOI can maintain a positive attentional attractiveness even if one of the two terms expectancy and value falls to 0. The multiplicative $E \times V$ computation predicts no scanning at all to the AOI in question, whereas the additive $E + V$ model assumes some scanning as long as either term is positive. Thus, for example, pilots may look at the blank or static OW (BW = 0), even if there is no movement there, simply to check whether something might be happening. Correspondingly, the pilot's attention may be drawn to an irrelevant but dynamic display (0 value, but >0 BW).

When correlations were calculated using this additive model, all values increased, suggesting that pilots scanned according to the additive rather than the multiplicative relationship. In addition, the additive model provided an especially improved fit for the two non-tunnel-separated conditions (DSV and DSI) relative to the multiplicative model. These two conditions, it will be recalled from Table 9.4, were the two that had the lowest level of model fit with the multiplicative model. The success of the additive model suggests that either expectancy or value can indeed retain its influence on scanning behavior even if the influence of the other term is negligible.

Coding the Model: Redundant Information

As noted before, modeling attention is challenged when a given task is served by more than one channel of information. The best example here is the task of *aviating* (maintaining stable lift for flight control), which, under VMC, is served by both the outside world horizon and the SVS horizon. The solution that we adopted, as shown in Table 9.2, was to code each AOI with a 0.5 coefficient for relevance to the *aviate* task—that is, to distribute relevance equally between these two redundant sources.

However, it is possible that pilots may choose to allocate attention more exclusively to one or the other source, leading to relevance weights of 1 and 0, respectively, for the two sources. The current application used the model fit as a tool to assess this possibility. More specifically, an alternative version of the model was run with the relevance of the outside world for aviating set to 0 for the four conditions in VMC and the relevance of the SVS AOI set to 1. The data produced by this version of the model fit the observed data well, especially for the separated data link VMC condition ($r =.75$, up from the previous value of 0.57 with no loss in model fit for other conditions; see Wickens, Alexander, Thomas, et al., 2004, for more details).

What these data suggest is that the pilots extracted little attitude information from the real horizon, even when it was visible in VMC; instead, they relied heavily on the synthetic horizon from the SVS display. As reported previously, in our discussion of off-normal event detection, this strategy of "attentional tunneling," nicely revealed by the modeling fit, put the pilots at risk when there was information in the outside world that was not revealed within the SVS panel (Thomas & Wickens, 2004; Wickens, 2005). This risk was captured by the less than perfect performance of these pilots in detecting either of the off-nominal events, whose detection depended upon OW scanning.

Dynamic Versus Static Model

All of the model fitting described here was carried out with the dynamic Monte Carlo simulation model. However, the finding that an effort parameter contributes little to the model fit suggests that this dynamic model, which predicts the course of transitions or attention switches between AOIs (the first-order Markov process), is unnecessary in the current application. This is because only transition frequencies, not the zero-order PDT values, are affected by the effort parameter. In contrast, both the static equation (Equation 9.2) and dynamic simulation model should make similar predictions about the PDT within each AOI (zero-order Markov). In order to validate this assertion, a static version of the additive model was implemented that, indeed, showed a profile of model fit identical to that shown in Table 9.4.

Summary: Model Adjustments

The model adjustments described earlier imply that effort inhibition had relatively little influence on scanning in either of the two SVS applications, suggesting that the simplified three-parameter version of either the dynamic or static model would suffice. It is important to note, though, that this conclusion is probably of somewhat limited generalizability. There are certainly circumstances where effort might play a more significant role. Effort might be expected to influence scanning, for example, if one or more areas of interest were located to the side or behind the pilot, such that larger head movement was required to access them; if long missions were modeled in which scanning became a fatiguing factor; or, in particular, if excessive scanning between the instrument panel and the outside world was required in order to monitor for relatively high-bandwidth information available only in the outside world.

Model Validation Against Individual Pilot Performance

The previous analyses indicate that the attention module of the A-SA model captures actual pilot scanning well. However, it says little about how this scanning supports Stages 2 and 3 SA, which underlie performance. We addressed this latter issue in two phases. First, we sought to determine whether individual pilots differed in the extent to which variance in their scanning behavior was captured by the model. Then, we assessed the extent to which those with higher model fits provided better flight task performance, under the assumption that those who better complied to the *optimal* (expected value) scanning model would be better performers.

In order to accomplish these two goals, for each pilot we plotted predicted versus obtained scan data across the eight conditions and five AOIs, just as was done for the average data in Figure 9.6. Each of these eight scatter plots (one per pilot) produced a *single correlation value* that could be described as that pilot's "adherence to the expected value model of scanning." We can refer to this as an "optimality score." The model version employed was a non-effort version in which expectancy was added to value $(E + V)$ based on our previously reported findings. Each of these eight pilot optimality scores was then associated with (a) the mean response time of that pilot for traffic detection, and (b) the mean flight control error measure (integrated lateral and vertical) of that pilot. This was done in order to provide two additional correlations: between the optimality score and SVS traffic detection time and between the optimality score and flight path tracking error. Both of these scatter plots and their associated correlations are shown in Figure 9.7.

Figure 9.7(a) reflects a nonsignificant inverse relation $(r = -.58)$ between optimality score and traffic detection latency $(p = .13)$. This effect is consistent with the possibility that pilots whose scanning was better captured by the optimal model

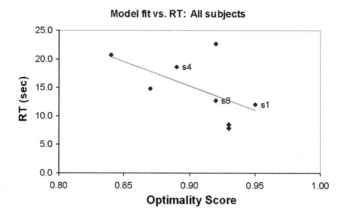

FIGURE 9.7a Scatter plot of pilots' scanning optimality score against the RT to detect traffic. Optimality score indicates the correlation between the pilot's PDT values across five AOIs and the model-predicted optimal PDTs. Each point is a single pilot. The correlation is $r = -0.58$.

were more rapid in detecting traffic, but it fails to provide firm evidence for this conclusion. Figure 9.7(b), however, reflects a very strong and statistically reliable relation between optimality score and integrated flight path tracking error ($r = -.88$; $p = .02$). *Thus, pilots who better conform to the optimal expected value model of scanning may be somewhat better at traffic detection and are reliably better at flight path tracking.* Additional analysis provided no evidence that the degree of model fit predicted the detection of off-nominal events. It is possible that this failure resulted because the estimation of individual pilot response to off-nominal events is of very low statistical power, since there is only one observation per pilot; therefore, there is no opportunity, in individual-pilot prediction, to increase power (reduce variability) via averaging.

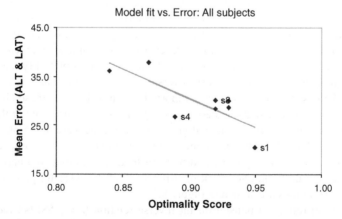

FIGURE 9.7b Scatter plot between optimality scores against combined lateral and vertical flight path tracking error. Correlation is $r = -0.88$.

Second, we reasoned that those pilots for whom inclusion of an effort parameter *improved* the model fit might actually generate poorer performance because effort would inhibit scanning to areas that should be visited to obtain optimal performance. Only two pilots (S2 and S6) showed higher model fits with the effort parameter than without it (and therefore may be characterized as "effort-burdened" in their information access). When all eight pilots were rank ordered in terms of the three indices of performance (traffic detection time, lateral error, and vertical error) and the rank orders were summed for each pilot across the three indices, the two effort-burdened pilots also had the two lowest rank orders of performance, an indication of the strong association between effort-based scanning inhibition and poor performance.

CONCLUSIONS

In conclusion, our analysis of Application 3 revealed that pilot's scan behavior was well captured by the attention (visual scanning) component of the A-SA model, both on average ($r = .93$) and individually (r values ranging from .84 to .93), whether computed by the analytic or the dynamic simulation version of the model. Importantly, we also found that those who better conformed to this optimal (without effort) expected value version of the model were better fliers and somewhat better traffic detectors and that those two whose scanning was inhibited by effort were poorer performers.

Our results also indicated three aspects regarding variations in the model. First, we improved the model fit by changing the assumptions, in this application, regarding the relevance to the pilot of perfectly correlated information (the horizon for aviating, visible both outside and on the SVS). When it was assumed that pilots only consulted one source (the SVS display), model fitting improved. Second, we found that an additive rather than a multiplicative relationship between expectancy and value better accounted for the data, an issue that needs to be pursued with further research (see chapter 5). Third, as in Application 2 (see Wickens, McCarley, et al., 2004), incorporation of an effort-inhibition parameter does not generally improve model fit, a conclusion that befits well-trained expert operators (Moray, 1986).

GENERAL CONCLUSIONS

In conclusion, we have described three applications of an attention-situation awareness model to important safety-critical aviation environments. All three applications were critical to the evolution and testing of the model, although only the third provided true statistical validation of the model's ability to predict operationally important performance levels: awareness and detection of traffic, and control of the airplane in a multitask simulation environment. In the following, we consider briefly the implications of the different forms of the model as they emerged across the three applications, and then the direct implications of the model for aviation safety.

Three important distinctions underlie the difference between Application 1 to taxiway errors, and Applications 2 and 3 to SVS performance:

- In Application 1 there was no direct assessment of SA, but rather an inference of how the modeled SA would influence choice performance. In contrast, in the SVS applications, Stage 1 SA was directly assessed via the allocation of visual attention.

- Emphasis in Application 1 was placed on modeling the allocation of attention to specific *events* in the environment, where the degree of attention capture produced by each event was based in part upon the event's salience of the event. Hence, the first application placed importance on salience and, because salience in turn is relevant to the *movement* of attention (i.e., first-order Markov process), it was essential to use the dynamic simulation version of the A-SA model to represent attentional behavior adequately. In contrast, for the two SVS applications, where interest was focused not on attentional responses to specific events but rather on adaptive changes in scanning in response to the long-term statistical properties of well-defined channels, the parameter of event salience became less relevant, and the key model outputs were the long-term statistical averages of attention transitions and percent dwell times used to estimate $P(A)$. This does not mean to imply, however, that salient events are unimportant in an SVS evaluation. It might be useful, for example, to examine the latency of attentional capture (noticing) by specific events (see Wickens, Sebok, Bagnall, & Kamienski, 2007).
- Application 1 was the only application that exercised the components of the belief module.

Within the two SVS applications, it became apparent that, to incorporate effort into the model, a simulation model involving movement of attention from one area to another was required. However, given that effort inhibition appeared to play only a minimal role, it was possible to model the most important aspects of attention solely in terms of expectancy and value, determining the frequency with which different areas were attended, not how attention moved between them. As a consequence, the simple, analytic equation could be effectively employed. Based on the results of Application 3, rather than the multiplicative version shown in Equation 9.1., the best form of this analytical equation appears to be the additive one:

$$P(A) = sS - ef\ EF + (exEX + vV)\qquad(9.3)$$

Where

$$v\ =\ \text{relevance } x \text{ task importance}$$

IMPLICATIONS

What does the model say to the aviation human factors professional? Data carry two strong messages. First, the linkage of expected value optimality to performance, seen in the third application, suggests that scanning strategies are themselves objectively and systematically tied to performance differences. This in turn suggests that the additive expectancy and value model might serve as a "gold standard" by which scanning training should be guided.

Second, the fact that it is possible to predict PDT on the basis of a task analysis provides a vehicle for predicting the occurrence of the safety-critical phenomenon of attentional tunneling (Thomas & Wickens, 2004, Wickens, 2005) as procedures are changed or new technology is added (Wickens et al., 2001, 2003). Within the SEEV model, it remains a little uncertain what should be the precise criterion for describing attentional tunneling, although an approach would be to decide some minimum PDT below which attention to a particular channel should not fall (e.g., attending to the outside world, at least 10%). If the model predicts less than that, there is an invitation for attentional tunneling. The output of the model run would be a distribution of lengths of time spent *away* from the critical area (this is sometimes called the "mean first passage time"; Moray, 1986), and this absent or neglect time would leave the pilot at risk for failing to notice safety-critical events that could only be observed there. Such a risk would be related to the frequency of occurrence of those events and their duration of visibility.

Finally, we see the need for continued validation of the full model. While the attention aspect has by now received considerable validation, what awaits further research is an empirical validation of the attention component coupled with its SA belief-updating companion.

ACKNOWLEDGMENTS

We wish to acknowledge the contributions of Bill Horrey, for assisting with eye-movement analysis; of T. J. Hardy, Ron Carbonari, and Jonathan Sivier, for developing the flight simulation used in Application 3; and of Roger Marsh, for developing algorithms for eye-movement scanning analysis.

REFERENCES

Alexander, A., Wickens, C. D., & Hardy, T. J. (2005). Sythetic vision systems: The efects of guidance symbology, display size, and field of view. *Human Factors, 47*, 693–707.

Bellenkes, A. H., Wickens, C. D., & Kramer, A. F. (1997). Visual scanning and pilot expertise: The role of attentional flexibility and mental model development. *Aviation, Space, and Environmental Medicine, 68*(7), 569–579.

Bundesen, C. (1990). A theory of visual attention. *Psychological Review, 97*, 523–547.

Carbonnell, J. F., Ward, J. L., & Senders, J. W. (1968). A queuing model of visual sampling: Experimental validation. *IEEE Transactions on Man Machine Systems, MMS-9*, 82–87.

Card, S., Moran, T., & Newell, A. (1986). The model human processor. An engineering model of human performance. In K. Boff, L. Kaufman, & J. Thomas (Eds.), *Handbook of perception and human performance* (Vol. 2). New York: Wiley.

Endsley, M. R. (1995). Toward a theory of situation awareness in dynamic systems. *Human Factors, 37*(1), 85–104.

Ericsson, K. A., & Kintsch, W. (1995). Long-term working memory. *Psychological Review, 102*, 211–245.

Foyle, D. C., & Hooey, B. L. (2003). Improving evaluation and system design through the use of off-nominal testing: A methodology for scenario development. *Proceedings of the Twelfth International Symposium on Aviation Psychology* (pp. 397–402). Dayton, OH: Wright State University.

Hogarth, R. M., & Einhorn, H. J. (1992). Order effects in belief updating: The belief-adjustment model. *Cognitive Psychology, 24*, 1–55.

Hooey, B. L., & Foyle, D. C. (2006). Pilot navigation errors on the airport surface: Identifying contributing factors and mitigating solutions. *International Journal of Aviation Psychology, 16*(1), 51–76.

Horrey, W. J., Wickens, C. D., & Consalus, K. P. (2006). Modeling drivers' visual attention allocation while interacting with in-vehicle technologies. *Journal of Experimental Psychology: Applied, 12*, 67–78.

Jones, D. G., & Endsley, M. R. (1996). Sources of situation awareness errors in aviation. *Aviation, Space, & Environmental Medicine, 67*(6), 507–512.

Klatzky, R. L. (2000). When to inspect? Recurrent inspection decisions in a simulated risky environment. *Journal of Experimental Psychology: Applied, 6*(3), 222–235.

Levison, W. H., Elkind, J. I., & Ward, J. L. (1971). *Studies of multivariable manual control systems: A model for task interference* (NASA CR-1746). Washington, DC: NASA.

Luce, R. D. (1959). *Individual choice behavior.* New York: Wiley.

Moray, N. (1986). Monitoring behavior and supervisory control. In K. R. Boff, L. Kaufman, & J. P. Thomas (Eds.), *Handbook of perception and performance* (Vol. II, pp. 40–51). New York: John Wiley & Sons.

Norman, D. A. (1981). Categorization of action slips. *Psychological Review, 88,* 1–15.

Parasuraman, R., Sheridan, T. B., & Wickens, C. D. (2000, May). A model for types and levels of human interaction with automation. *IEEE Transactions on Systems, Man, & Cybernetics, 30*(3), 286–297.

Peterson, L. R., & Peterson, M. J. (1959). Short-term retention of individual verbal items. *Journal of Experimental Psychology, 58,* 193–198.

Prinzel, L. J., III, Comstock, J. R., Jr., Glaab, L. J., Kramer, L. J., Arthur, J. J., & Barry, J. S. (2004). The efficacy of head-down and head-up synthetic vision display concepts for retro- and forward-fit of commercial aircraft. *The International Journal of Aviation Psychology, 14*(1), 53–77.

Reason, J. (1990). *Human error.* New York: Cambridge University Press.

Sarter, N., & Woods, D. D. (2000). Team play with a powerful and independent agent: A full-mission simulation study. *Human Factors, 42*(3), 390–402.

Sarter, N. B., & Woods, D. D. (1995). How in the world did we ever get into that mode? Mode error and awareness in supervisory control. *Human Factors, 37*(1), 5–19.

Sklar, A. E., & Sarter, N. B. (1999). Good vibrations: Tactile feedback in support of attention allocation and human–automation coordination in event-driven domains. *Human Factors, 41*(4), 543–552.

Spence, C., & Driver, J. (2000). Audiovisual links in attention: Implications for interface design. In D. Harris (Ed.), *Engineering psychology and cognitive ergonomics.* Hampshire, England: Ashgate Publishing.

Thomas, L. C., & Wickens, C. D. (2004). Eye-tracking and individual differences in off-normal event detection when flying with a synthetic vision system display. *Proceedings of the 48th Annual Meeting of the Human Factors and Ergonomics Society.* Santa Monica, CA: HFES.

Wickens, C. D. (2000). The trade-off of design for routine and unexpected performance: Implications of situation awareness. In M. R. Endsley & D. J. Garland (Eds.), *Situation awareness analysis and measurement* (pp. 211–225). Mahwah, NJ: Lawrence Erlbaum Associates.

Wickens, C. D. (2005). Attentional tunneling and task management. *Proceedings of the 13th International Symposium on Aviation Psychology.* Dayton, OH.

Wickens, C. D., Alexander, A. L., Horrey, W. J., Nunes, A., & Hardy, T. J. (2004). Traffic and flight guidance depiction on a synthetic vision system display: The effects of clutter on performance and visual attention allocation. *Proceedings of the 48th Annual Meeting of the Human Factors and Ergonomics Society.* Santa Monica, CA: HFES.

Wickens, C. D., Alexander, A. L., Thomas, L. C., Horrey, W. J., Nunes, A., Hardy, T. J., et al. (2004). *Traffic and flight guidance depiction on a synthetic vision system display: The effects of clutter on performance and visual attention allocation* (AHFD-04-10/NASA(HPM)-04-1). Savoy, IL: University of Illinois, Aviation Human Factors Division.

Wickens, C. D., Gempler, K., & Morphew, M. E. (2000). Workload and reliability of predictor displays in aircraft traffic avoidance. *Transportation Human Factors Journal, 2*(2), 99–126.

Wickens, C. D., Goh, J., Helleberg, J., Horrey, W., & Talleur, D. A. (2003). Attentional models of multitask pilot performance using advanced display technology. *Human Factors, 45*(3), 360–380.

Wickens, C. D., Helleberg, J., Goh, J., Xu, X., & Horrey, B. (2001). *Pilot task management: Testing an attentional expected value model of visual scanning* (ARL-01-14/NASA-01-7). Savoy: University of Illinois, Aviation Research Laboratory.

Wickens, C. D., Helleberg, J., & Xu, X. (2002). Pilot maneuver choice and workload in free flight. *Human Factors, 44*(2), 171–188.

Wickens, C. D., & Hollands, J. (2000). *Engineering psychology and human performance* (3rd ed.). Upper Saddle River, NJ: Prentice Hall.

Wickens, C. D., & McCarley, J. S. (2001). *Attention-situation awareness (A-SA) model of pilot error* (Final Tech. Rep. ARL-01-13/NASA-01-6). Savoy: University of Illinois, Aviation Research Lab.

Wickens, C. D., McCarley, J. S., Alexander, A. L., Thomas, L. C., Ambinder, M., & Zheng, S. (2004). *Attention-situation awareness (A-SA) model of pilot error* (AHFD-04-15/NASA-04-5). Savoy: University of Illinois, Aviation Human Factors Division.

Wickens, C. D., Sebok, A., Bagnall, T., & Kamienski, J. (2007). Modeling situation awareness supported by advanced flight deck displays. In *Proceedings of the Human Factors and Ergonomics Society 51st Annual Meeting*. Santa Monica, CA: Human Factors and Ergonomics Society.

Wiegmann, D. A., & Shappell, S. A. (1997). Human factors analysis of postaccident data: Applying theoretical taxonomies of human error. *The International Journal of Aviation Psychology, 7*(1), 67–82.

Woods, D. D., Johannesen, L. J., Cook, R. I., & Sarter, N. B. (1994). *Behind human error: Cognitive systems, computers, and hindsight* (State-of-the Art Rep. CSERIAC 94-01). Wright–Patterson AFB, OH: CSERIAC Program Office.

Part 3

Implications for Modeling
and Aviation

10 A Cross-Model Comparison

Kenneth Leiden and Brad Best

CONTENTS

INTRODUCTION

This chapter compares and contrasts the human performance modeling capabilities developed and demonstrated by the five modeling teams during the National Aeronautics and Space Administration Human Performance Modeling (NASA HPM) project. The teams are referred to by the name of their corresponding human performance model (HPM) tool:

- Adaptive Control of Thought-Rational (ACT-R);
- Improved Performance Research Integration Tool/ACT-R (IMPRINT/ACT-R);
- Air Man–machine Integration Design and Analysis System (Air MIDAS);
- Distributed Operator Model Architecture (D-OMAR); and
- Attention-Situation Awareness (A-SA).

The cross-model comparison is intended to provide insight into each HPM's level of fidelity for representing a diverse range of modeling capabilities commonly associated with human performance:

- Error prediction (surface operations modeling only);
- External environment;
- Crew interactions;
- Scheduling and multitasking;
- Memory;
- Visual attention;
- Workload;
- Situation awareness;
- Learning; and
- Resultant and emergent behavior.

The comparison was performed by studying the HPM teams' respective final reports and chapters in this book and then contacting the authors with clarifying questions. In some instances, this comparison identifies interesting aspects of the modeling that are not covered in the respective chapters (chapters 5–9). The reason for this is that the HPM teams were not required to follow the same format as represented by this chapter in discussing the predefined set of modeling capabilities listed earlier.

It is important to emphasize that the scope of this comparison is discussed with respect to the synthetic vision system (SVS) modeling effort rather than the surface (taxi) operations modeling effort because, in general, more time and effort were spent on the former, resulting in higher fidelity modeling capabilities being developed. The one exception is the Error Prediction section, which applies to the surface operations modeling effort. Due to this limitation in scope, it is possible that a particular HPM is under-represented for a modeling capability under the SVS modeling effort, but the capability has been exercised in higher fidelity in other modeling efforts. For example, a team may have decided not to develop a situation awareness (SA) capability for the SVS modeling effort, but has done so on other projects unrelated to the NASA HPM project. In these instances, corresponding references are provided by the HPM teams as a service to the reader as well as to acknowledge this research.

ERROR PREDICTION

Modeling capabilities for error prediction come in many forms. In Leiden et al. (2001), error taxonomies and the error prediction capabilities of a range of HPM tools are presented. In general, error prediction can be characterized by the capabilities of the modeling tool. For example, memory-related errors require a modeling tool that represents memory interference and decay. Errors due to flight crew interactions require high-fidelity modeling of the captain and first officer (FO), and scripted behavior would not be effective. Moreover, error prediction in human performance modeling almost always results in some type of priming to allow the error to occur. The reason for this is that most complex environments such as aviation have procedures in place to deal with commonplace human errors. For example, the read-back of a clearance ensures both that the pilot understood the controller correctly and that the controller (while listening to the read-back) issued the intended clearance. It is still possible that the pilot would perform the clearance incorrectly, but there needs to be a more complex layer in the error chain to manifest itself.

Thus, in order to prime the model to produce errors, the modeler must have an understanding of realistic events and activities (and their frequency of occurrence) so that the types of errors produced represent realistic errors beyond the commonplace errors mitigated by checks in routine procedures. Otherwise, these errors may never occur at all in simulation or occur so frequently as to lack the face validity necessary to be considered meaningful. This priming aspect is most likely the crux of predicting accurate error rates by the models since the information is often difficult to quantify and is therefore often crudely estimated. A summary of the representation of error prediction is shown in Table 10.1.

TABLE 10.1
Error Prediction Capability Synopsis

HPM Tool	Error Prediction Representation
ACT-R	Developed five primary decisionmaking strategies that a pilot employs when approaching a taxiway intersection to decide which taxiway to take. Hypothesized that the different strategies (each with different probability of error) would have a probability of selection based on the decision time horizon.
IMPRINT/ ACT-R	Represented errors due to time-based decay or activation noise that resulted in a failure to recall a chunk holding turn information. The resulting error would be a missed turn. Also, represented errors that resulted in the wrong chunk being recalled. This would cause the model to schedule a turn on the wrong taxiway or in the wrong direction on the correct taxiway.
Air MIDAS	Represented three types of error mechanisms: (1) Pilots forgetting to complete a procedure as a result of multi-tasking between too many procedures at a given time. (2) Loss of information in working memory. (3) Worldview inconsistency between two operators. For example, this error could occur when air traffic control (ATC) erroneously assumed the pilots had accurately received taxi clearance information.
D-OMAR	Represented errors based on: (1) Episodic memory—an element of autobiographical memory and a source for habit-based actions (2) Context-based expectation based on partial knowledge—the captain knows the location of the destination concourse gate and, based on this knowledge, may reasonably have an expectation that the next turn will take them on the shortest route to that gate. (3) Remembrance—the captain remembers a correct or incorrect taxi sequence issued by the ground controller.
A-SA	Error prediction is provided in terms of errors of SA. The value of SA (between 0 and 1) was assumed to reflect the probability that the pilot when reaching a decision point at a taxiway intersection would select the correct response. If the pilot failed to recognize the correct response, then the default responses (modeler-specified), such as "when in doubt turn toward the terminal" or "when in doubt go straight," would represent the default decision.

ACT-R

For the surface operations modeling effort, the ACT-R team (using subject matter expert [SME] input) developed five primary decision-making strategies that a pilot employs when deciding which taxiway to take when approaching a taxiway intersection. They hypothesized that the different strategies would have a probability of selection based on the decision time horizon—the time available until a turn must be executed at an intersection. For example, making the decision to turn based on guessing would be fast, but inaccurate. Thus, the pilot is likely to choose this strategy only when there is insufficient time available for the other strategies. The strategies are summarized in order of the decision time horizon required:

1. Guess randomly—this is a very fast, but inaccurate strategy.
2. Remember the correct clearance—while fast, this strategy is increasingly inaccurate as memory decays.
3. Make turns toward the gate—this is somewhat slower than the first strategy, but 80.7% accurate (based on their internally generated analysis for nine heavily used airports in the United States; see chapter 5 for details).
4. Turn in the direction that reduces the larger of the X or Y (cockpit-oriented) distance between the aircraft and the gate—this is about 93% accurate based on the ACT-R team's analysis.
5. Derive from map/spatial knowledge—this is the slowest strategy available. The accuracy of this strategy is dependent on remembering the clearance, which decays with time, but can be enhanced by cues on the map.

Rather than running the ACT-R model to predict specific instances of error, the ACT-R team took a more global approach to the problem. Production rules were generated for each of these strategies where each strategy had a minimum time horizon (based on the preceding order) that was required for it to be selected. Furthermore, the probability-of-success component of the production utility was set to how accurate the strategy was deemed to be. For Strategy 1, the accuracy was based solely on the taxiway geometry. For example, if the intersection had three choices weighted equally, the probability of success would be 0.33. For Strategies 3 and 4, the values corresponded to 0.807 and 0.93, as indicated earlier. For Strategies 2 and 5, the probability of success was based on the declarative memory component of ACT-R to recall these chunks.

The ACT-R model was run in Monte Carlo fashion to predict the probability of selection of these strategies for decision time horizons varying from 0.8 to 10.6 sec in 0.2-sec increments. Thus, the error rates were a function of decision time horizon such that:

$$\textit{Probability of error}_k = 1 - \sum_j St_{jk}P_j$$

where

k is the decision time horizon (in 0.2-sec increments)
j is the decision strategy
St_{jk} is the probability of strategy selection for each j and k
P_j is the probability of a strategy's success

For example, their results indicated that for a 4.2-sec decision time horizon, the probabilities of selecting Strategies 3 and 4 were each about 50% (see Figure 5.3 in chapter 5). Thus, the probability of error is about 13%—that is, equal to $1 - (0.93 * 0.5 + 0.807 * 0.5)$.

IMPRINT/ACT-R

For the surface operations modeling effort, the IMPRINT/ACT-R team predicted taxiway errors based on memory decay and interference. The taxiway turns were represented by two chunks each, one chunk indicating the name of the taxiway and the other holding the direction to turn. At the appropriate time in the simulation, the ACT-R pilot was called for the initial memorization of the list of taxiways. Later in the sequence, the aircraft approached a taxiway intersection and the ACT-R pilot attempted to recall the list of taxiways. If the attempt was successful, the correct taxiway was followed. If unsuccessful, the following error types could be produced by the model:

- For errors due to time-based decay or activation noise that resulted in a failure to recall a chunk holding turn information, the resulting error would be a missed turn.
- Errors due to interference, similarity-based partial matching, priming, or activation noise that resulted in the wrong chunk being recalled would cause the model to schedule a turn on the wrong taxiway or in the wrong direction on the correct taxiway.

Air MIDAS

For the surface operations modeling effort, the Air MIDAS team represented three types of error mechanisms. The first error type represented the pilots forgetting to complete a procedure as a result of multitasking between too many procedures at a given time. The occurrence of this error was modeled by causing events in the simulation to occur at approximately the same time such that postponed and interrupted activities would be forgotten because there were too many procedures competing in the scheduler (see the Scheduling section).

The second error type represented errors due to loss of information in working memory. These two error types could be combined to replicate a wrong turn. For example, if the FO had his or her head down while approaching a required turn due to task interruption and forgot to resume the task in a timely manner, the captain could make an erroneous turn based on his or her failure to recall the correct turn from working memory. The third type of error represented worldview inconsistency between two operators. For example, this error could occur when the ATC erroneously assumed the pilots had accurately received taxi clearance information.

Two manipulations were used to test the impact of memory demands on crew performance: the working memory decay rate and capacity. Decay rate was set for information loss at either 12 or 24 sec from the time of perception or recall and the memory capacity was either 5 chunks or 10 chunks.

D-OMAR

For the surface operations modeling effort, the D-OMAR team represented taxi-way errors by priming the model such that reasonable scenarios were constructed in which the FO was occupied with event-driven tasks and unable to provide the prompt to the captain to turn at an intersection. Without the explicit prompt, the decision of the captain to turn at an intersection was based on other intentions:

- *Episodic memory*—this is an element of autobiographical memory and a source for habit-based actions. Experience from similar situations provides a response that has worked well in the past.
- *Context-based expectation based on partial knowledge*—the captain knows the location of the destination concourse gate and based on this knowledge may reasonably have an expectation that the next turn will take them on the shortest route to that gate.
- *Remembrance*—the captain remembers a correct or incorrect taxi sequence issued by the ground controller.
- *Explicit prompt* by the FO based on written taxi instructions—nominally, the intention of the captain and the prompt from the FO are in agreement.

The priority assigned to these intentions and timing of events determined the intention selected by the model. The timing and priorities were varied by hand to explore their impact on producing errors in the taxi sequences.

A-SA

For the surface operations modeling effort, error prediction is provided by A-SA in terms of errors of SA. (See also the Situation Awareness section). In other words, the value of SA (between 0 and 1) was assumed to reflect the probability that when the pilot reached a decision point at a taxiway intersection, he or she would recognize the correct response option and behave appropriately as a result. If the pilot failed to recognize the correct response option, then the default responses (modeler specified) such as "when in doubt turn toward the terminal" or "when in doubt go straight" would represent the default decision. If more than two responses applied to a decision point, the response selected would be chosen at random. The equation for the probability of error can be expressed as:

$$Probability\ of\ error = 1 - [SA + (1 - SA)P_{avg}]$$

where P_{avg} is the average probability of all default responses being correct.

Because these default responses tend to be reasonably accurate, even with low SA the probability of making a wrong decision is lower than one would first suppose. For example, the results in chapter 9 show the probability of correct behavior is never lower than 80% despite the value of SA falling below 0.5. It is important to note that the error prediction capability within A-SA (and all the other models for that matter) is quite application specific.

Lastly, A-SA can characterize specific predictions of attentional errors (i.e., failures to notice) when pilots do not allocate attention according to optimal prescriptions. Such errors may be characterized with attentional tunneling both in aviation and driving (Horrey, Wickens, & Consalus, 2006).

EXTERNAL ENVIRONMENT

From a human performance modeling perspective, the external environment is the environment in which the simulated operators interact. The external environment can represent something simple (e.g., an automated teller machine) or something quite complex (e.g., a nuclear reactor and its control room). In the aviation domain, the external environment typically represents an aircraft's state (position, velocity, attitude) and performance characteristics, dynamic responses to pilot actions, some representation of flight deck displays and instrumentation, and the atmosphere and surroundings through which the aircraft travels.

The level of fidelity at which a complex external environment should be represented is often a difficult decision for the modeler to make. On the one hand, a high-fidelity environment most closely represents the closed-loop behavior between a pilot's actions (or inactions) and the aircraft's response. Thus, if a pilot performs a task poorly, the aircraft flight dynamics should exhibit the appropriate response, which in turn may require a corrective action by the pilot. This feedback mechanism is often a crucial aspect of human performance modeling, particularly with respect to workload. Without it, the modeler risks concealing important human performance issues.

On the other hand, a high-fidelity model of the external environment can be difficult, time consuming, and prohibitively expensive to develop. Fortunately, there are commercial off-the-shelf (COTS) flight simulators readily available to provide a high-fidelity external environment, so the issue becomes one of integration rather than development. Of course, integration does have its own issues: The process of passing information between simulations can also be difficult and time consuming, and require specialized expertise. Thus, when considering the fidelity of a complex external environment, the modeler must carefully consider these issues, particularly since the decision is made early in the project and midcourse corrections may not be feasible.

It is important to note that ACT-R, Air MIDAS, and D-OMAR have all implemented high-level architecture (HLA), developed by the U.S. Department of Defense to address simulation interoperability issues, during previous modeling projects. However, unless specifically required by a particular project (which is sometimes the case for military applications), most HPM teams prefer to use other methods for simulation interoperability due to HLA's complexity and overhead.

For the SVS effort, a Boeing-757 commercial air transport carrier performing an area navigation (RNAV) approach was assumed. During the RNAV approach, the flight path was determined using lateral and vertical navigation (LNAV, VNAV) modes in which waypoints (latitude, longitude, and altitude) entered a priori defined the lateral and vertical paths of the aircraft. Under this autoflight mode, the aircraft essentially flies itself. The only control inputs needed from the pilots were speed settings via the mode control panel (MCP), flap settings, and lowering the landing gear. In addition, there were no significant winds, wind shear, or other off-nominal events that would alter the aircraft's flight path significantly from a nominal path prior to breaking through the cloud ceiling or reaching decision altitude, whichever came first. (Pilots typically take manual control of the aircraft at this point, but the modeling teams did not attempt to model these motor tasks.)

These assumptions are stated here to emphasize that there is reasonable justification for representing the aircraft dynamics by simple kinematic equations and scripted events (as was done by the IMPRINT/ACT-R, D-OMAR, and A-SA teams) without a significant impact on the pilot–action/aircraft–response feedback mechanism. *For this particular modeling effort,* it is the belief of this chapter's authors that this lower fidelity modeling of the external environment provides an adequate rendering for the HPM component predictions. That said, the primary drawback to these simpler external environments is that the models' *extensibility to future work* may be hindered if the future work must demonstrate a closely coupled feedback mechanism between pilot–action and aircraft–response (and vice versa).

On the other hand, the ACT-R and Air MIDAS teams chose to integrate their HPM with higher fidelity flight simulators, avoiding this potential drawback. Furthermore, by coupling to a higher fidelity flight simulator, some aspects of model verification become readily apparent. For example, if the speed brakes were not retracted on landing and a tail strike occurred, it would be obvious to the modeler that the pilot experienced some type of distraction during the run that prevented him or her from retracting the speed brake. In contrast, it would be unlikely that the lower fidelity models would capture the dynamics of a tail strike. (Of course, if the modeler had the foresight to include a model flag to identify that the speed brake was not retracted, the effect of identifying the problem would be the same.) A summary of how each modeling team represented the external environment is shown in Table 10.2.

ACT-R

The ACT-R team connected their model to X-Plane, a PC-based flight simulator software package. X-Plane is available for a small fee and includes 40 aircraft types including props and jets, both subsonic and supersonic. X-Plane scenery is worldwide and includes 18,000 airports in its database.

The default control for flying an aircraft is the keyboard/mouse/joystick. To enable inputs from the HPM, the ACT-R team built a two-way network interface

TABLE 10.2
External Environment Capability Synopsis

HPM Tool	External Environment Representation
ACT-R	X-Plane, a commercially available PC-based flight simulator package
IMPRINT/ACT-R	Simple aircraft kinematic equations using IMPRINT discrete event simulation
Air MIDAS	PC Plane, a NASA-developed PC-based flight simulator
D-OMAR	Simple aircraft kinematic equations internal to D-OMAR
A-SA	Does not require an integrated external environment

using user datagram protocol (UDP) to allow information to pass between ACT-R and X-Plane such that the simulated ACT-R pilot is actually flying the aircraft. X-Plane information from the navigation display (ND), primary flight display (PFD), mode control panel (MCP), and out-the-window (OTW) view is passed to ACT-R so that, within the ACT-R environment, this information could be accessed and "viewed" by the simulated pilot during a scan. X-Plane information that is numeric is passed numerically to ACT-R. Information that is based on interpretation such as alignment with a runway or the visibility OTW must first be translated into numeric values. The combined ACT-R/X-Plane simulation runs in real time. Reasonably accurate time synchronization is accomplished implicitly by requiring each simulation to synch to its internal clock.

For the SVS modeling effort, the ACT-R team estimated that about 50% of the total model development time was required for modeling the external environment. This is further decomposed into about 15% of the time for the communication link between ACT-R and X-Plane and 35% to represent the information depicted on the displays.

IMPRINT/ACT-R

The IMPRINT/ACT-R team represented the external environment using IMPRINT, which was developed by Micro Analysis & Design through direction of the U.S. Army. Like ACT-R, IMPRINT is itself a human performance modeling tool, but IMPRINT simulates human performance at a larger level of granularity as compared to the cognitive level of ACT-R. In addition, IMPRINT operates in a discrete event simulation framework that facilitates modeling both human operators and the external environment through use of a graphical tool for rapid model generation.

For the SVS modeling effort, the delineation between ACT-R and IMPRINT is that ACT-R was utilized to model the captain at high fidelity whereas IMPRINT was utilized to model the FO at lower fidelity, ATC communication, the aircraft dynamics (using relatively simple kinematic equations), and the flight deck displays and controls. IMPRINT did not represent the aircraft's heading or distance to waypoints on the ND, which meant that the simulated captain did not scan this display as frequently as the subject pilots. This did have an impact on the visual attention allocation, as discussed later in the Visual Attention section.

The connection between ACT-R and IMPRINT was provided by the component object model (COM) programming interface, which supports a relatively low-bandwidth connection. The elements passed across the COM connection from IMPRINT to ACT-R allowed updating the various ACT-R chunks corresponding to the pilot's knowledge and awareness of the displays and controls. Conversely, information flow from ACT-R to IMPRINT consisted of the various actions the ACT-R simulated pilot is capable of performing (e.g., lowering the landing gear, setting flaps).

For the SVS modeling effort, the IMPRINT/ACT-R team estimated that 70% of the total model development time was required for modeling the external environment. This is further decomposed into about 20% of the time for the communication link between ACT-R and IMPRINT and 50% to develop the aircraft dynamic model and represent the aircraft state information on the displays.

Air MIDAS

Similar to the ACT-R team, the Air MIDAS team connected their model to a flight simulator software package. In this case, the Air MIDAS team selected PC Plane, a NASA-developed, PC-based flight simulator. Dynamic link library (DLL) functions generate aircraft inputs and provide time synchronization. Microsoft Access, a database application, serves as middleware between Air MIDAS and PC Plane to hold numeric information representing the ND, PFD, MCP, and OTW views. One notable difference between the Air MIDAS team's approach and that of the ACT-R team is that the Air MIDAS approach provides more precise time synchronization, since both PC Plane and Air MIDAS are controlled by an independent scheduler (i.e., the DLL functions).

For the SVS modeling effort, the Air MIDAS team estimated that 30% of the total model development time was required for modeling the external environment. This includes the PC Plane time control mechanism, model of displays, and the communication link between Air MIDAS and PC Plane.

D-OMAR

The external environment for D-OMAR was represented within D-OMAR itself. D-OMAR simulated the aircraft dynamics (using relatively simple kinematic equations), the flight deck displays and controls, runway configuration, and ATC communication. The external environment developed for the surface operations modeling effort was leveraged extensively for the SVS modeling effort. The simulated pilot gathers numeric information from the D-OMAR representation of flight deck displays and controls through fixations specified by the pilot's background scan pattern or goal-directed scans. The D-OMAR team estimated that less than 25% of the total model development time was required for modeling the external environment.

A-SA

For the attention module, the salience, effort, expectancy, and value (SEEV) model within A-SA requires coefficient estimates for salience (in the surface modeling effort only), effort, bandwidth (frequency that data are changing on a display), relevance, and task priority. Salience, effort, and bandwidth characterize the external environment. As such, the external environment needs to be represented only to the extent that reasonable estimates of these coefficients are possible. Three examples can clarify this point.

First, the effort coefficient is dependent on how the flight deck displays are situated with respect to each other (as can be depicted by a simple mock-up of the flight deck) where transitions between separated displays require more effort than adjacent displays. Second, the bandwidth coefficient can be estimated (and held constant) for a segment of flight in which the bandwidth of a key parameter on the display remains nearly constant. Alternatively, for a more accurate estimate, the bandwidth coefficient can be derived from the actual bandwidth of parameters extracted from flight simulator data without requiring direct integration between A-SA and the flight simulator. The bandwidth of discrete events can be generated by any time-based script. Third, salience of events can be coded in a dynamic simulation (as was done for the surface modeling effort) using simple heuristics and rules.

For the other module of A-SA—namely, the belief-updating module—no environmental modeling is required, since the effect of the external environment is entirely mediated by the attention module.

CREW INTERACTIONS

This section discusses the method for modeling interactions between the captain and FO as well as their interactions with ATC. The level of fidelity at which the crew interactions should be represented is similar to the external environment issues discussed earlier. In fact, from a single-operator perspective, the crew interactions are essentially a subset of the external environment. By identifying what question the modeling effort is trying to answer, insight can be gained into level-of-fidelity requirements.

If, for example, new procedures are being developed specifically to address crew resource management issues, then the modeling effort should emphasize high-fidelity crew interactions because poor performance of one crew member could require corrective actions by the other crew member—similar to the feedback mechanism between pilot and aircraft for the external environment mentioned earlier. On the other hand, if it is a new display that does not directly affect crew interaction procedures, then lower fidelity modeling, such as scripting the interactions of one of the crew members, should be sufficient. A summary of the representation of crew interactions is shown in Table 10.3.

ACT-R

The ACT-R team only modeled the captain. Interactions with the FO (and ATC) were scripted.

IMPRINT/ACT-R

The IMPRINT/ACT-R team represented crew interactions using IMPRINT. As mentioned earlier, IMPRINT is itself a human performance modeling tool. Thus, the IMPRINT/ACT-R team chose to use ACT-R to model the captain and IMPRINT to model the FO. This allowed interactions between the crew to occur in a nonscripted

TABLE 10.3
Crew Interactions Capability Synopsis

HPM Tool	Crew Interaction Representation
ACT-R	Modeled the captain. Interactions with the FO (and ATC) were scripted.
IMPRINT/ ACT-R	Used ACT-R to model the captain and IMPRINT to model the FO (lower fidelity).
Air MIDAS	Both the captain and FO were simulated at equivalent levels of fidelity by Air MIDAS.
D-OMAR	Both the captain and FO were simulated at equivalent levels of fidelity by D-OMAR.
A-SA	Modeled the captain. Interactions with the FO (and ATC) were scripted.

manner, although the FO's behavior modeling was lower fidelity by comparison. In any case, this proved to be a good compromise between the required fidelity of the captain and perhaps unnecessary complexity for the FO.

AIR MIDAS

Recent developments in Air MIDAS (Corker, 2005; Corker, Fleming, & Lane, 2001) have emphasized multiple-operator environments so that modeling studies can investigate the closed-loop behavior (e.g., between pilots of the same aircraft as well as pilots of different aircraft) that would be essential to the understanding of new operational concepts such as the self-separation of aircraft. These developments have been leveraged for the SVS modeling effort to simulate both the captain and FO at equivalent levels of fidelity. This effort required detailed model input for both pilot positions in terms of external environment, procedures, task priorities, etc. The resulting interactions between the pilots thus provided realistic closed-loop behavior, with latency being the most common effect, while adhering to attentional and workload constraints.

D-OMAR

The modeling framework of D-OMAR lends itself quite well to modeling multiple human operators. For both NASA modeling efforts, the D-OMAR team modeled the captain and FO at equivalent levels of fidelity. Similar to Air MIDAS, this effort required detailed model input for both pilot positions in terms of external environment, procedures, task priorities, etc. The resulting interactions between the pilots thus provided realistic closed-loop behavior.

A-SA

Crew interactions in A-SA are scripted and typically represented as an auditory event in an event timeline. The salience of an auditory event (which is typically assigned maximum weights in A-SA) captures the pilot's attention. Whether pilot SA is improved or degraded depends on the relevance of the crew interaction to tasks at hand. Interestingly, an irrelevant interaction would degrade SA faster than if there had been no interaction at all because of the interference of that interaction with memory-dependent maintenance of SA.

SCHEDULING AND MULTITASKING

This section discusses the modeling representation of pilot task scheduling as determined by task priorities, human resource limitations, and system resource constraints. Task priorities refer to both knowledge based (e.g., lower landing gear prior to landing) and data driven (e.g., an auditory alert captures the pilot's attention). Human resource limitations refer to the visual, auditory, cognitive, and motor resources available to the pilot versus those required by concurrent task demands. System resource constraints refer to constraints of the external environment (e.g., only one person can speak at a time on the radio) that can impose limitations on what tasks can be performed by a pilot at a given moment.

TABLE 10.4

Scheduling and Multitasking Capability Synopsis

HPM Tool	Scheduling and Multitasking Representation
ACT-R	Scheduling is a stochastic process performed by production rules. In practical terms, the production rules encapsulate both basic and domain-specific human behavior.
	Multitasking was not implemented for this modeling effort (the ACT-R architecture does permit it with certain restrictions).
IMPRINT/ACT-R	Same as above.
Air MIDAS	Scheduling is deterministic and encapsulates both basic and domain-specific human behavior. Uses visual, auditory, cognitive, and motor channel resources on a numeric scale developed by McCracken and Aldrich (1984).
	Multitasking is possible based on resources available versus resources required (e.g., structure allows two visual tasks to be performed concurrently provided sufficient resources are available).
D-OMAR	Similar to Air MIDAS in some ways. The primary differences between D-OMAR and Air MIDAS are that the types of resources do not map directly and are quantified differently. The resources in D-OMAR represent five Boolean values: the eyes (visual attention), ears (listening), mouth (speaking), and a dominant and nondominant hand (motor movements). As such, two visual tasks cannot be performed concurrently as they can with Air MIDAS.
A-SA	There is no explicit scheduler or task shedding mechanism in A-SA. However, task priority (value) provides implicit guidance for when more relevant or less relevant information is sampled.

Scheduling is a key component of any HPM effort because it determines under what conditions interruptions and multitasking can occur, if at all. Multitasking can be modeled as either true concurrent activities (Air MIDAS and D-OMAR) or as rapid interleaving between multiple, serial activities (ACT-R). Perhaps more important than the theories themselves are the practical modeling considerations of the two, as rapid interleaving is more difficult to implement (discussed in the ACT-R section below). Although the SVS modeling effort is *not* characterized by a strong need for interruptions and multitasking (it is primarily a visual attention study), other topical aviation-related questions such as the impact of controller–pilot data link communication in reducing ATC workload would require solid modeling capabilities in these areas to provide an answer. A summary of the representation of scheduling and multitasking is shown in Table 10.4.

ACT-R AND **IMPRINT/ACT-R**

The scheduling of memory retrieval, visual attention, manual actions, and other skill-based tasks is performed by the central production system in ACT-R. The production system is intended to represent procedural memory, which stores associations between stimuli and response (in ACT-R parlance, condition and action) that indicate how to do something without consciously thinking about it. The production system detects patterns in the ACT-R buffers to determine what to do next. The production system is also capable of learning, which is discussed in the Learning section.

The production system is a rule set that operates by default in 50-ms time steps, though the timing of individual productions can be extended by retrievals or actions that take longer. A production rule (or simply "production") is an "if *condition,* then *action*" pair. The *condition* specifies a pattern of chunks that must be present in the buffers for the production rule to apply. The *action* specifies some action to take. Multiple production rules can satisfy their respective conditions, but only one of those, the one with the highest utility, can execute its corresponding action. The production utility U_i is expressed as:

$$U_i = P_i G - C_i + \varepsilon \qquad \text{Production utility equation}$$

where

> P_i is the probability of success that the production rule, if executed (or "fired" in ACT-R parlance), can achieve the current goal
>
> G is the value of the goal (G is independent of the production rule, default value is 20)
>
> C_i is the cost (typically an estimate of the time from when the production is selected until the objective is finally completed)
>
> ε is a noise parameter (e.g., if several production rules match the chunks in the buffers, there is a stochastic influence as to which production rule would be fired)

The values of C_i and P_i can be set for each production rule or can be learned from experience.

In practical terms, the production rules encapsulate both first-principle and domain-specific human behavior and thus represent task scheduling and priorities of the pilot. Thus, it is important that these two types of behavior be well understood by the modeler to reflect realistic task scheduling and interruptions of pilots. As should be evident by this section and the Memory section, there is an art to ACT-R modeling (as there is with all HPMs). It would be unlikely that two modeling teams would represent the production rules for a complex model such as the SVS modeling effort in the same manner. In fact, this was exactly the case for the ACT-R and IMPRINT/ACT-R teams as they each chose a unique production rule design for modeling visual attention (see the Visual Attention section).

Historically, multitasking has not been an emphasis of ACT-R and most models created with ACT-R (including the SVS models) do not attempt to represent multitasking. However, with the release of version 5.0, which incorporates certain aspects of the perceptual and motor modules in the executive-process interactive control (EPIC) model, ACT-R is capable of modeling parallel activity *with certain restrictions imposed by limitations observed in human performance.*

First, modules in ACT-R, such as the visual and manual modules, operate independently of each other so that, for example, visual and manual tasks can be done in parallel as long as their initiation is staggered. This is due to a bottleneck in the central production system that results in a production rule for a visual task being delayed while a production rule for a manual task is being fired (or vice versa). Since there is evidence for direct module-to-module interaction, ACT-R researchers acknowledge that ACT-R may require a modification to the framework to allow for this direct interaction. Second, each module can only do one job at a time due to the constraint on each module's buffer to hold only one chunk at a time (so two visual tasks cannot be done concurrently, regardless of their resource demands).

The preceding restrictions, which are intended to represent human limitations, nonetheless result in a cumbersome process for multitasking between higher level tasks (e.g., making an MCP entry while listening to an ATC communication). Multitasking at this higher level is better described as rapid interleaving between tasks as it requires ACT-R to switch continually between goals (for instance, in the preceding example, ATC communication must be broken into chunks to store in memory and each of these reflects a new goal); each switch requires the previous goal to be stored and then later retrieved from memory. Fortunately, there have been some techniques developed to facilitate this process, but it is still rather difficult compared to Air MIDAS and D-OMAR.

Air MIDAS

In Air MIDAS, pilot activities and tasks can be triggered for scheduling under two conditions. The first condition occurs when a value or event is encountered (typically through a monitoring scan or as an auditory event) in the environment that has significance to the pilot. The second condition occurs based on the modeler-specified task decomposition (e.g., Task 2 necessarily follows Task 1). Once triggered, activities and tasks are then scheduled based on priorities, resources required, and resources available. The priorities and resources required for each activity are assigned by the modeler (often with subject matter expertise input) whereas the resources available are calculated during run-time.

The scheduling mechanism in Air MIDAS considers both an estimate of the time it takes to complete a task and the availability of visual, auditory, cognitive, and motor channel (VACM) resources on a scale developed by McCracken and Aldrich (1984) and shown in Table 10.5. This theory assumes that the VACM channels are independent of each other. The scale for each channel ranges from 0 (no workload) to 7 (very high workload). The visual and auditory channels refer to the signal detection required for a task. The cognitive channel consists of the information processing and the motor channel is the physical action required to accomplish a task. One advantage to the VACM method is that it quantifies whether a task can be scheduled in a straightforward fashion by assuming the workload levels are additive within channels and independent between channels. For example, assume $M = 2$ and $V = 4$ for task A and $V = 2$ and $C = 2$ for task B. If task A is being performed, task B can only be scheduled to be performed concurrently with task A if each channel's total workload is less than 7, which it is for this example since $M = 2$, $V = (4 + 2) = 6$, and $C = 2$.

If the time or resources required to perform a task are not currently available, task priorities determine the outcome. Scheduling is accomplished through a set of queues of current, interrupted, and postponed activities to keep track of the activity order. Postponed and interrupted activities can be forgotten if there are too many procedures of the same type competing in the scheduler. The modeler can specify the number of active goals that can be kept in working memory at any given time. The active goals have procedures associated with their performance. By limiting the active goals, the modeler can influence the number of procedures that can be kept

TABLE 10.5
VACM Values and Descriptors

Value	Visual Scale Descriptor	Value	Auditory Scale Descriptor
0.0	No visual activity	0.0	No auditory activity
1.0	Register/detect image	1.0	Detect/register sound
3.7	Discriminate or detect visual differences	2.0	Orient to sound, general
4.0	Inspect/check (discrete inspection)	4.2	Orient to sound, selective
5.0	Visually locate/align (selective orientation)	4.3	Verify auditory feedback (detect occurrence of anticipated sound)
5.4	Visually track/follow (maintain orientation)	4.9	Interpret semantic content (speech)
5.9	Visually read (symbol)	6.6	Discriminate sound characteristics
7.0	Visually scan/search/monitor (continuous/ serial inspection, multiple conditions)	7.0	Interpret sound patterns (pulse rates, etc.)

Value	Cognitive Scale Descriptor	Value	Motor Scale Descriptor
0.0	No cognitive activity	0.0	No motor activity
1.0	Automatic (simple association)	1.0	Speech
1.2	Alternative selection	2.2	Discrete actuation (button, toggle, trigger)
3.7	Sign/signal recognition	2.6	Continuous adjusting (flight control, sensor control)
4.6	Evaluation/judgment (consider single aspect)	4.6	Manipulative
5.3	Encoding/decoding, recall	5.8	Discrete adjusting (rotary, vertical thumbwheel, lever position)
6.8	Evaluation/judgment (consider several aspects)	6.5	Symbolic production (writing)
7.0	Estimation, calculation, conversion	7.0	Serial discrete manipulation (keyboard entries)

Source: McCracken, J. H., & Aldrich, T. B. (1984). Technical note ASI479-024-84. Fort Rucker, AL: Army Research Institute, Aviation Research and Development Activity.

current or pending based on first-in, first-out queuing process. If a higher priority goal is activated, which causes the available goal queue size to be exceeded, then the first goal to have entered the queue (i.e., the least recent) is aborted. In fact, this is how one of the error mechanisms in the surface operations effort was represented (see the Error Prediction section). Note that for output purposes, the "forgotten" tasks are recorded even though they are never acted upon.

Multitasking is made possible through the same scheduling process with the requirement that the resource demands to perform parallel tasks do not exceed the human or system resources available. Numerical VACM scores above the threshold of seven are typically not seen in Air MIDAS because the scheduler accounts for the excessive demand and postpones or sheds tasks accordingly. Practically speaking, the visual demand of the SVS modeling effort inhibits multitasking since a high value of visual resource is required for most tasks. For example, the Air MIDAS team assigned a value of 5.9 to the visual channel during the task of monitoring (i.e., "visually read (symbol)" from Table 10.5). This means the remaining visual resource available is only 1.1 (i.e., $7 - 5.9 = 1.1$) for other tasks to be performed in parallel. As can be seen in Table 10.5, most tasks have visual resource requirements higher than 1.1. Thus, the monitoring scan must be suspended or interrupted in order to initiate most other tasks, thereby inhibiting multitasking.

Lastly, it should be noted that Air MIDAS differs in three ways from ACT-R in regard to multitasking. First, there are no architectural bottlenecks in Air MIDAS analogous to the central production system bottleneck of ACT-R (see the ACT-R sub-section for explanation). Second, multiple tasks using the same channel resource are permitted provided the sum of their demands does not exceed the channel threshold (e.g., two low-demand visual tasks can occur concurrently). Third, the Air MIDAS scheduler is deterministic rather than stochastic.

D-OMAR

In D-OMAR, scheduling and multitasking share some similarities to Air MIDAS. For example, pilot activities and tasks are triggered for scheduling under essen-tially the same two conditions from before: (1) in response to values or events in the environment that have significance to the pilot, or (2) based on the sequence of tasks resulting from normal pilot procedures and basic human behavior. If the resources required for the task are available, the triggered activity or task can begin immediately, possibly in concurrence with the current task. If the resources required are not available, task priorities assigned by the modeler determine if the current task should be interrupted to start the new task or if the new task should be postponed. In some instances, task priorities are based on well-accepted policy (e.g., intracrew conversation should be interrupted when an ATC communication is heard over the radio).

The primary differences between D-OMAR and Air MIDAS are that the types of resources do not map directly and are quantified differently. In Air MIDAS, the visual, auditory, cognitive, and motor resources are scaled numerically from 0 to 7 for each task element so that two low-demand visual tasks, for example, can occur concurrently. In contrast, the resources in D-OMAR represent five Boolean values: the eyes (visual attention), ears (listening), mouth (speaking), and dominant and non-dominant hands (motor movements). In other words, each resource can only be allo-cated to one task at a time (e.g., an operator can only listen to one conversation at a time). For example, with regard to the eyes as a resource for concurrent activity, the model must interrupt its background scan to perform another visual task (e.g., set a lever) and then resume the background scan. Lastly, it is assumed that the cognitive channel is implicitly coupled with the other resources and is therefore not resource limited.

A-SA

There is no explicit scheduler or task-shedding mechanism in A-SA. However, task priority (value) provides implicit guidance for when more relevant or less relevant information is sampled. A higher priority task will increase the likelihood of sam-pling information sources that are relevant to that task (better preserving SA for that task) at the expense of lower priority tasks supported by other information sources (degrading SA for those tasks). The direct links between task priority as explicitly varied, scanning in the attention module, and multitask performance are well docu-mented in a highway-driving simulation by Horrey et al. (2006).

MEMORY

Modeling memory provides some unique capabilities to HPM efforts. First, memory models enable an HPM tool to produce certain types of human errors. For example, when a wrong memory chunk is retrieved and then used in a decision or calculation, the outcome can be a human error such as making the wrong turn on a taxiway. Second, memory models enable the retrieval of a memory chunk to fail occasionally (i.e., forgetting). Forgetting may require the operator to reacquire the missing chunk (e.g., the pilot attends to the display to read the value again or calls ATC and asks them to repeat a clearance), which can increase task load. Alternatively, forgetting can result in the operator making a decision or calculation without this piece of information (perhaps trying to infer or guess the information), which can lead to human error. Lastly, memory models, depending on the amount of memory retrieval required, can be useful in estimating task time using a bottom-up approach, particularly if the task is mostly cognitive in nature.

Notwithstanding the preceding, there are reasons that some HPM teams choose not to incorporate memory models. Memory models add significant complexity to HPMs. The ability to incorporate the unique experiences of one pilot compared to another over a lifetime of flying is virtually impossible. Thus, the memory models must assume some type of average knowledge, which may not reflect the real-world experiences of a pilot who has flown into Chicago O'Hare regularly for 20 years versus the pilot who has flown there only a handful of times over 20 years.

Furthermore, for routine pilot tasks that have been performed thousands of times, the correct retrieval of a memory chunk is virtually guaranteed and the retrieval time becomes a function of noise only (which can be incorporated without a full memory model). Thus, for routine pilot tasks, there may be no change in predicted pilot performance between HPM results in which the memory model is enabled versus disabled. As such, one may ask if incorporating the complexity of a memory model is necessary if it does not impact pilot performance results. As with many aspects of human performance modeling, the trade-off among depth, breadth, and plausibility of the memory model must be carefully weighed. A summary of the representation of memory as implemented by each modeling team is shown in Table 10.6.

ACT-R AND IMPRINT/ACT-R

Two types of memory, declarative memory and procedural memory, form the foundation of ACT-R. Declarative memory stores facts and events (i.e., chunks), whereas procedural memory stores associations between stimuli and response that indicate how to do something without consciously thinking about it. Procedural memory is more closely associated with the Scheduling section of this chapter, and thus is discussed in more detail there.

It is worth noting that ACT-R does not explicitly make a distinction between declarative memory and working memory as do some of the other modeling tools. In ACT-R, working memory is simply a subset of declarative memory that is active (or "highly activated" in ACT-R parlance). If a chunk is not activated sufficiently, it cannot be retrieved. The higher the chunk's activation is, the higher the probability is

TABLE 10.6
Memory Modeling Capability Synopsis

HPM Tool	Memory Representation
ACT-R	Model for declarative memory including terms for base-level activation and context activation
IMPRINT/ACT-R	Similar to above, but includes terms for partial matching (for number retrievals only) instead of context activation
Air MIDAS	Working memory capacity and the time duration that memory chunks remain in working memory—the functions associated with interference and decay—are user-specified parameters
D-OMAR	Assumes perfect memory for the SVS modeling effort; see Error Prediction section for surface modeling effort
A-SA	A-SA does not attempt to represent individual chunks of memory; however, overall SA is assumed to decay with time

that the chunk can be retrieved and, if retrieved, the faster the chunk's retrieval will be. Activation (i.e., the value of chunk) is the sum of the following components:

- Chunk's usefulness to past experiences, referred to as *base-level activation*;
- Chunk's relevance to the current context, referred to as *context activation*;
- Chunk's similarity to other chunks, referred to as *partial matching activation*; and
- A noise term.

Discussion of the equations associated with these four components is outside the scope of this comparison. However, at a high level, they are discussed in more detail because the two ACT-R modeling teams had different approaches to incorporating them into their respective models.

Base-Level Activation

In general, base-level activation is the most successfully and frequently employed component of the ACT-R theory in that it represents the activation of a chunk growing with frequency of use (i.e., learning) and decaying with time (i.e., forgetting). Applied to a specific domain such as aviation, populating the frequency of use and decay of a chunk can be a time-consuming and imprecise effort, particularly if emulating realistic experiences that vary from pilot to pilot. The ACT-R and IMPRINT/ACT-R teams address this issue as follows.

The ACT-R and IMPRINT/ACT-R teams applied different techniques for base-level activation in the SVS modeling effort. The IMPRINT/ACT-R team completely disabled base-level learning. In its place, they set the base-level activations of the knowledge chunks to a *constant* value that they believed was high, but reasonable. Their reasoning was that knowledge pertaining to the pilot task (e.g., how to set each control based on the values of various state variables) was so well practiced that it should be retrieved quickly and reliably such that the usefulness of the chunk to past experiences would remain constant over the course of the simulation. In contrast, the ACT-R team used a hybrid base-level learning equation (outside the scope of this discussion) and populated the initial base-level activation to very high values. As such, the base-level activation would vary slightly during the course of a simulation based on frequency of use and decay, but it would remain high at all times, ensuring correct and quick chunk retrieval. Practically speaking, both team approaches are similar in that they model quick and successful chunk retrieval for highly practiced tasks.

The preceding discussion addresses retrieving declarative memory for highly practiced tasks. The other aspect of base-level activation is the working memory component, which was exercised in the surface operations modeling and is discussed in the ACT-R portions of the Error Prediction section.

Context Activation

In ACT-R, context activation (or spreading activation) refers to how the chunk of memory in the goal buffer activates (or "excites") chunks in declarative memory that have a direct connection to that chunk. For example, assume one is asked to name a type of animal that is small and brown. Chunks that are activated may include skunk (for being small), bear (for being brown), or squirrel (for being both small and brown). Using ACT-R defaults, squirrel would be the chunk retrieved out of those three choices because it receives activation from two sources whereas skunk and bear only receive activation from one source each. On the other hand, if both squirrel and prairie dog are stored in declarative memory, then they both receive activations from two sources. In this tie-breaker condition and assuming ACT-R defaults, the one with higher base-level activation would be retrieved (assuming no activation noise).

In addition, context activations must also consider the strength of association between a source in the goal buffer ("small" or "brown" in the preceding example) and a fact or chunk in declarative memory. As the number of associations increases, the strength of association decreases. From an experimental psychology perspective, context activation in ACT-R is feasible to investigate because an experiment can be designed specifically to limit the number of associations between sources and facts to manageable amounts. However, from a domain-specific perspective such as aviation, context activation is less useful, not so much for what is modeled, but rather for what is *not* modeled.

In other words, the ACT-R modeler can readily incorporate the minimum associations between sources and the corresponding facts, but it would be extremely difficult if not impossible to incorporate all, or even most, of the other associations that exist based on a pilot's operational experience (e.g., the number of runway configurations experienced by a commercial air carrier pilot over a 30-year career must be in the hundreds). In terms of ACT-R modeling performance, the impact of this would most likely be that the time to retrieve facts is faster than actually occurs in real pilots. (Recall that for the SVS modeling effort, the base-level activation was set to high values by both teams so this effect is probably not noticeable.)

For the SVS modeling effort, the IMPRINT/ACT-R team disabled the context activation component for the preceding reason. On the other hand, the ACT-R team employed context activation using the minimum associations and ACT-R default values.

Partial Matching Activation

Partial matching activation addresses the similarity between chunks in declarative memory and is the mechanism responsible for simulating erroneous retrievals such as memory slips and lapses. In ACT-R, a declarative memory retrieval request specifies both the type of chunk being requested and the attributes or slots of that chunk. The partial matching component addresses the similarity between an attribute or slot in the retrieval request specification and chunks in declarative memory. The IMPRINT/ACT-R team used partial matching only for numerical values, which provided the possibility for erroneously retrieving nearby numbers (in the ordinal sense) from memory. The IMPRINT/ACT-R team used it for all number retrievals in their model. In contrast, the ACT-R team did not incorporate partial matching in their model.

Summary of ACT-R Memory Modeling

An interesting aspect of the ACT-R memory model discussion is to compare its performance as represented in the SVS modeling effort to simpler memory models used by other HPM tools. Both ACT-R teams chose to assign high values for base-level activation. The performance impact of this is that declarative memory corresponding to highly practiced pilot tasks is quickly and reliably retrieved. The time to retrieve the chunks would vary slightly based on activation noise, but this would be similar to a simpler memory model that assumes the retrieval accuracy is perfect and retrieval time is based on a mean time and standard deviation.

Context activation was disabled by the IMPRINT/ACT-R team. In contrast, it was enabled by the ACT-R team, using the minimum number of associations. However, compared to the high values assigned to initial base-level activation, the context activation would be of little or no consequence in the chunk retrieval process. Thus, a simpler memory model that ignores this effect would perform similarly. Lastly, for number retrievals only, the IMPRINT/ACT-R team enabled partial matching, which allows for retrieval errors between similar numbers. This effect would not be readily incorporated into a simpler memory model and thus is the only factor that really stands out compared to simpler memory models used by other HPM tools.

AIR MIDAS

The memory structure implemented in Air MIDAS is referred to as the updateable world representation (UWR) and models both long-term memory and working memory. Long-term memory is represented by a semantic network and embodies the declarative knowledge of the pilots. The time to retrieve a chunk from long-term memory is assumed to be constant and specified by the fetch time parameter.

Working memory is the subset of UWR that can be interfered with, forgotten, and replaced. A chunk in working memory becomes active when it is referenced by a task or activity or when it is updated by perception. A chunk leaves working memory when the activity that references it is completed, working memory capacity is reached, or duration of time is exceeded. In terms of implementation, working memory chunks are subject to decay by default. (The modeler can override the default settings such that individual chunks do not decay). Activation level for a chunk is determined by the decay rate on each simulation time step. When a chunk's activation level falls below a retrieval threshold, the chunk is marked unretrievable and is no longer available for information retrieval.

Working memory capacity and the time duration that chunks remain in working memory—the functions associated with interference and decay—are user-specified parameters. Air MIDAS assumes the "model human processor" of Card, Moran, and Newell (1983) for these parameters. Decay rates are assumed to be linear and a single retrieval threshold applies to all working memory chunks. Lastly, Air MIDAS has standard values that specify how long it takes to store visual and auditory stimuli in working memory.

D-OMAR

For the SVS modeling effort, the D-OMAR team assumed perfect memory and did not explicitly model memory retrieval. The surface operations model included the influence of episodic memory on action selection. In D-OMAR, episodic memory represents an element of autobiographical memory and is a source for habit-based action by having experiences from similar situations (that worked well in the past) guide the current response.

A-SA

The memory component of A-SA is straightforward and does not attempt to represent individual chunks of memory. Recall that A-SA predicts a value of SA between 0 and 1, with 1 representing perfect SA. The memory component focuses on the decay of SA over time. Qualitatively, it represents the notion that preservation of SA is a resource-demanding process requiring rehearsal and is not likely to be maintained at a high level, particularly when events irrelevant to SA are encountered. Situation awareness is assumed to decay faster (time constant of 5 sec) when irrelevant information is encountered because of interference with this rehearsal. Situation awareness is assumed to decay more slowly (time constant of 60 sec) in the absence of irrelevant information and when no new relevant information is available, representing uninterrupted decay of working memory.

VISUAL ATTENTION

Developing a capability to predict visual attention was a key aspect of the SVS modeling effort for most of the teams. Visual attention describes how the pilot moves attention between displays and between items on a display to extract information from the external environment. During the monitoring task, the pilot periodically views each display in a manner that we refer to in this chapter (using D-OMAR parlance) as a *background scan pattern*. Interestingly, each modeling team took a unique approach to modeling the background scan pattern.

During the background scan, certain values encountered from the external environment trigger reactive or procedural tasks (as do certain auditory events), which take precedence over the background scan pattern and direct visual attention elsewhere. Once the reactive or procedural tasks are finished, the background scan is resumed. The resultant visual attention allocation is thus the combined effects of the background scan pattern and the reactive or procedural scans, typically discussed in terms of fixations and dwells. A "dwell" is defined as the time period during which a fixation or series of continuous fixations remain on a particular display.

Note that none of the teams attempted to emulate "perception" or "pattern recognition." Text and numeric values were simply assumed to be read into memory when a fixation of sufficient duration occurred. For nontextual and non-numeric information, each of the modeling teams devised ways to convert the context or content of the information into text or numeric parameters within the model. For example, the OTW representation of "runway in sight" was typically converted to a Boolean variable. Furthermore, none of the teams attempted to address visual workload issues related to data fusion. For example, the SVS display presents runway alignment rather easily via a three-dimensional rendering of the runway and flight path predictor, whereas the traditional display requires piecing together several display items to develop an equivalent mental picture. A summary of the representation of visual attention is shown in Table 10.7.

ACT-R

For the SVS modeling effort, the primary goal of the ACT-R team was to model visual attention as thoroughly as practical with minimal tuning of the default ACT-R parameters. In order to do this, they needed to model bottom-up (data-driven) and top-down (knowledge-driven) factors. The primary bottom-up factor was the layout of the displays and the associated effort of acquiring information from the displays. This was particularly pertinent when predicting the background scan pattern when SVS was available since the SVS display contained redundant information from the

---

TABLE 10.7
Visual Attention Capability Synopsis

HPM Tool	Visual Attention Representation
ACT-R	Background scan based on the information needs of the pilot as determined by the team's internally generated task analysis and the effort of acquiring information from the displays
IMPRINT/ACT-R	Background scan employed the learning mechanism in ACT-R to learn both the information needs of the pilot and the effort of acquiring information from the displays
Air MIDAS	Background scan derived from eye-fixation data and augmented with scan policy for certain segments of flight
D-OMAR	Background scan consists of three scan types—basic PFD scan plus heading, ND scan, and OTW scan; frequency of these scan types determined by calibrating with eye-fixation data
A-SA	Visual attention is represented by the SEEV model, which asserts that pilot attention is influenced by two factors: (1) bottom-up capture of salient or conspicuous events, and (2) top-down selection of valuable event information in expected locations at expected times. In contrast, attention is inhibited by the bottom-up level of effort required, which encompasses the effort of moving attention elsewhere and the effort required due to concurrent cognitive activity

PFD (e.g., altitude, airspeed) and the ND (e.g., heading). With SVS, pilots can expend less effort (minimize cost in ACT-R parlance) if they attend to the nearest redundant information. Although ACT-R has a mechanism for updating the cost/effort during run-time (used by the IMPRINT/ACT-R team following), the ACT-R team estimated the cost/effort a priori based on the physical layout of the displays and saccade (rapid voluntary eye movements used to move from one fixation to another) latency and accuracy.

The key top-down factor in the model was the information needs of the pilot based on the ACT-R team's internally generated task analysis. This task analysis was decomposed into three high-level tasks (*aviate, navigate,* and *manage systems*) and their corresponding monitoring loops:

- The *aviate* monitoring loop consisted of monitoring the speed, altitude, flap settings, and vertical speed.
- The *navigate* monitoring loop consisted of monitoring the flight trajectory, heading, location along flight path, and distance to next waypoint.
- The *manage systems* monitoring loop consisted of monitoring the MCP to confirm entries and monitoring the modes of the flight management system (FMS) to confirm LNAV and VNAV.

How frequently the three monitoring loops are executed relative to each other identifies the information needs of the pilot for the monitoring task. The ACT-R team calibrated the frequencies heuristically until they achieved a good match with visual attention allocation data from *one flight segment of one SVS run* from the NASA human-in-the-loop (HITL) simulation. The complete process is as follows: Calibrating the frequency of "aviate," "navigate," and "manage systems" monitoring loops (with respect to each other) causes an adjustment to the information needs of the pilot, which causes an adjustment to the background scan pattern, which, combined with event-driven, reactive, or procedural scans, results in the overall allocation of visual attention. The calibration process is repeated until a good fit of visual attention is achieved.

Note that this was the only calibration performed. Model predictions for the other conditions were generated without additional visual attention calibration. Assuming that the model predictions are indeed good for these other conditions, minimizing calibration is desirable because it indicates that the underlying methodology is robust and ultimately more useful than an approach that requires continuous tweaking for each new condition. Furthermore, the model's applicability to other aviation studies can be conducted without significant model "redevelopment" time.

IMPRINT/ACT-R

Compared to the ACT-R team, the IMPRINT/ACT-R team took a fundamentally different approach to modeling visual attention by enabling learning in ACT-R. As mentioned previously (chapter 4), to accomplish this type of learning with the standard version of ACT-R 5.0 was not possible. Instead, many aspects of the ACT-R 4.0 hierarchical reinforcement scheme, which was based on the goal stack from ACT-R 4.0, were implemented in a variant version of ACT-R 5.0.

The learning mechanism had two components to it. The first component learned the *information needs* of the pilot during the background scan by assigning "success" to rewarding information. The modelers defined rewarding information on two levels: first, that the information was changing and, second, that the information resulted in the pilot taking some sort of action (e.g., setting the flaps). (By comparison, the ACT-R team used calibration to identify the information needs.) Information on the display that is neither changing nor results in action is a "failure." The pieces of information (e.g., altitude, airspeed, etc.) were initialized identically at the beginning of the run in terms of success (i.e., no successes or failures yet recorded). The pieces of information selected for the pilot to attend to early in the run were purely random.

Over time, the successes and failures of the subgoals (whether the information was changing and whether it triggered pilot actions) became more important in terms of guiding the selection toward rewarding information. However, since the successes decay according to a power law process, the recent successes and failures carried more weight in guiding the current selection. Indeed, the IMPRINT/ACT-R team believed it was critical to use the decay form of the probability of success and cost equations (see chapter 6) because this enabled rapid adaptation to changing information needs.

The second component is the *choice of display* component and only applied when SVS was available. This component was specifically designed to address why some of the pilots from the NASA HITL simulations relied mostly on the ND and PFD, whereas other pilots relied on the SVS (which in addition to the three-dimensional rendering of the terrain also contains an overlay of information, such as airspeed and altitude, that are redundant with the PFD). The IMPRINT/ACT-R team hypothesized that once visual attention was directed toward a particular display, the availability of redundant information (due to the presence of SVS) would increase the probability that the next display item needed would also be found on that display. Thus, a pilot could make his or her background scan pattern more efficient by choosing the next display item (assuming it is one of the redundantly available pieces) from the currently attended display since less time and effort are required to move a small distance within the current display than to move to a different display altogether.

To implement this, production rules were developed to levy a "cost" corresponding to the time and effort of moving attention to the next display item—the further the distance, the higher the cost. Two production rules were necessary to apply the cost mechanism. One production rule specified the traditional displays (ND and PFD) as the location of redundantly available information; the other specified SVS. Initially, the production rule selected was determined purely randomly (with each production rule having a 50% chance of being selected) corresponding to which display type, either traditional or SVS, would win the pilot's attention. However, over time, as one display randomly received more attention (analogous to a coin toss resulting in multiple "heads" in a row), the time, effort, and cost of the scan would incrementally decrease since the next display item needed would more likely than not be found on that same display. The reduced cost would then incrementally increase the probability of that particular display type being selected in the future, resulting in convergence to an all-or-nothing use of the display type—it would be all traditional and no SVS or vice versa. Of course, this all-or-nothing convergence refers to the background scan patterns. The total visual attention allocation is the combined effects of the background scan pattern and the reactive or procedural scans.

A key benefit to this learning methodology is that it is possible to model and evaluate behavior in an environment for which no human performance data exist. The allocation of attention and consequent use of the interface is a *predictive* model of human behavior, rather than a descriptive model that simply emulates existing data.

As mentioned in the External Environment section, the IMPRINT representation of the ND in the external environment did not capture the heading of the flight. In the real world, a pilot would scan the aircraft's progress with respect to waypoints and attend to the ND more frequently prior to and during a heading change to ensure that the aircraft is maintaining its flight path. Without this effect modeled, the simulated pilot would attend to the ND less frequently than in the real world.

Air MIDAS

The Air MIDAS team took a completely different approach to visual attention compared to the two ACT-R teams. The modeling of visual attention in their Simulation 3 started with four representative scan patterns of the pilot task based on eye-fixation data collected from the HITL simulations. This data were represented in terms of the percentage of time the HITL pilots fixated on the various flight displays. The displays where the subjects spent significant percentages of time were the PFD, ND, OTW, and SVS (when available). The four patterns represented visual meteorological conditions (VMC) without SVS, VMC with SVS, instrument meteorological conditions (IMC) without SVS, and IMC with SVS. In addition, the NASA SVS simulation scan patterns were slightly modified to account for small differences in display mapping between the NASA SVS simulation layout and the layout of the Boeing-757 simulated by Air MIDAS.

Next, a scan "policy," based on an aircraft operating manual for the captain and FO, specified how the scan pattern should be modified to include the OTW view in the scan, depending on whether the runway was in sight and whether the aircraft had approached decision altitude (DA). The scan policy assumptions were the last step necessary to construct the background scan pattern prior to run-time.

As compared to the ACT-R team's approach, an interesting aspect of the Air MIDAS approach is that the behavior of the background scan pattern closely replicates the required input frequency of the "aviate/navigate/manage systems" monitoring loop components of the ACT-R model (recall that this was the calibration mechanism used by the ACT-R team). Ignoring SVS for the sake of discussion, the aviate monitoring loop in the ACT-R model primarily scans the PFD, the navigate monitoring loop scans the ND, and the manage-systems monitoring loop primarily scans the MCP.

By comparison, the background scan pattern of Air MIDAS specifies directly what percentage of time the PFD, ND, and MCP should be scanned. Thus, the background scan pattern in Air MIDAS acts as a surrogate for the monitoring loop task decomposition structure by *implicitly* representing the "aviate/navigate/manage systems" frequency of execution. However, the Air MIDAS approach is more dependent on eye-fixation data to define the background scan pattern (the ACT-R team needed only one flight segment from one scenario) and might not as accurately generalize to a different cockpit setup or scenario.

Another aspect of visual attention that the Air MIDAS team modeled was the failure to acquire information during fixation if the duration of the fixation was below a threshold. The threshold was determined heuristically to provide a 10% failure in fixation and applied to background scans, but not reactive or procedural scans. Thus, 10% of the fixations from the background scan acquired no information from the external environment, while directed fixations always succeeded.

Lastly, the Air MIDAS team assumed that the speed and altitude values from the SVS were not used to trigger any reactive or procedural task. This is due to the original concept of operations for SVS that stated that SVS was *not* intended for flight directive or commanding purposes. The Air MIDAS team chose to embed this concept of operations into their model. In contrast, the other modeling teams explored the utility of violating the concept of operations by using the SVS for flight directive or commanding purposes to demonstrate the efficiency of attending to redundant information overlaid on the SVS display.

D-OMAR

The D-OMAR team took yet another approach to modeling the background scan pattern compared to the other teams. The D-OMAR team developed a "basic" scan that read speed, altitude, vertical speed, and heading. In the baseline condition (no SVS), the first three display items were read from the PFD and the latter item was read from the ND. (It is worthwhile to note that although the location of the heading display item was on a separate display, the effort to acquire it was minimal when scanning the PFD because it was situated directly below the PFD at the top of the ND.) In addition to the basic scan, a "horizontal situation" scan of the ND was developed to look at the current flight segment, next flight segment, heading, display range setting, and distance to the active waypoint from the ND. Lastly, the OTW view was scanned periodically, but at a much lower frequency.

With SVS, the basic scan was assumed to be read from either the PFD/ND or SVS in an alternating fashion. In addition, the view of the runway was added to the scan of SVS prior to descending below the cloud cover. The horizontal situation scan and OTW scan were unchanged from the baseline, although their frequency was impacted by the addition of the scan of the SVS.

The intervals between scans for each of the three scan types (basic, horizontal situation, and OTW) were controlled by three user-specified variables, respectively. These variables were tuned against the NASA HITL simulation data such that the three scan types collectively utilized all the pilot's available time (assuming a segment of the flight when no other tasks were being performed). This defined the background scan pattern. As with the other models, scanned display items were used to trigger reactive or procedural tasks, which took precedence over the background scan pattern and directed visual attention elsewhere. Lastly, transitions from display to display were not specified by the modeler, but rather determined based on trying to maintain the desired intervals while interleaving between the background scan and reactive or procedural scans.

A-SA

Visual attention in A-SA is represented by the SEEV model. SEEV asserts a pilot's attention is *influenced* by two factors: (1) bottom-up capture of salient or conspicuous events (visual or auditory), and (2) top-down selection of valuable event information in expected locations at expected times. In contrast, attention is *inhibited* by the bottom-up level of effort required, which encompasses both the effort of moving attention elsewhere and the effort required due to concurrent cognitive activity. Under ideal conditions, a skilled pilot should not be influenced by salience (unless this is explicitly correlated with relevance) or inhibited by the level of effort required.

The probability of attending $P(A)$ to a particular flight deck display or OTW view is represented by the following equation:

$$P(A) = sS - efEF + (exEX + vV)$$

where coefficients in upper case (S for *salience*, EF for *effort*, EX for *expectancy*, and V for *value*) describe the properties of a display or environment, while those in lower case describe the weight assigned to those properties in directing an operator's attention. The values for these coefficients were estimated by the A-SA team, although in the surface (taxi) operations model, expectancy was defined by the frequency of discrete events in the time-based script.

For the SVS modeling effort, visual attention allocation was predicted for four flight segments of the approach. SEEV parameters were estimated for each of the four segments. They modeled the *effort* of moving attention from one display to another by assuming that such effort is monotonically related to the distance between displays. The *value* of a display was equal to the value of the task served by the display multiplied by the relevance of that display to the task in question. The value of the task was based on the priority of the *aviate* and *navigate* high-level tasks to the flight segment. For example, when the pilot took manual control of the aircraft in the last segment, these priorities increased dramatically. The *expectancy* for information contained on the display was determined by bandwidth (rate of change) of the display. Note that the A-SA team did not model the *salience* of events in the SVS modeling effort because there were few salient events to capture during approach.

WORKLOAD

This section discusses the capabilities of the modeling tools for predicting workload. The purpose of having a workload metric is to gain an understanding of how busy the operator is over some time period of interest. Workload predictions could be used to assess new cockpit technology and procedures in an HPM environment prior to any prototype development to assist in the design process. Perhaps the ultimate goal for HPM workload predictions is to have the capability of replicating objective task load measurements and/or subjective workload ratings so that the number of HITL

TABLE 10.8
Workload Capability Synopsis

HPM Tool	Workload Representation
ACT-R	No explicit measures of workload available as an output, but workload constraints are still present in the system (e.g., if task demands exceed available resources, some tasks or subtasks will not be completed).
IMPRINT/ACT-R	See above.
Air MIDAS	Workload estimates based on the VACM resource theory of workload. Average workload over a flight segment was calculated to compare baseline condition to SVS.
D-OMAR	Estimates of workload are not currently computed, but trace data are available to support the assessment of workload.
A-SA	No attempt to model the workload of action-related phenomena. Does model the effect of cognitive load on attention and SA.

experiments could be reduced in favor of less costly HPM experiments. A summary of the representation of workload is shown in Table 10.8.

ACT-R AND IMPRINT/ACT-R

For the SVS modeling effort, there were no explicit measures of workload available in the models developed by ACT-R or IMPRINT/ACT-R other than the visual attention allocation already discussed. The lack of explicit workload metrics is more of a bookkeeping issue than a theoretical one. In the course of a run, ACT-R essentially characterizes the human operator's state every 50 ms, including the resources used. Thus, it would be possible to add the required code to generate some type of workload metric based on the percentage of time human resources are allocated to tasks and activities versus the percentage of time these resources are unused and available. Workload constraints are still, however, present in the system. If task demands exceed available resources, some tasks or subtasks will not be completed. Thus, excessive workload will produce performance errors in ACT-R models.

AIR MIDAS

Workload in Air MIDAS is based on the scale in Table 10.5 to provide workload estimates compatible with the VACM resource theory of workload. Numerical VACM scores above the threshold of seven per channel are typically not seen in Air MIDAS because the scheduler accounts for the excessive demand and postpones or sheds tasks accordingly.

Average workload is a useful parameter for characterizing relative differences between a baseline condition and a condition that represents new procedures or new displays such as SVS. Average workload over a time period of interest (such as a flight segment) is expressed by:

$$WL_{i\ avg} = \sum_{j=1}^{n} \frac{WL_{i,j}\Delta t_{i,j}}{t_{total}}$$

where

    i is the VACM channel

    j represents each task element executed by the model (e.g., move hand to mode control panel, read value from display)

    n is the number of task elements executed over the total time duration

    $WL_{i,j}$ is the VACM workload of each task element

    $\Delta t_{i,j}$ is the duration of the task element

    t_{total} is total time duration to average across

Note that the $WL_{i,j}$ values are VACM channel inputs assigned by the modeler for every possible task element to be performed, whereas $\Delta t_{i,j}$ is calculated during run-time using micromodels of human performance such as Fitt's law or speech duration based on the number of words in the message.

A similar equation was used to categorize workload by type of procedure or decision such as the "land or go around" decision, performing the landing checklist, and performing the background scan pattern.

D-OMAR

Explicit workload assessment is not an integral part of a D-OMAR model. D-OMAR does provide trace data that could readily support a variety of approaches to work-load assessment, leading to a comparison between conditions or scenarios. However, this was not done for the SVS modeling effort.

A-SA

A-SA makes no attempt to model the workload of action-related phenomena. How-ever, it does attempt to model the effect of cognitive load on attention and SA. For example, for the surface operations model, cognitive load for an event is incurred for some time after the event is first processed. As a new event occurs, the attentional resources available to process it are inversely proportional to the sum of residual attentional weights from previous events, which effectively enables cognitive load to modulate current attentional control. In addition, cognitive load of distracting events degrades SA with a more rapid decay (see Memory section).

SITUATION AWARENESS

Situation awareness is of considerable importance in commercial aviation because a study of accidents among major airlines revealed that 88% of those involving human error could be attributed to problems with SA as opposed to problems with decision making or flight skills (Endsley, 1995). Situation awareness is often discussed in terms of Endsley's (1988) three levels:

- Level 1 SA—perception of the elements in the environment;
- Level 2 SA—comprehension of the current situation; and
- Level 3 SA—projection of future status.

For the SVS modeling effort, the background scan patterns developed by the modeling teams address Level 1 SA. One can assume that if the pilot interleaves his background scan with other tasks requiring attention, Level 1 SA would be maintained. With the exception of A-SA, the modeling teams did not attempt to predict Levels 2 and 3 SA as an explicit metric in their models. However, the ACT-R team did attempt to model the go-around decision, based on an integration of information that included trajectory alignment components as well as human predispositions. This decision could be treated as an implicit form of SA.

For the surface operations modeling effort, both captain and FO were modeled in Air MIDAS and D-OMAR. Again, although no explicit measures of SA were generated, the activity of the FO had an impact on the crew interactions, which had an impact on the SA of the captain, which had an impact on whether the pilot made the correct turn at each intersection encountered. This again could be treated as an implicit form of SA. See Error Prediction section for more discussion on this topic.

A-SA

A-SA was specifically designed to address SA. The SA component of A-SA is discussed in the belief-updating module (see chapter 4 and chapter 9). The model structure is designed such that the value of SA is used to guide subsequent attentional scanning, as indicated by the following equation from the surface operations version of the model:

$$W = S + SA * V$$

where

W is the attentional weight of a stimulus/event
S is the salience of the event
SA is the current value of SA (between 0 and 1)
V is the information value

Thus, good SA guides attention toward relevant information, whereas poor SA guides attention toward conspicuous, but less relevant, information. A-SA also predicts errors of SA; see the Error Prediction section.

LEARNING

Learning or adapting is a key feature of certain HPM tools. Typically, these learning models are academic in nature and attempt to represent the "change in behavior" process a human experiences due to interactions with the environment under a controlled set of conditions. Learning applications include arithmetic, skill acquisition, and problem solving.

IMPRINT/ACT-R

Of the HPM tools discussed in this book, only ACT-R features learning as a core capability. There are dozens of references available to describe the many facets of learning explored with ACT-R (see http://ACT-R.psy.cmu.edu/publications).

The IMPRINT/ACT-R team was the only team that employed learning for the NASA HPM project. Learning was developed as a means to predict visual attention for the baseline and SVS conditions to compare with the pilot eye-fixation data collected from the NASA SVS simulations. Although one could argue that there may have been some learning involved with the SVS display (since the subject pilots had limited training with SVS prior to the runs in which data were collected), this was not the primary intention of the IMPRINT/ACT-R team when they developed their learning approach. Rather, they were trying to develop a robust capability for predicting visual attention under both the baseline and SVS conditions that could be applied:

- In a scenario-independent manner;
- In the absence of detailed information needs of the pilot; and
- Without requiring calibration against human performance data.

In addition, their approach could be extended to other phases of flight including takeoff and cruise with minimal modification to the learning/visual attention methodology.

The full discussion of IMPRINT/ACT-R team's approach is provided in the Visual Attention section, but there is another point of discussion worth presenting here. The IMPRINT/ACT-R team's approach is different from that of the ACT-R team, which used an "information needs of the pilot" approach to predict visual attention. To develop a *robust* modeling capability, a learning approach is a less arduous and time-consuming process than an information-needs approach, and this certainly has many advantages, particularly since many HPM projects are conducted under tight funding constraints.

However, using the information-needs approach can improve the efficiency of the background scan pattern during the time period around discrete events because certain pieces of information can be explicitly assigned more weight during these time intervals. For example, if an aircraft is descending and is required to level out at its assigned altitude, an information needs approach can more realistically represent a human pilot's scan pattern, which would more frequently scan the display for altitude and vertical speed as the aircraft nears the assigned altitude to see that the rate of descent is slowing and then stops. If the aircraft were to "bust" the assigned altitude, the information-needs approach would more likely result in a quicker response for corrective action than the learning approach.

Of course, the learning and information needs approaches are not mutually exclusive, so these explicit adjustments to the background scan pattern could be made in the IMPRINT/ACT-R model. However, it would somewhat defeat the purpose of having implemented learning in the first place. This issue aside, the learning approach offered by the IMPRINT/ACT-R team provides a fresh perspective to the prediction of visual attention.

A-SA

Learning is represented indirectly in A-SA by its assumption that expertise brings about selective attention that is more closely driven by the expectancy and value components of the SEEV model than by salience or effort. Adhering to this "expected value" model provides optimal scanning that is found to be a predictor of better multitask performance (see Application 3 in chapter 9).

RESULTANT AND EMERGENT BEHAVIOR

This section describes the capabilities of the modeling tools for demonstrating resultant and emergent human behavior. Resultant and emergent behaviors are often confused in this context and the difference between the two is worth mentioning. Resultant behavior is that which could be "reasonably expected" from a collection of objects and the interactions between them based on an understanding of their individual behavior (Auyang, 1999). In contrast, emergent behavior is that which is "truly unexpected." For both the SVS modeling and surface operations efforts, it is the opinion of this chapter's first author that the behavior predicted by the HPM tools demonstrates resultant rather than emergent behavior.

In most HPM efforts, by the time the modeling teams have constructed their models and understood the implications of the underlying assumptions (for both domain-specific and first-principle behaviors), the behavior predicted should be reasonably expected. If this is not the case, then it is very likely that there is a bug in the model. Indeed, in the early rounds of this cross-model comparison effort, we noticed unexpected pilot behavior predicted by two of the five teams' SVS models, both of which were traced back to bugs in their models that required fixing. Resultant rather than emergent behavior is most readily apparent in the Error Prediction section, where "priming" of the models was necessary to induce the types of errors the modelers hoped to see. Thus, to an outside observer, what may appear to be unexpected behavior is in fact reasonably expected behavior to the informed analyst.

That said, we do believe truly emergent behavior can be produced through human performance modeling if the HPM considers multiple agents, perhaps with different strategies, that are interacting in a complex system. For example, airlines that run hub-and-spoke operations have very different strategies for their aircraft fleet compared to airlines that run point-to-point operations. An HPM effort that considers how these strategies interact in response to uncertainties in the National Airspace System (e.g., ATC or aircraft equipment failure or diminished capacity due to thunderstorm activity or special use airspace) could certainly produce emergent behavior in terms of both pilot and airline decision making, although the models to do so would be significantly more complex than the models discussed here.

For the surface operations modeling effort, the resultant behaviors predicted by all the teams' models were types of taxiway errors (wrong turns or missed turns) made by the simulated flight crew. The different team approaches to modeling these errors are discussed in the Error Prediction section. In contrast, for the SVS modeling effort, the teams had more leeway in choosing problems to study based on their interest and expertise (e.g., emphasis on workload vs. visual attention allocation). Thus, the resultant behaviors predicted by the modeling tools are more varied and are discussed further next.

ACT-R

The key resultant behaviors demonstrated by the ACT-R model were allocation of visual attention allocation and the transition of attention from one display to another. In particular, the ACT-R model revealed that pilots use SVS as a proxy for the PFD and ND, at least in the early phases of flight, entirely as a result of the symbology overlaid on the SVS.

IMPRINT/ACT-R

The IMPRINT/ACT-R model demonstrated pilot monitoring behavior adapting to the frequency of changing information on the various displays. Another resultant behavior demonstrated that some pilots adopted SVS as their primary display while others chose still to rely on the PFD. Both of these resultant behaviors are discussed extensively in the Visual Attention section of this chapter. Lastly, the IMPRINT/ACT-R model demonstrated decisions by the pilot to go around or land based on the aircraft's offset from the runway and the pilot's ability to see the runway at decision altitude.

AIR MIDAS

Key Air MIDAS resultant behaviors include pilot response times in reaction to key events such as "runway in sight," "go around due to an ATC instruction," and "go around due to inability to see the runway at decision altitude." The addition of SVS to the flight deck revealed that the response times were longer compared to the baseline display conditions (no SVS).

D-OMAR

The D-OMAR team found that the SVS essentially created a second attitude instrument (i.e., an integrated display showing horizon, pitch, altitude, altitude rate, and speed) on the flight deck that resulted in pilots devoting more time to the attitude instruments and less time to navigation. In response to this, they modeled an enhanced SVS that combined the functionality of the PFD and SVS in a single instrument and showed that the pilot's attention could be redistributed among the displays to maintain "pre-SVS" scan behavior. Another resultant behavior demonstrated was the decision by the pilot to go around or land based on the aircraft's offset from the runway, distance to the runway threshold, and a personal attribute of the pilot related to his or her priority to land.

A-SA

The key resultant behavior revealed by the A-SA model was that a pilot's adherence to an optimal scan pattern (strong contributions from the expectancy and value components of SEEV) does impact multitask flight path tracking error and could impact traffic detection latency. In addition, the effort required to scan between more distant displays plays only a small role in inhibiting those more distant scans.

VERIFICATION AND VALIDATION TECHNIQUES

Human performance models and modelers must have credibility with decision makers in order to influence the design and development of new cockpit technologies. Verification and validation are necessary steps in achieving this credibility. During the *verification* phase, the complete HPM software package (including integration with the external environment) should be tested as extensively as possible to establish that it performs as intended by the modeler. For example, verification should demonstrate that the scheduler schedules correctly, the random number generator produces truly random numbers, and the communication link between the HPM and the external environment works as designed. In the strictest sense, verification means the model is bug free. In reality, there may be bugs in the model that do not affect the performance or output of the model to the extent that they can be detected through methodical analysis. Proof of verification is usually not explicitly required in a modeling effort such as the NASA HPM project, but rather assumed as a normal part of the model development and analysis process. However, sometimes lack of verification is evident when model results are counterintuitive.

The *validation* phase focuses on the ability of the model to provide sound predictions within certain bounds. Of course, "sound predictions" and "certain bounds" are subjective. The bounds could include phase of flight, weather conditions, flight deck configuration, or types of procedures. The classic approach to validation (often referred to as *results validation*) is to model some baseline scenario in which some type of human performance data is available (e.g., fixation percentages) for comparison. If the differences between the model predictions and the baseline data are unacceptable, the model is tuned or modified by trial and error as needed (often referred to as *calibration*), rerun, and compared against the data once again. This process is repeated until the comparison between predictions and data is deemed acceptable. The model is then run for another scenario or condition where data are available but have not been previously utilized for comparison or calibration. If the comparison between calibrated model predictions and the new data is acceptable, the model is considered validated. At this point, the validated model can be used to make predictions within certain bounds.

Results validation is not always feasible to perform due to the difficulty of acquiring human performance data, particularly for systems that do not yet exist. It can be a Catch-22: Decision makers want validated models to guide future technology development, but since it is a future system, no directly applicable human performance data exist. To address these issues, the Defense Modeling and Simulation Office (http://vva.dmso.mil) has outlined alternative forms of validation specifically to tackle human behavior representation issues that, while not as rigorous as results validation, do present a process for attaining model credibility.

The following subsections discuss the validation techniques of the five HPM teams' efforts based on results presented in their respective chapters. It should be noted that validation based on only three subject pilots should be treated with care since HPM tends to focus on average human performance and, with only three subjects, it would be statistically feasible to have all pilots either perform above or below the average. Thus, the modeler may assume that there is a problem with the model when in fact the model may be behaving quite well.

ACT-R

The validation techniques used by the ACT-R team were to calibrate the model at a high level of abstraction (i.e., visual attention allocated to each of the six types of displays—PFD, ND, etc.) and then validate it at a lower level of abstraction (i.e., transitions matrix—transition of visual attention from one display to another display, 6×6 matrix for baseline, 7×7 matrix for SVS). Specifically, they used one flight segment of one condition of the NASA SVS data to calibrate the information needs/visual attention component of their model (see the Visual Attention section for details). They chose the SVS condition of the IAF to FAF flight segment for calibration, which resulted in an r-squared fit of 98% for the fixation percentages (i.e., percentage of time that the pilot fixated on each of the displays) spread over the six types of displays.

Next, for validation purposes, they executed the model for three flight segments to predict fixation percentages and transition matrices. The resulting r-squared fit for the fixation percentage data explained about 85% (baseline, start to IAF), 68% (SVS, start to IAF), and 93% (baseline, IAF to FAF) of the variance, respectively. The resulting r-squared fit for the transition matrices explained 79% (baseline, start to IAF), 42% (SVS, start to IAF), 77% (baseline, IAF to FAF), and 69% (SVS, IAF to FAF) of the variance.

IMPRINT/ACT-R

The IMPRINT/ACT-R team had three steps in their validation process. In the first step, they demonstrated at a high level that the ACT-R model achieves the same outcomes as the human pilots for the 10 scenarios tested in the NASA simulation. In other words, if the outcome of a scenario for a human pilot was a go-around, then the ACT-R outcome was a go-around as well. The outcome possibilities were land, sidestep and land, or go around.

Next, the IMPRINT/ACT-R team looked at detailed model traces of the pilot activities during the course of the ACT-R simulated approach and compared them with the human pilot activities collected from the NASA SVS simulations. To do this, they transcribed the videotaped scenarios described in chapter 3 and the corresponding timestamp. They focused on the following approach procedures: (1) responding to ATC instructions for approach and initial configuration of aircraft for approach, (2) descending from 3,000 ft and performing flap and speed settings, (3) performing landing checklist procedures, and (4) performing final approach procedures including decision to land at decision height. Note that this step does not meet the classical quantitative definition of *results validation*, but instead is an example of qualitative validation techniques sometimes used by the HPM community when desired quantitative data are not available.

The final step of validation was comparing visual attention allocation from the IMPRINT/ACT-R model with the subject pilots. They used dwell percentage as the visual attention metric. The IMPRINT/ACT-R data corresponded to approximately 280 sec, representing the latter half of the approach. This was compared to a weighted average of Phases 3 and 4 of the NASA simulation data representing the flight segment from final approach fix to trial end. Comparing the model to Pilot 1 across all scenarios, the *r*-squared fit explained 58% of the variance. Comparing the model to Pilot 2 across all scenarios, the *r*-squared fit explained 48% of the variance. Comparing the model to Pilot 3 across all scenarios, the *r*-squared fit explained 47% of the variance. Aggregating Pilots 1–3, the *r*-squared fit explained 55% of the variance. It is important to emphasize that there was no calibration required for visual attention in the IMPRINT/ACT-R model because of the learning mechanism the team developed (see the Visual Attention and Learning sections).

AIR MIDAS

The Air MIDAS team applied several verification techniques to their SVS modeling efforts. First, they stepped through the flight profile and task sequences for nominal approaches and those approaches that resulted in a go-around. This included timing for key events such as flaps, speed setting, and communication. Next, they provided a detailed task sequence corresponding to three events: (1) the timing of "runway-in-sight" callout, (2) go-around due to ATC, and (3) go-around due to pilot decision.

Before discussing additional verification techniques used by the Air MIDAS team, it is worthwhile reviewing a particular aspect of their visual attention modeling. The scan pattern in Air MIDAS is an input to the model (as opposed to being derived by the model as was done by the other HPM teams) based primarily on eye-fixation data provided, with some minor modifications to account for scan strategy. In the initial SVS effort (corresponding to Simulation 2 in chapter 7), the scan pattern was based on eye-fixation data from Mumaw et al. (2000). In contrast, for the most recent SVS effort (corresponding to Simulation 3 in chapter 7), the scan pattern was based on eye-fixation data from the NASA SVS simulations discussed in chapter 3.

For Simulation 2 using Mumaw's (2000) data for input, verification of Air MIDAS was performed by comparing the model's prediction of fixation percentages against the Mumaw data. The *r*-squared fit explained 99% of the variance for both baseline and SVS conditions, which was evidence that the input data for the scan pattern was interpreted correctly by the model. In addition, this demonstrated that the input scan pattern was essentially identical to the output visual attention predictions of Air MIDAS (Mumaw data input = Air MIDAS output). Next, the Air MIDAS team validated this model by comparing their visual attention predictions for Simulation 2 (Mumaw input) against the NASA SVS simulation data. However, to do so required the modeling team to make several assumptions to extend the model from the conditions tested in Mumaw to the new conditions tested in the NASA SVS simulation. (Mumaw had neither an SVS condition nor a sidestep condition.) These assumptions were made based on the information needs of the pilot that were required to carry out the necessary procedures.

The Air MIDAS team also made the assumption that SVS was to be used only as a secondary display. Using fixation percentage as the metric, the *r*-squared fit for the baseline IMC (no SVS) condition explained 58% of the variance. The *r*-squared fit for the IMC–SVS scenario explained 43% of the variance. The *r*-squared fit for the SVS with sidestep condition explained 31% of the variance. Lastly, it should be noted that the Air MIDAS team did not compare their visual attention predictions developed during their Simulation 3 against any data set. But, recall that in this effort they were using the NASA SVS simulation data as the primary input to the scan pattern, so the NASA SVS simulation data could not also be used for validation.

D-OMAR

The D-OMAR team developed four graphical display tools to assist in the verification process. As detailed model behaviors were developed, these tools were used to assure that the behavior represented reasonable, human-like aircrew performance. First, the event recording subsystem captured data to display in the simulation control panel's trace pane. Second, a plan view display of the aircraft's progress along its flight path provided a high-level view that the pilot was flying the aircraft as intended. Third, a Gantt chart display provided a lower level view into the pilots' goals and procedures as orchestrated in time to produce actions. Lastly, an event timeline provided detailed insight into the crew's behavior.

For model validation, D-OMAR predictions of visual attention were compared against the NASA SVS simulation data using percentage dwell time as the metric. No r-squared statistics were provided, although the model predictions look reasonable when compared to the three subject pilots (see Table 8.1, chapter 8).

A-SA

The A-SA team took an approach to model validation different from those of the other teams. Instead of relying on the NASA SVS simulation data for validation, they used it as a stepping stone for validation against data collected at the University of Illinois. The Illinois simulation data were collected for eight pilots flying eight experimental conditions (four display configurations conducted in both IMC and VMC), providing a larger set of data to provide statistical evaluation. In addition, the simulation afforded the capability of assessing traffic detection time and flight path tracking error as a function of the pilot's adherence to an optimal scan pattern. The A-SA team determined the coefficients of the model a priori and did not calibrate these parameters to achieve a better fit. The predictions of visual attention allocation were compared against the Illinois simulation data. Using dwell percentage as the metric, the r-squared fit across all eight conditions and all eight pilots explained 86% of the variance. For individual conditions, the r-squared fit explained between 33 and 99% of the variance; six of the conditions had an r-squared fit of 80% or higher.

SUMMARY

This chapter is intended to provide comparative insight into each HPM tool discussed in this book. A comprehensive range of modeling capabilities provides a discussion platform to highlight similarities and differences in the HPM tools and approaches while providing the reader with an up-to-date perspective of human performance modeling in aviation. The authors of this chapter hope that this cross-model comparison advances the state of HPM by allowing a cross-fertilization of ideas and techniques for future modeling efforts.

As should be evident to the reader, the architecture of the modeling tools does have bearing on how readily core capabilities such as workload, visual attention, crew interactions, procedures, situation awareness, and error prediction can be modeled and predicted. Some representative examples are discussed next:

- Multiple operator models (e.g., captain and FO) are more easily accommodated by Air MIDAS and D-OMAR. Thus, if flight crew or pilot–ATC interactions are expected to be significant drivers to a future modeling effort, then the Air MIDAS and D-OMAR tools would be more straightforward to apply.
- The A-SA and ACT-R models focused specifically on what drives visual attention from a bottom-up as well as top-down perspective. Hence, if visual attention allocation needs to be understood for a particular technology, the ACT-R and A-SA modeling tools would more easily facilitate the analysis.
- Only the Air MIDAS model provides an explicit workload metric as an output of the model. However, the D-OMAR and two ACT-R models have workload constraints (via human and system resource limitations) embedded in their models, so metrics reflecting the underlying resource utilization information could be output from their models with some modification.
- Multitasking is explicitly supported by the frameworks of Air MIDAS, D-OMAR, and the attention module in A-SA. On the other hand, it is a cumbersome process to model in ACT-R.
- Complex memory modeling and learning are available in the IMPRINT/ ACT-R model, which can be useful in modeling memory errors and processes where little human performance data exist, respectively.
- A-SA is the only HPM tool to provide an explicit situation awareness metric. Indeed, a worthwhile objective to pursue would be to incorporate the A-SA framework into the other modeling tools.

Each of the HPM tools discussed in this chapter has extended its capabilities significantly to support the SVS and surface operations modeling efforts. An evaluation of new aviation technology beyond these efforts would most likely require modifications to the models to account for the specific details of the new technology, associated procedures, and perhaps different phases of flight such as departure or cruise. However, this would be a smaller effort compared to the work already expended during the NASA HPM project for the development of the external models to represent the aircraft flight dynamics, flight deck displays, and the communication link between external environment and HPM tool (recall that the teams expended between 25 and 70% of their effort to represent these functions). The teams that connected to higher fidelity flight simulators (the ACT-R and Air MIDAS teams) should be acknowledged as they are poised to tackle problems in which the closed-loop behavior between the pilot's action and the aircraft's response (and vice versa) is a key factor.

Of course, the respective modeling tools and capabilities are dynamic. Each successive modeling effort in a complex environment such as aviation most likely adds to a tool's repertoire of capabilities. Human performance modeling has come a long way, but still has a long way to go as well. It is likely that in a few years the HPM tools discussed in this document will be able to further demonstrate enhancements applicable to aviation modeling. Lastly, it is useful to keep in mind that HPMs do a better job of predicting relative measures rather than absolute measures. Thus, the question, "Is cockpit A better than cockpit B?" is easier to answer than "Is cockpit A ready for full-scale development?"

ACKNOWLEDGMENTS

We wish to thank Michael Byrne, Kevin Corker, Stephen Deutsch, Alex Kirlik, Christian Lebiere, Dan Schunk, and Chris Wickens for their assistance in answering our many modeling questions during the course of this cross-model comparison effort.

REFERENCES

Auyang, S. Y. (1999). *Foundations of complex-system theories: In economics, evolutionary biology, and statistical physics*. Cambridge, England: Cambridge University Press.

Card, S. K., Moran, T. P., & Newell, A. (1983). *The psychology of human–computer interaction*. Hillsdale, NJ: Lawrence Erlbaum Associates.

Corker, K. (2005). Computational human performance models and air traffic management. In B. Kirwan, M. Rodgers, & D. Shafer (Eds.), *Human factors impacts in air traffic management* (pp. 317–350). London: Ashgate.

Corker, K., Fleming, K., & Lane, J. (2001). Air–ground integration: Dynamics in exchange of information for control. In L. Bianco, P. Dell'Olmo, & A. Odoni (Eds.), *New concepts and methods in air traffic management* (pp. 125–142). Berlin: Springer.

Endsley, M. R. (1988). Design and evaluation for situation awareness enhancement. *Proceedings of the Human Factors Society 32nd Annual Meeting* (pp. 97–101). Santa Monica, CA: HFES.

Endsley, M. R. (1995). A taxonomy of situation awareness errors. In R. Fuller, N. Johnston, & N. McDonald (Eds.), *Human factors in aviation operations* (pp. 287–292). Aldershot, England: Ashgate Publishing Ltd.

Horrey, W. J., Wickens, C. D., & Consalus, K. P. (2006). Modeling driver's visual attention allocation while interacting with in-vehicle technologies. *Journal of Experimental Psychology: Applied, 12*, 67–78.

Leiden, K., Laughery, K. R., Keller, J. W., French, J. W., Warwick, W., & Wood, S. D. (2001). *A review of human performance models for the prediction of human error* (Tech. Rep.). Boulder, CO: Micro Analysis & Design.

McCracken, J. H., & Aldrich, T. B. (1984). *Analyses of selected LHX mission functions: Implications for operator workload and system automation goals* (Tech. Note ASI479-024-84). Fort Rucker, AL: Army Research Institute, Aviation Research and Development Activity.

Mumaw, R., Sarter, N., Wickens, C., Kimball, S., Nikolic, M., Marsh, R., Xu, W., & Xu, X. (2000). *Analysis of pilot monitoring and performance on highly automated flight decks* (NASA Final Project Rep.: NAS2-99074). Moffett Field, CA: NASA Ames Research Center.

11 Human Performance Modeling
A Virtual Roundtable Discussion

David C. Foyle, Becky L. Hooey,
Michael D. Byrne, Alex Kirlik,
Christian Lebiere, Rick Archer, Kevin M. Corker,
Stephen E. Deutsch, Richard W. Pew,
Christopher D. Wickens, and Jason S. McCarley

CONTENTS

INTRODUCTION

In this chapter, a series of questions are posed to each of the five modeling teams regarding their individual models, the challenges they faced, and their reflections on the National Aeronautics and Space Administration Human Performance Modeling (NASA HPM) project. The reader has a unique opportunity to join a virtual round-table discussion with members of the five modeling teams, gaining some perspective on the take-home lessons and insights that the modeling team members had after completing the NASA HPM project. Thirteen questions are discussed in this "virtual roundtable" covering five areas. These five areas of discussion are: general modeling issues, the impact of model structures and architectures on the problem being addressed, the use of models to address aviation issues, model results and validation, and a look back at the NASA HPM project efforts and a look ahead to the future of human performance modeling in aviation.

These five general topic areas and themes of each of the detailed questions are:

- General modeling issues
  - Why model?
  - Who is the model user?
- Model structures and architectures
  - Impact of model architecture and structures
  - Representation of the external environment
- Using models in aviation
  - Domain information requirements for modeling
  - Modeling error and performance
  - Discovering latent errors, and rare unsafe conditions
  - Addressing aviation issues
  - Reusing models
- Model results and validation
  - Model assumptions and "scripting"
  - Model validation
- Models: The past and the future
  - Successes and difficulties
  - Future directions and challenges

GENERAL MODELING ISSUES

WHY MODEL?

Modeling a problem requires detailed inputs (e.g., subject matter experts [SMEs], task analyses, and detailed data). Critics have argued that this rich data set might be sufficient to understand the human operator's behavior, and have questioned the added value of the modeling efforts. What did you learn from the model effort that was not already knowable from SMEs, task analyses, and human-in-the-loop (HITL) data? In other words, to what extent did your model:

- Accurately predict behavior versus replicate HITL data;
- Produce emergent behavior; and,
- Allow for extrapolation to other nontested display/procedural conditions?

Byrne and Kirlik (ACT-R)

We know of no examples in science in which a data set, however rich, is itself sufficient to "understand" a phenomenon without the use of a theory or model to impose structure on the data. Even simple curve fitting and analysis of variance (ANOVA) represent the use of a model and many assumptions to move from data to conclusions, which are thereby necessarily tentative and model-dependent. The question of what we learned that was not actually "knowable" is actually quite peculiar. Knowable by whom? An omniscient being? If that is the context, then, of course, we could not have learned anything new. For mortal creatures lacking omniscience, the task of moving from SME verbalizations, task analyses, and behavioral data to conclusions about these scenarios required some sort of conceptual structure or model.

Given the richness and high dimensionality of these data, we decided to bring to bear a similarly rich modeling structure on them in order not to trivialize the dynamics of the situation. While our results are clearly tentative and model-dependent, by making our model (a logical device for moving from data to conclusions) public and explicit we allow others to inspect our reasoning and find fault where they may. The alternative would seem to be to provide all of these data to some so-called expert and ask for his or her intuitive conclusions. That is a lot cheaper, but those conclusions are tentative as well, and there is no possibility of publicly examining the *method* by which the expert reached those conclusions. This is a crucial problem, since science is inherently a public, error-correcting social enterprise.

Lebiere and Archer (IMPRINT/ACT-R)

Much of our modeling work was based on predicting rather than "postdicting" data. While we did use some data to constrain the model, much of our results are direct predictions of the architecture applied to the task.

The core of our model was based on the ACT-R learning mechanism for production selection. The first emergent behavior was a round-robin type of monitoring loop that was sensitive to the frequency of change of the various displays. This learning approach replaced a hardwired round-robin monitoring loop in a previous version of the model. The second emergent behavior was a primary reliance on either the primary flight display (PFD) or the synthetic vision system (SVS), depending upon the course of learning. This replicated the empirical results, showing a dichotomy between some crews adopting the SVS as their primary display while others chose still to rely on the PFD. These two emergent effects are particularly noteworthy because, while they arise from the same architectural mechanism, their effects are in many ways opposite: The first one makes sure that all individual displays get their proper share of attention while the second is a winner-take-all competition between bundled sources of information. But they arise from the constraint of each task (maximizing the amount of new information vs. minimizing the time taken to access it, respectively) as reflected through a general-purpose architectural learning mechanism.

Applying the model to new display and procedural conditions is relatively straightforward. The task network model has to be modified to reflect the new display and procedural conditions, and the cognitive model has to be updated to the extent that the changes introduce new elements not present before (e.g., a new maneuver or a new display component). Beyond that, as described earlier, the model should be able to adapt to the new conditions. A major benefit of the principled separation between task network model and cognitive model is to create a clear separation between assumptions about the environment and the cognitive processing, allowing each model to be changed without affecting the other. As a measure of the generality of our model, the learning approach to attention management and display selection can be viewed as a general solution to a much broader multitasking problem of great current interest to cognitive modelers.

Corker (Air MIDAS)

Why model—especially if you have access to analytical and human-in-the-loop data for the phenomena you are modeling? The standard answer to that has been documented throughout the history of human performance modeling. Systems are modeled because in so doing you understand the system in a way that observation does not provide. A system model allows you to make explicit statements about the causal structure in your model and to manipulate that causal structure in ways that are unavailable in the full systems (especially the case in human/system models). You model the system in order to understand the limits of generalizability of your model. You model a system because, once you have established some level of validity and generalizability, you can manipulate the model to explore procedures and operating modes that were not otherwise available.

Emergent behavior is a difficult term to define in computational models. Is there anything that can be considered "emergent" behavior from computational models? In some sense all behavior (including learning) is determined by the mechanisms, rules, weightings, heuristics, algorithms, and decision processes represented in the model. However, it is the case that when multiple agents are set in interaction with each other and those algorithms and heuristics interact, it is possible that the complexity of that interaction in time is not able to be anticipated by the modeler. In this sense, emergent or new behavior is exhibited by the model.

Models of human performance in large-scale and complex systems have long served engineers in prediction of system performance. They have also been used to predict requirements for aiding systems to augment human performance and assure safe system operation. Human performance models (HPMs) have also served the human factors and cognitive sciences by establishing a platform for the embodiment of architectural and functional representations of key human operator characteristics. In order to describe the new process and procedures for this evolving concept safely and effectively, the human operator's performance must be clearly and consistently included in the design of the new operation and of any automation aiding proposed to help the operators in their distributed activities.

Deutsch and Pew (D-OMAR)

It is striking just how different the answers to this question are for the taxi versus the SVS modeling efforts. For the SVS modeling effort, we would readily argue that the model results were predictive: They produced results similar to the HITL trials. The data used to *build* the HITL experiment were sufficient to support the building of the HPM experiment; HITL experiment *results* did provide further insight into subject behaviors.

The story with respect to the taxi modeling effort was very different. Nominal aircrew taxi procedures, like all aircrew procedures, are very robust. As implemented in the model they led to essentially error-free performance that was in no way predictive of the high error rates generated in the HITL trials (or, as we understand it, in the real world). In response to this striking mismatch, we then probed the robust behaviors of the model to look for the windows of opportunity through which the errors observed in the HITL experiment might emerge. For any given action, there can be one or more sources for the intention for that action, some leading to correct responses and others leading to errors. In one case that we examined (trials in which subjects incorrectly turned *away* from their designated gate) we were able to identify habit-capture as a likely source of the error. Theoretical foundations established to build the D-OMAR HPMs supported the analysis and understanding of the sources of the errors seen in the HITL trials. By probing the model in this manner, we were able to create situations that in fact led to errors similar to those observed in the HITL experiment trials.

As modeled in D-OMAR, the aircrew's approach, landing, and taxi procedures have the flexibility to address a reasonable range of event sequences as seen in the various taxi and SVS scenarios, but the models do not produce new behaviors or address unforeseen events.

A central finding from our early SVS model analysis was the subject's shift in time allocation between monitoring aircraft *attitude* and *position* when the SVS was added to the flight deck. Based on the HPM trials, we found that a second attitude instrument on the flight deck led the subjects to devote more time to attending to aircraft attitude. When we looked back at the HITL data, we found a similar adjustment in subjects' scan patterns. We formed the hypothesis that the shift in scan pattern was the product of having two attitude instruments and that combining the functionality of the PFD and SVS in a single instrument would lead the subjects to return to their baseline scan pattern. We modeled an enhanced SVS that combined the functionality of the PFD and SVS in a single instrument. In the enhanced SVS trials, the model did in fact resume the baseline scan pattern as predicted. The D-OMAR model trials led to a finding not previously noticed in the HITL experiment data and thereby led to a new line of investigation that quickly and efficiently developed a solution to the observed problem.

Wickens and McCarley (A-SA)

The key elements revealed by our modeling that would have been far more difficult to discern from the raw data provided were:

- The strong contributions of the expected value components of the model to visual scanning (attention allocation), demonstrating the optimality of scanning behavior and its relation to performance; and
- The relatively low role of effort conservation in inhibiting scanning between more distant displays.

WHO IS THE MODEL USER?

Some argue that for human performance models to be a useful tool, they must be put in the hands of those doing the work, including designers, researchers, subject matter experts, and acquisition specialists—not just those of computer scientists. Others, however, argue that human performance models are tools designed to be used by computer scientists or human performance modelers who understand the limitations and assumptions of the model, and should not be widely distributed. That is, one would not put a computation fluid dynamics model in the hands of a NASA human factors researcher and expect him or her to design an airplane. Which position do you support and why? Ideally, what skills, experience, and knowledge should one have (or be trained in) to develop valid models and interpret their output?

Byrne and Kirlik (ACT-R)

Scientific and engineering models are tools for well-trained scientists and engineers. In fact, very little science or engineering can be done without models. But we would disagree that computer scientists will remain the end users of human performance models in the future. Initially, it was only computer scientists who used JAVA, HTML, or even the Internet itself. As tools and technologies mature, they will increasingly become used by professionals who find that it is cost effective to solve their problems with models, as computational fluid dynamicists have done. The real problem for the future of human performance modeling, we believe, lies not in the possible benefits such models could provide—which are significant—but rather in whatever economic forces are currently operating to render technology-driven, rather than need-driven, design nearly the only viable option. Perhaps we need our models to aid technologists in synthesizing designs (this is the market), rather than merely evaluating designs before they are used (there is little market for this).

Lebiere and Archer (IMPRINT/ACT-R)

As usual in dichotomic debates, the right position often ends up being somewhere in the middle. While it is unreasonable to expect users without any computational or modeling experience to be able to develop highly complex human performance models, the current state of the art where human performance modeling tools are only accessible to highly trained users with a Ph.D. in cognitive science or human factors is a crippling barrier to the practical success of the field. One would expect the domain of application of human performance models to be much broader than that of fluid dynamic simulations because HPMs should be relevant to most complex artifacts being designed rather than just a small, specialized class.

A key development in HPM tools would be to embed their assumptions in such a way that only working knowledge of the system, rather than a theoretical knowledge of its underpinnings, is required for proper use of the tool. For instance, to use a cognitive architecture requires basic knowledge not only of its operations, something that would be accessible to most people with computational or modeling experience, but also of complex psychological theories as to which cognitive operations allowed by the software implementation are actually in keeping with the theory.

In that approach, overlays to very general tools can be designed to make them more accessible to domain practitioners for each given domain. This approach has been used, for instance, in the development of a family of task network modeling tools such as IMPRINT based on the discrete event MicroSaint simulation engine; it is also the underpinning of our GRBIL tool that allows any interface to be evaluated against a cognitive model by simply defining it and letting the cognitive model itself be generated automatically.

Corker (Air MIDAS)

In an earlier review of models (Baron, Kruser, & Huey, 1990), the National Research Council noted the problem of providing human performance models to individuals who may not have the background to understand the implications of those models (i.e., the implications of the model structures and limitations in use). In my experience, in the recent reviews of operational concepts focused on capacity enhancements, a combination of reduction of the complexity of the controller tasks associated with the required delimitations of the modeling used to represent those tasks has resulted in a significant overestimate of the effectiveness of specific technologies.

I feel that the human performance modeling effort should be conducted by those who have both background and experience in representing human performance as part of a larger team of analysts. This team approach assures that the technologies under examination as well as the human operators that are intended to use them are appropriately represented. This is not to say that there is no value to reducing the complexity of the models and improving the speed of their implementation, as well as making the output of the models more transparent with respect to the questions they are being asked to answer. However, I think that the dream of a turnkey human performance model usable by someone without training in human performance is a misguided pursuit.

Deutsch and Pew (D-OMAR)

Each step in adapting a human performance model for a particular application can be quite difficult and is probably best done in collaboration with someone from the model building team. First, the establishment of reasonable goals for the model-based experiments can best be done by taking advantage of an insider's knowledge of a model's current and potential capabilities. Second, each problem space will demand additions to the model (e.g., the ability to model and process SVS data) that are best done by the modeling team. Third, the interpretation of model results should also have input from the model team; some of a model's numbers will be "better" than others. We would suggest that the level of effort put in by an "outsider" working with a model to contribute adequately in a meaningful way would at the same time qualify that person as an "insider"; we know of no alternate way to acquire the necessary skills to contribute meaningfully.

Theory building is a necessary prerequisite to model building; one has to start from a theory covering perceptual, cognitive, and motor capabilities; their constraints; and how they combine to produce human multitask performance. Unfortunately, this has not been the goal of any one discipline. Hence, there can be many points of approach to developing functional models: experimental, development, social psychology, cognitive science, cognitive neuroscience, philosophy, and so on. With a theory in place, one then needs some simulation skills to develop an architecture in preparation for modeling. One might be lucky and find a model that one likes as a starting point, or one might find today's offerings less satisfying and initiate a new modeling effort.

Wickens and McCarley (A-SA)

Compromise between the two positions is both feasible and desirable. Modelers should endeavor to consider and produce what we call "user-friendly primitives." These are terms in the model that nonmodeling professionals will find both meaningful and easy to calculate. For example, in our model, the parameter of expectancy bandwidth for a particular area of interest can be calculated by anyone with access to the event stream. At the same time, it is unrealistic for human factors practitioners to expect all models to be "walk up and use" operations. Prospective users should be prepared to invest substantial time and effort in understanding and becoming competent with more complex and powerful models.

MODEL STRUCTURES AND ARCHITECTURES

IMPACT OF MODEL ARCHITECTURE AND STRUCTURES

To what extent do you believe the specific architectures and structures in the various models impact the user's choice of a modeling tool, the ability to describe or predict the data, and the validity of results?

Byrne and Kirlik (ACT-R)

Ideally, a description of a human performance modeling tool or technique (including a cognitive architecture) should include a clear specification of the nature of the problem or situation it is designed to address. This is necessary to allow a potential user to choose the right modeling tool for the job at hand. We are clearly many years away from this level of maturity in the realm of human performance modeling. In part, this is due to a conflict between the goals of architecture developers to describe the widest possible range of behavior within a unified framework and the needs of human-machine systems researchers to represent behavioral situations and tasks at very high levels of specificity.

While some of the models used in this NASA research program may apply to certain problems or situations more naturally than others, one conclusion from this project would seem to be that a wide variety of models can be configured to address essentially the same phenomena, albeit in different ways. Because the insights that will be gained from applying the various modeling approaches cannot be known until situation-specific models are created and evaluated, there is as yet little support for a user to work backwards from a problem or situation to the choice of an appropriate model or architecture. So the theoretical constraints on model selection are currently quite weak. We believe that practical constraints are instead the current drivers of model selection, such as expertise with a particular modeling approach.

Because developing this expertise in most modeling formalisms is time intensive and difficult, we are not convinced that the specific features of various architectures have a lot to do with the user's choice of modeling tool. That choice is primarily driven by each user's history.

However, the choice of modeling architecture can affect the ability to describe or predict data and the validity of the results (though it does not necessarily do so). In some cases, these differences are fairly obvious; for example, if the architecture does not model down to the level of eye movements, it cannot predict or describe eye-movement patterns. In point of fact, however, in more applied contexts there are many aspects to "the data" and the choice of architecture that will tend to drive which aspects of the data are considered most central.

Validity is a more difficult animal (see also the Model Validation section on this). Architectures that constrain possible models and have had those constraints validated in a wide variety of contexts are probably more likely to produce more "valid" results in some senses, but not necessarily in all senses.

Lebiere and Archer (IMPRINT/ACT-R)

While all architectures are designed to attain a large degree of generality, in practice they also tend to focus on specific areas that reflect their origin, the interest of their creators and their most common applications. Some architectures focus on cognitive mechanisms, others on artificial intelligence reasoning and inference capabilities, and yet others on perceptual and motor mechanisms. All tend to have some functionality across the entire spectrum but are particularly strong in one or more areas. Ultimately, the field might converge on an integrated architecture representing the best of each area, but in the meantime, the best tool for any given application will reflect the architecture's chosen focus.

Within their focus area, architectures can strongly constrain the resulting model and results to the extent that knowledge engineering and parameter manipulations are kept to a minimum. Ideally, the architecture would be given nothing more than the description of the task and would learn a performance model of the task without any additional inputs, yielding parameter-free predictions of performance. In practice, especially in domains involving significant training and expertise like aviation, a representation of the cognitive agent's knowledge and strategies still has to be specified, together with some parameterization of the model to reflect individual differences and factors not yet represented in the architecture. However, modelers should strain toward the ideal of relying as much as possible on the architecture's mechanisms and constraints, especially its learning capabilities, as a way of decreasing the importance of data in the process and moving away from "postdicting" and toward predicting human performance.

Corker (Air MIDAS)

I will assume that, by "user," you mean the person who is going to select and use a model for some analysis.

In addressing the impact of architecture and structures, I will make distinction between the two. The architecture of a model, we feel, determines the methods of functional coordination for the processes that the model seeks to represent. The architecture of the model provides the process by which time is represented, by which knowledge is represented, by which the external world and its systems are represented, and by which the human performance functions are represented, and the mechanisms by which these elements interact with each other. The structures of a model are the elements that are chosen to be significant to the human-system function that the model is attempting to represent. Structures, then, are the way knowledge is stored and accessed or the way perception of communication occurs. "Structures" also refers to decisions as to what is to be represented about the world.

Both of these should influence the user's choice of modeling tool, as they both have an immediate impact on what can be analyzed by the model. If, for instance, the architecture of the model is one that represents events (e.g., a discrete-event simulation) as opposed to continuous time (at some representational granularity), then any processes of human-system performance that occur between the defined events will not be able to be represented. The model of the human performance in such systems is usually represented as a series of states and the probability of transition among them. Events in human performance that might be expected to occur between those transition events are not presented. As another example of the differences among models, an architecture that assumes stochastic inputs will require multiple runs of the model to determine a distributional characteristic for behaviors, while an architecture that assumes a static task network will not require these Monte Carlo analyses and may be used to ask questions about "what if" a specific configuration might occur.

The structures that are chosen determine the functional analyses that can be undertaken. If there are, for instance, no input or perceptual mechanisms represented, then some other process is required to get information from the world to interact with those model structures that represent cognition. Both Soar (Laird, Rosenbloom, & Newell, 1986) and EPIC (Meyer & Kieras, 1997) models, for some part of their early development, had no processes that represented either input or output in connection with a simulation world. They were thereby restricted in the sorts of questions an analyst could approach.

The validity of the model, however, may not be affected by these choices because, presumably, the modeler has chosen a phenomenon to represent that is compatible with the architecture and function chosen for the model. So, a model that concentrates on learning (that represents the input to that learning process in a fairly minimal way) may accurately predict the cognitive process of learning and can be validated against human learning in a similar situation. Prediction of large-scale dynamic system behavior with assumed feedback and closed-loop response requires that the architecture for the model be able to represent the system dynamics and that the structures be able to represent the modification of current model output influenced by prior system performance. Such a model may not have an explicit process representing the "learning" of the significance of the feedback, but could represent its impact and make accurate predictions of the closed-loop system behavior that could be validated against large-scale, closed-loop system behavior.

Deutsch and Pew (D-OMAR)

Potential users of human performance models are in a very difficult position. Less is known about the sources of skilled human behaviors than we like to admit. At best, models built from this limited base perform "well" within a not well-defined narrow band of situational stimuli. At the same time, very different architectures can yield very similar results; hence, good "numbers" are not necessarily indicative of "correct" models. Model validation remains an open problem with no easy solutions on the near horizon. For a model, its architecture and structure should be the instantiation of a theory concerning the source of human performance. They are factors that potential users should evaluate; they do drive the predictive capabilities of the model, but particular architectures or structures have not led to easier paths to model validation.

Wickens and McCarley (A-SA)

The response to this question strikes us as a trade-off between the ability of the model to describe the specific data provided (i.e., the model's validity), on the one hand, and the ease and generic flexibility of use that make the model accessible to other users, on the other hand. A model that is closely tailored to a particular application may show high validity, but may run the risk of requiring so much modeler expertise as to inhibit its use by others.

REPRESENTATION OF THE EXTERNAL ENVIRONMENT

Describe how your model represented the external environment and how this representation affected (i.e., either improved or limited) the model results.

Byrne and Kirlik (ACT-R)

We recognized the need very early to describe cognition and behavior within its dynamic, closed-loop context. Trying to model pilot behavior without detailed descriptions of the aircraft cockpit, aircraft dynamics, and the operational environment would have been about as successful as trying to model the behavior of a bicycle rider without models of the bicycle and terrain. In the first taxi modeling scenarios, we did not have access to computational models of the NASA simulator dynamics, although we were provided with a visual scene database. So we created our own aircraft model in a somewhat ad hoc way and were thus able to create a functioning model of closed-loop, pilot-aircraft-taxiway interaction.

Our experiences in taxi modeling, viewing errors as arising from an interaction between atypical taxiway layouts and default knowledge, only emphasized the importance of detailed environmental modeling. This prompted us to begin the second phase, SVS approach and landing scenarios, using the commercially available X-Plane (http://www.x-plane.com/FTD.html) flight simulator as an environmental model. Although we had to deal with a number of low-level programming issues associated with communication and coordination between ACT-R and X-Plane, having this closed-loop capability was crucial to our SVS modeling, especially the interdependencies between visual scanning behaviors, scenarios, and cockpit display design.

Lebiere and Archer (IMPRINT/ACT-R)

The external environment, including the airplane, its systems, and its controls (including the synthetic vision system), as well as the world outside it, including air traffic controllers, is represented using the IMPRINT task network-modeling tool. Task network modeling is well suited for representing complex systems and missions as a series of decomposing tasks triggered by events. This approach has the advantage of tractability: The system can be decomposed to the optimum level of complexity versus fidelity. Only the level of details relevant to the particular model needs to be represented. Moreover, the approach is naturally composable, allowing independent pieces to be assembled in modular fashion. The main drawback of the approach is that it is not based upon any given principles or theories. Therefore, the validity of the model developed has to be determined rather than inherited from an existing, validated simulation.

Corker (Air MIDAS)

Our model is a continuous time representation of the external world. The behaviors that a model chooses to take—that is, the action of the model—depend on several factors that can be roughly divided into "internal" and "external" with respect to the human operator model itself. The internal factors are the mission goals that the operator's model is assumed to "know"—that is, that have been defined by the modeler. Other internal factors are the priority and resource requirements to perform the action required. In the first phase of the SVS work, we linked these internal generators of model behavior to an event-based script; that is, there were no external aircraft or airspace factors driving the model. We found that the script had a limiting effect on the model's response. That is, independent of what the model did in response to a scripted event (e.g., extend flaps at a specified altitude and speed), the behavior of the model did not affect the next scripted event. This led to a condition in which the model's performance could be tracked, but other than noting overloads or action later than expected, no consequence on the simulation sequence was represented; more importantly, no consequence of the model's behavior was propagated downstream into subsequent performance demands.

In our opinion, the value of our model comes from the interaction of the external drivers of the model behavior with the internal processes representing perception, cognition, and motor behavior in response to those external requirements. By including an active representation of the aircraft (PC Plane; Palmer, Abbott, & Williams, 1997) into the simulation, the model in the later phases of our development needed to adapt its function to the interaction of the world requirements (represented by the aircraft in descent); the information generated by that aircraft (provided on the standard and SVS displays); and the internal capacities and time-constants of the perceptual, cognitive, and motor functions of the human operator. The result of the model's goals interacting with the physical dynamics of the aircraft provides a way for behaviors to have consequence and propagated effect. We feel that validation for a model of a closed-loop dynamic system depends on the system dynamics being properly represented.

Deutsch and Pew (D-OMAR)

The D-OMAR event-based simulator and representation languages enabled us to model the external environment much as we modeled the aircrews and the controllers. The principal elements included the aircraft/flight deck, ATC workplaces, airport runways, taxiways and signage, and airspace navigational aids and fixes. We modeled selectively, using more or less fidelity for individual elements as required by the particular research objectives for the designated scenario trials. We were easily able to add an SVS to the flight deck and subsequently explore aircrew use of an enhanced SVS that combined the functionality of the PFD and SVS in a single instrument. The ability to run all elements of the scenarios in a single simulator operating in fast time led to significant time savings in model development.

Wickens and McCarley (A-SA)

The external environment was represented by a series of events, each coded by the analyst in terms of the *value* of the information provided by the event, the *location* of the event in the visual field (if it was visual), and, in some applications, the *salience* of the event. The rate at which the events occurred in a real-time simulation of the scenario (video tape or computer-generated model) determined the key *expectancy-bandwidth* properties of the model. It is also possible for our model to function without an external event generator (simulation, video, etc.), if the bandwidth is simply specified as a single parameter. In this case, the analytic form of the model is used.

USING MODELS IN AVIATION

DOMAIN INFORMATION REQUIREMENTS FOR MODELING

The NASA HPM project provided extensive data sets, including task analyses and performance and eye-tracking data from human-in-the-loop simulations. However, some would argue that if you have access to this level of information and data, you do not need to build a model. To what extent was this rich data set necessary to answer questions for the aviation community versus building model capabilities that can now be reused? How would your results have differed with lesser data? For those in industry or academia who lack the resources and availability of data, what information is required at a minimum?

Byrne and Kirlik (ACT-R)

We have stated our position on the "you do not need to build a model" criticism earlier. For our team, the answer to how things would have been different with lesser data depends greatly on which effort (taxiing or SVS) is under discussion. Since the SVS had the denser data provided, we will focus on that. Frankly, without the eye-tracking data, we would have been seriously hamstrung. There is not enough variance in the rest of the data to have very much of interest to explain from an ACT-R perspective.

As for what information is required of the data, there is no single answer to that question, because what data are required depends entirely on what kinds of questions you need to answer with the model. That said, from the ACT-R perspective, the more detailed the data are, the better. Since ACT-R actually performs the task, rather than producing a metric or global estimate, the denser the data are, the better.

Lebiere and Archer (IMPRINT/ACT-R)

As mentioned previously, many of our results arise from the nature of the task combined with the constraints of the architecture. Data were used to guide the design of the model but the primary goal of our modeling effort was to generate the data rather than merely replicate it. The emergent behaviors described in our answer to the "Why Model?" question are an instance of that. The model might have differed in minor quantitative ways without the data guidance, but the primary qualitative findings would have been preserved regardless. Our general position is that the same information made available about the task to human subjects should be given to the modeler, and no more. For expert users in a sophisticated domain such as aviation, that can mean a lot of expertise to reproduce, but fortunately much of that is codified.

Our primary source of guidance in developing our model was the 767 flight manual, which codified the procedures to be used and provided the basic data used in populating the instance-based decision-making model. In tightly controlled domains with no prerequisite specialized knowledge, such as cognitive psychology experiments, an increasingly popular approach has been to give the architecture the very same instructions as the human subjects and to let it learn the task in the same way that subjects do (Anderson et al., 2004). For generic interface manipulation tasks, we have developed a general tool called Graph-Based Interface Language (GRBIL) that can generate a cognitive model of the task from just the description of the interface and a demonstration of the task to perform and then yield a priori predictions of performance on that task (Lebiere, Archer, Hamilton, & Schunk, 2005).

Corker (Air MIDAS)

Regarding the data sets supplied, our model used data from one eye-movement study (Mumaw et al., 2000) to parameterize the information-seeking behavior of the pilot. We then predicted the NASA HITL data. In addition, we generalized the model to include sidestep maneuvers that were not in the original data set (also see the "Why Model?" section).

Deutsch and Pew (D-OMAR)

The basic elements necessary to support model development and experiment design to address a target problem are information on flight deck and ATC workplace instrumentation, aircrew and ATC procedures, and airport and airspace data. The process is very much like that for the design of HITL experiments. From this point on, model building provides what has proven to be a very important contribution to the analysis process. Much of the process of building generative models is about building an understanding for the sources of human behaviors. Much of what we learn about a problem space is learned in the iterative process of building and refining a model that, ironically, simply produces the expected performance. At the points at which problems are uncovered (e.g., an altered scan pattern when the SVS is added to the flight deck), ideas put forward as solutions can be explored using the model that is then on hand. At the points at which the resultant performance is not the expected performance (e.g., the very low error rates in the modeled taxi scenarios), the modeling environment (i.e., the theory and the model itself) can provide a base from which to support further analysis to better understand actual aircrew behaviors.

The rich data set that NASA provided enabled us to move rapidly on the immediate target problems and led to models that have been reused on related aviation domain problems. The processed HITL subject data, while not essential to model development, served to help validate some model results and pointed to problems with other model outcomes (e.g., taxiway error prediction). The HITL subject data could be further exploited to refine model elements, particularly for basic perceptual and motor capabilities.

Wickens and McCarley (A-SA)

The data sets provided were absolutely essential for us to build and validate the model. However, our particular model could have been constructed and validated in reasonably accurate form even with criterion data of poorer resolution (e.g., percentage time fixated on each area of interest rather than second-by-second scan paths). In this case, as noted in our answer in the "Representation of the External Environment" section, the analytic form of the model would have been used for validation, as it has in other applications (Wickens, Goh, Helleberg, Horrey, & Talleur, 2003).

The model could likewise have been tested using predictor variables less detailed than those provided. As noted previously, stimulus events were represented by a small set of parameters (information value, location, salience, expectancy bandwidth). Values for most of these were determined by the analyst using a heuristic lowest ordinal integer strategy. The model thus requires only minimal information about stimulus parameters as input. This sparse method of coding input data also provides the corollary benefit of making the model accessible to nonspecialist users who might wish to apply it to domains beyond those tested here.

Modeling Error and Performance

The first modeling effort focused on error prediction. Why is modeling error, as opposed to performance, so difficult? Is it inherent in the nature of the problem (i.e., human error) or is it that human performance models are not suited to answer this question? Is there or is there not inherently a difference between a human "error" model and a human "performance" model?

Byrne and Kirlik (ACT-R)

It is not clear that difficulties with error modeling are intrinsic to the modeling, but rather intrinsic to the data. Errors are generally (hopefully) low-frequency events. This makes having clean error data especially difficult. Reluctance to model error data is (we think) often largely driven by issues with data collection and analysis. Field data are fraught with lack of control issues, so causality can be very difficult to establish, while experimental (or simulator) data on errors require impractically large sample sizes. Thus, most error data are less than ideal. It is thus reasonable that there is reluctance to model such data.

As for whether error models are performance models, in very important senses a good error model has to be a performance model; it would be a mistake to try to model ONLY the errors in the data. The same cognitive/perceptual/motor mechanisms that produce correct performance also produce the errors. How can separating them make sense?

Lebiere and Archer (IMPRINT/ACT-R)

Our position is that errors are just one facet of performance. Basically, error is what happens when the combination of task, system, and scenario overwhelms the capacity of our cognitive, perceptual, and motor capacities. As such, errors should not be modeled separately, but instead as part of a general model of performance. The frustration with modeling errors specifically is that they are low-probability events whose occurrences usually involve a high amount of variability. As such, they provide fairly unreliable data around which to build a model, and a frustrating target whose predictability is likely to be highly limited on an individual basis. When error is modeled as part of overall performance, however, and the inherent limitations of the dimension are taken into account, a more productive approach can result. For instance, Lebiere (2005) developed an ACT-R model of a synthetic air traffic control task for the AMBR modeling competition that predicted each type of error that occurred based on the constraints on the task and the other performance dimensions.

Corker (Air MIDAS)

Models of error take two forms, both of which have shortcomings. The first is the behaviorally emergent model of error. In this case, to make a reference back to the previous questions, error is the quintessential emergent behavior that occurs at the intersection of human performance capability and operational demand. The varieties of models that predict performance have mechanisms for variation of that performance. The application of that variety to conditions not anticipated in the model's world does not lead to the purposive processes of error. So what is lacking is a cohesive theory of error. Our models are capable of representing the processes of error as they are understood—for example, frequency gambling or confirmation bias. They are also capable of having those erroneous behaviors occur under conditions that are suitable for their performance (e.g., time- and resource-constrained performances). So in these senses, models of human error do exist and can be exercised.

However, what is lacking from the state of knowledge is when these errors will manifest; that is, in what convergence of world and model state will a specific type of erroneous behavior occur? In fact, there is so little consensus on this that some (cf. Dekker, 2001) maintain that to speak of human error at all is to miss the point—that humans are interactively engaged with their environment in adaptive behaviors that are only maladaptive through systems design inadequacy. Or to use Hollnagel's (1993) terms, human operators are excellent at coping and muddling through—that is, absorbing the impact of our propensity to error and still managing to move toward a goal. So it is difficult to envision a model of error that successfully captures this process, because we do not have good understanding of the processes of satisficing and goal adaptation.

The second method that purports to represent human error processes is discrete event simulation in event tree analyses. Here all possible errors or hazards and the points in the simulation in which they are available to occur need to be identified prior to the simulation. Then a large number of stochastic simulations are run to identify the impact of the occurrence of the error or hazard at various points in the simulation evolution. Recovery and impact as well as numerical assessment of overall system risk are the focus of these models. They can provide these estimates, but require an extensive analysis of all possible risks in advance of the system being analyzed; thus, underdefinition of the system (especially in early conceptual design) is a pitfall of such analyses.

There are several ways in which "safety risk" can be produced by a computational model. These include running the human-system model beyond a performance point that would normally be considered safe (and beyond what you would ethically perform in a simulator). You can also introduce intentional failure and see how the system recovers (if it recovers) from that failure and what the propagated effects of that failure are; these are essentially modeler-introduced stressors on system performance

There is also a set of safety risk behaviors that are unlikely to occur in simulation or in operations. Although unlikely, they can still occur in actual system performance. These are the incidents that require the convergence of a number of systems states and human responses to those states. In the rare cases when they do occur in the simulation (i.e., multicause risk incidents), safety risk can be identified.

Deutsch and Pew (D-OMAR)

We found the error modeling a unique challenge. Our approach to modeling human error was the same as that for modeling human performance: build models of aircrew and controller procedure execution representative of the robust behaviors seen in aircrews and controllers. As expected, our models for the taxi scenarios produced robust performance; unfortunately, the error rate was well below that seen in the HITL trials. Our recourse was to explore a series of scenarios that focused on a subset of the errors seen in the HITL trials. At critical points we identified potentially competing sources for action selection—some leading to correct outcomes and others leading to the errors as seen in the trials. Through this approach, we were able to reproduce some of the errors seen in HITL trials and provide insight into the reasons for those errors.

For the present, HPMs can be useful tools in helping to understand the sources of human error. With more experience in the problem space, we may find ways to refine the models to become tools better predictive of error. The high incidence of errors in the HITL trials yielded a range of particular situations that lead to error. As such, it provided a rich source of human subject data that it would be useful to further probe to better understand the vulnerabilities to error in aircrew procedures that, on the surface, appear very robust.

Wickens and McCarley (A-SA)

In our opinion, the biggest challenge for modeling error is that errors are (fortunately in most aerospace applications) very low-frequency events. As a consequence, the sample of errors from any particular scenario (e.g., the video data provided) does not provide a terribly reliable estimate of the population tendency of the circumstances precipitating performance failures. Efforts at modeling errors thus will often be frustrated by a sparsity of empirical data. This is particularly the case to the extent that models are challenged to predict the occurrence of errors at particular points in time, rather than differences in mean error rates.

Two caveats should be added to the preceding points. First, we view errors as one component of "performance," rather than something separate from performance. But predicting continuously available aspects of performance (e.g., tracking deviations) is generally easier because data are more plentiful and thus more reliable. Second, in light of the concerns regarding unreliable error data, we were gratified that the database for the first validation (taxiway errors) was remarkably well populated (high N) compared to most other forms of aerospace errors. (Voice communications errors would be another well-populated example.)

Discovering Latent Errors and Rare Unsafe Conditions

One argument that people have made is that an advantage of modeling—especially, fast-time modeling—is that it allows one to test a large number of scenarios and environmental conditions to uncover potentially unsafe conditions that occur very infrequently in the real world. To what extent are the current state-of-the-art HPMs able to identify error vulnerabilities that were not known by task analysis or HITL simulation? What is required for models to be able to uncover latent error conditions? Do we have sufficiently detailed models of human behavior to predict responses to these rare, unsafe conditions?

Byrne and Kirlik (ACT-R)

While there is likely some promise that models may ultimately be able to help identify latent error conditions that are currently not addressed by task analysis or HITL studies, we are not yet convinced that architectures are complete enough to do this in the general case. Part of the problem is scientific coverage; not all architectures are complete enough to really capture everything. However, another part of the problem is the underconstrained nature of the beast. Humans are highly flexible and adaptive and will often find multiple ways to achieve the same set of goals. Many of those ways will be idiosyncratic and infrequent, thus making them unlikely to be implemented in whatever the architecture in question is. (However, task analysis will not uncover them and they are not likely to appear in HITL studies at just the right time to generate errors.) On the other hand, there is perhaps potential to gain some leverage on slightly more common errors derived from the interaction of common approaches taken by operators to particular task-environment combinations. For some of these we probably have sufficient detail presently, but for others almost certainly not.

Lebiere and Archer (IMPRINT/ACT-R)

Current methods for identifying error vulnerabilities have significant weaknesses: Task analysis is notoriously prone to blind spots, and human-in-the-loop simulations are significantly limited in their ability to reproduce a large number of situations. Theoretically, computational cognitive models remedy both limitations: The formalization required in developing a cognitive model uncovers (most) blind spots and fast-time simulations allow the exploration of orders of magnitude more situations than human-in-the-loop simulations. This should allow for the detection of the most common error conditions that arise from the relatively normal operation of the system (e.g., increasing the workload will at some point result in an error).

However, for cognitive models to be able to detect the rarest error occurrences, additional capabilities might be required. Some factors that often play a role in catastrophic failures (e.g., fatigue or poor training) need to be represented in a principled fashion in cognitive models. A guided, rather than random, exploration of the space of possible situations might be necessary to whittle down the combinatorial space of situations. The cognitive model itself should provide the guidance necessary to uncover the situations that are likely to give rise to low-probability errors.

Corker (Air MIDAS)

We have an approach to risk assessment in which the full high-fidelity human performance model representing the controller behavior (in this case, in operations associated with time-based metering) is run. The runs are subject to variations of "stressors"—for example, changes in the direction of the wind, density of traffic, or visibility. The full model performance is then examined to identify the functions in the model that are critical in their contribution to risk performance (measured as points of closest approach, for example). These critical functions are then made efficient in code and embedded in a high-speed parallel simulation cluster. Runs are then performed in the cluster to look for the "rare event" in model and system performance environmental "stress" contexts. The large number of runs allows the estimate of the very low probability event.

Deutsch and Pew (D-OMAR)

The low frequency of errors and the even lower frequency of errors combining to produce a consequential outcome render a simple stochastic exploration of the behaviors space impractical. The robustness of aircrew procedures that employ checking and cross-checking of critical actions means that most errors will be caught, further compounding the search task. Estimating error frequency for error types can also be a problem. For some errors (e.g., discrimination of taxiway signage), their frequency might be reasonably estimated; for others (e.g., the onset of a particular intention to act), it is more difficult.

Timing is also critical. Very small variations in timing can open or close the window in which an error might occur. Timing was particularly critical in the taxi scenario in which the habit-induced error occurred. The combination of the demand on the part of the first officer for head-up time to monitor the approaching traffic and the precise moment of the ground controller's interruption of the first officer's prompt for the upcoming turn was necessary to open the window to error. It might well have been possible to generate many hundreds of runs slightly varying several of the timings and never have opened the window for the habit-induced error.

To address this problem, we employed a heuristically guided search of the space in which forced sequences of errors were generated looking for those that led to a consequential outcome. The errors currently reproduced by the model are initial examples of the product of such a process. We have identified several novel potential sources of errors and worked to create situations in which they might reasonably be expected to occur. We have taken advantage of the effect of time pressure by manipulating the timing of events to uncover sequences of errors that do in fact lead to consequential errors. For the present, this heuristically guided exploration of the error space has been manipulated by hand. In the future, it would be possible to develop a more automated heuristically guided exploration of the error space.

Wickens and McCarley (A-SA)

It is not clear that current (or even near-term future) modeling will be able to predict errors that could not be predicted as well by careful task analysis. This being said, models of visual scanning or attention allocation (such as our A-SA model) can provide compelling evidence for vulnerabilities in certain situations. For example, such a model could provide evidence that the "mean first passage time" of looks outside the cockpit (or looks to the roadway in a driving application) was above N seconds. This, in turn, would suggest that there are N seconds of total vulnerability to outside hazard, a measure that could be directly translated into a collision risk measure (Horrey & Wickens, 2004).

Addressing Aviation Issues

In terms of supporting the aviation research and development community, what issues and research questions are HPMs best able to address? What issues are HPMs not yet capable of addressing? What will it take to address those issues?

Byrne and Kirlik (ACT-R)

Without trying to duck the question, we have to say that the area is too immature to be able to draw the boundary between what is feasible and what is not. We do think it is safe to say that modeling human cognition and behavior in response to rare or unpredictable events is clearly beyond current capability; in the case of unpredictable events, it may even be impossible. Cultural, social, emotional, and physiological factors are insufficiently understood in many cases to capture their effects within a computational framework.

Lebiere and Archer (IMPRINT/ACT-R)

The issues and research questions that HPMs are best able to address at this point fall into the general category of decomposable factors. Human factors, including perceptual and motor processing, have been integrated in recent years into cognitive architectures. While their accuracy is still subject to improvement, those factors can now be accessed in a relatively easy, straightforward fashion by the modeling process. Similarly, factors such as the impact of scenarios can be explored in tractable fashion through the use of simulation. Issues that HPMs cannot currently address fall into the general category of nondecomposable factors. Those include the role of expertise and high-level cognition and errors arising from complex conditions in dynamic systems. Addressing those issues will require fundamental advances such that these aspects of human performance gain the sort of compositionality and reproducibility afforded to other factors.

Corker (Air MIDAS)

Impact of procedure and procedural change is likely the best use of our type of model. Second to that would be application of the model to understand the communication and mutual understanding requirements. Finally, for our type of closed-loop system model, time-sensitive changes and stability requirements can be effectively explored.

As noted earlier, our framework is not well suited at this stage to represent specific error emergent behavior, and is not currently adapted to assess emotions and stress. The problem is not so much that the model structures are inadequate to this task, but rather that the state of knowledge as to the impact of these processes on behavior is not well identified.

Deutsch and Pew (D-OMAR)

The evaluation of the use of new flight deck and controller equipment and procedures is perhaps the best candidate for using HPMs to support aviation research goals. Good target applications are those related to support for increased airspace capacity and improved safety. At a more basic level, models can also be used to gain insight into aircrew and controller capabilities and thereby contribute to the design of equipment and procedures. The recent research effort that provided additional insight into error vulnerabilities can support further work on improved aviation safety. Error *prediction* proved to be a more difficult problem. Additional research that drew on the National Transportation Safety Board (NTSB) accident reports and the NASA Aviation Safety Reporting System (ASRS) database would lead to model development with better predictive capabilities. Additionally, there is a wealth of human performance data that have been captured in the many NASA HITL experiments and related research, particularly with respect to perception. These data can be used as the basis for improving the representation of perceptual and motor capabilities in our HPMs; improvements would then enable the models to support finer grained decisions related to equipment design and operational procedures.

Wickens and McCarley (A-SA)

Human performance models provide scenario-based performance and error metrics that can *suggest*:

- Advantages of one display or interface configuration or format over another (e.g., SVS vs. conventional instrumentation);
- Vulnerabilities (e.g., error likelihood increases) that are imposed by certain systems or procedures; and,

- Parameters or features of a system-environment configuration whose variance is most likely to alter performance or errors (i.e., a sensitivity analysis) and, therefore are candidates for efficiently defining conditions in a pilot-in-the-loop (PITL) simulation.

We do not believe that HPMs are yet capable of fully and accurately predicting levels of performance or errors within complex scenarios. Good, reliable validation data remain too scarce.

REUSING MODELS

To what extent can the models developed in the HPM project be reused for other aviation applications? What level of effort would be required to reapply the models to answer different research questions or evaluate different systems? For example, now that you have developed an approach-and-landing scenario, what would it take to modify your model to address a slightly different question (for example, a change in crew procedures on approach and landing, or a different kind of display) during approach and landing (same phase of flight, different question)? What would it take to modify your model to address a different phase of flight, say takeoff? What is it that characterizes a new problem as a modeling "tweak" as opposed to a whole new problem? How can model reuse be made more efficient? What should the HPM community be striving for along the lines of data-input tools, data structures, scenario definitional standards, data protocols, and others that would make HPM more efficient?

Byrne and Kirlik (ACT-R)

We think there may be some possibility of reuse, but it would be piecemeal and perhaps somewhat limited in the case of a new task like takeoff. The biggest limiting factor is the knowledge engineering problem; if we had to model takeoff, we would have to do a lot of it from scratch because we do not know how to take off. However, we would not have to start completely from scratch because we would be able to reuse the simulation software (X-Plane) and the communications. We could probably reuse some of the model, too—particularly some of the lower level visual access routines. However, all the knowledge in the model about what tasks to do and when to do them and when to expect certain state variables to take certain values and such would have to be built anew.

Modifications to the existing task would be easier; we would get a lot of reuse. Different displays, in particular, would not be that hard (we have actually experimented with modifying the SVS overlay ourselves). New procedures would be slightly more difficult because that would require new knowledge engineering, but certainly of lesser scope than a change to a new task.

As for what defines a new task versus a tweak, we would again go back to knowledge engineering. If what the pilots have to know is substantially different, then it is a new task. We think this is a limiting factor to which there is no obvious clear solution, though there are a few things that might help.

There are a few things that could be done to assist reuse in general. One thing that would help us be more efficient in modeling would be a standardized task analysis representation and level of detail; a good detailed hierarchical task analysis (HTA) would be great. Some kind of standardized representation of the data would be helpful. One thing that would aid reuse would be a standardized aircraft simulation package with well-defined communication links and a standardized and well-defined language for representing what is on the display. It would also be nice if it could run faster than real time, allowing for external control of the clock. We custom engineered this for the SVS task, and while what we ended up with was good, it was a fair bit of work and there were hassles along the way.

Lebiere and Archer (IMPRINT/ACT-R)

Model reuse is a general problem, especially for cognitive models. While models are often built on a general architectural framework and with generality in mind, direct reuse by the modeling community has been relatively rare. Part of it might be due to technical factors and part to social factors, but it seems that the primary culprit is that models are designed in such a way that task-specific details tend to get hardwired in the logic of the model so as to make its adaptation to another task, sometimes with little difference from the first, as difficult as creating another model entirely. While our experience developing our model and applying it to other scenarios indicated that it would be generally quite tractable to retarget it to other scenarios, phases of flight, or research questions, it is worth emphasizing the general characteristics that underlie that conviction.

One general principle to apply in modeling is modularity, both in terms of separating cognitive model from simulation of the environment and of designing each model as a set of independent functional blocks rather than a large, monolithic wad of interdependencies. This principle might seem rather obvious and has yielded great dividends in allowing very large software engineering systems to be constructed in a relatively tractable manner, but it is by and large not the norm in current modeling and simulation practice, especially cognitive modeling. Large, cross-platform repositories of freely available models would be a first step toward their integration into larger, more complex models.

A significant barrier to the efficient use of modeling is the cost of integration into existing systems and simulations, which has been estimated as being larger than the cost of developing the models itself. Progress toward general integration protocols not just at the technical level, such as the High Level Architecture (HLA), but also at the conceptual level, in the form of modeling ontologies and other such general representations would seem to be a key enabling step toward making the practice of human performance models more directly applicable to real-world problems.

Corker (Air MIDAS)

The model framework can and has been used in a range of aviation operations. Once you have invested in the development of the knowledge-based procedures and other SME-developed characteristics of air traffic management and pilotage, the specific variations for representing a particular technology of a particular procedural change are relatively easy to implement. New particularizations of the basic airmanship and ATC function can be implemented in the knowledge base of our model, while the structures (scheduling, load, activity management, etc.) of the model remain intact.

In a rough estimate, new techniques or new scenarios in aviation require efforts on the order of person-months to implement and to conduct an analysis.

We have seen a number of efficiencies beginning to have some traction in this field. The most significant of these is the ability of various specialized models to be tied together to take advantage of distributed expertise and parallel development. Integration languages (HLA, for instance) play an important role in this model integration process. There is a cost for assuring coherence and allowing integration. The second area of improvement of efficiency is the ongoing development of libraries of capability and action—for example, conflict detection and resolution as a common and shareable module for ATC simulation. Similarly, pilotage tasks that have a common structure can also be reused by several modeling teams. Scenario definitional standards and a common data dictionary would also help the development of modular efficiency.

Deutsch and Pew (D-OMAR)

Model reuse is beginning to yield noticeable economies. The approach, landing, and taxi procedures developed for the O'Hare taxi scenarios formed the basis for the model for the SVS modeling at Santa Barbara Airport (SBA). There was a new airport with new runways, taxiways, and signage; the procedures were modified to execute an RNAV rather than an ILS approach; the SVS was added to the flight deck; and scan procedures were adjusted to include the new instrument. For another NASA project, the model was adapted to simulate the 1994 wind shear accident at Charlotte-Douglas Airport, North Carolina. Coping with the weather cells was a new task addressed by the models for both the aircrews and the controllers, and once again there was a move to another airport and airspace. We are currently modeling new en route controller equipment and procedures to evaluate controller workload under increased sector loading.

The same basic aircrew and controller models operate in each of these scenarios. It is not too hard to envision stitching the model procedures together and filling the gaps to handle each of the major flight phases from takeoff to landing. In a bit more of a stretch, we are building models for a pilot and sensor operator for a UAV—essentially an aircraft with a flight deck that remains on the ground and a unique set of scanning devices whose images must be evaluated.

The efficiency of information capture drives the efficiency in getting scenarios operational and trials executed. Standard definitions on airport and airspace entities would certainly help. Information on flight deck and controller workplace configurations and procedures is usually more difficult to obtain. The modeling efforts associated with the taxi and SVS trials went as well as they did, in part, because of the excellent work that NASA did in assembling the information to facilitate model building. Finding ways to convey this basic information in more standardized form would make it more likely that this needed information would be more readily available across more projects.

Wickens and McCarley (A-SA)

Our A-SA model is something of an exception in terms of modifying it for other environments. This is because we do not model performance per se, but rather the attentional scanning required to sustain performance. Such scanning is useful, but not sufficient for predicting performance. As shown in our Application 1 (taxiway errors), additional assumptions that are not a central part of the model may be needed to actually predict performance errors. To be truly comprehensive, the A-SA model would need a series of "performance-resource functions" to translate measures of attention (visual scanning resource allocation) into measures of performance (Alexander & Wickens, 2005). We note also that our model remains a single resource model and does not have the capability of predicting the concurrent performance between auditory and visual channels, between voice and manual outputs, or between action and perception. An account of performance under these conditions would require a model including multiple resource functions.

MODEL RESULTS AND VALIDATION

MODEL ASSUMPTIONS AND "SCRIPTING"

One criticism of the human performance modeling field is that it is difficult for others to assess the validity of the model, or even know what data and theories were used to populate the model, and what behavior was programmed ("scripted") and what emerged. Arguments have been made for the need to state assumptions of a model explicitly. In your opinion, what should be included in a statement of assumptions? Is it possible to capture all of the relevant assumptions in a meaningful way? How, as a modeling community, can we address this?

Byrne and Kirlik (ACT-R)

While the problem of hidden assumptions is certainly a problem in modeling work, it is just as much a problem in any other human factors approach. Task analyses always contain assumptions, hidden or otherwise, about what the operators really do. HITL studies make assumptions about the representativeness of scenarios used and where fidelity to reality is and is not required. Psychologists find modeling assumptions more mysterious because they have less practice identifying them, but they are not especially more problematic than assumptions made other places.

That said, there are a lot of systems where the assumptions are more deeply buried than others. There are numerous formalisms where we still have no idea how the simulation actually generates the output it generates because the authors fail to provide the relevant equations (either continuous equations or state-transition information) that underlie their models. At the bare minimum, those should be laid out. A basic outline of how the task analysis is transformed into something the architecture understands should also be required. However, it is probably unlikely that we will ever have full disclosure of all assumptions because we are still at a point at which it is hard to remember every single one made in the course of constructing a complex model.

We think a lot of the "scripting versus emergent" debate is misguided. Anyone who has seriously programmed (or paid any attention when using Microsoft Word) knows that results of large programs are, in fact, often highly unpredictable. For a slightly less tongue-in-cheek treatment of this issue, see Simon's (1969) *The Sciences of the Artificial.* On the other hand, there are real differences in degree of emergence. These are not easy things to address; journal editors are not going to tolerate huge appendices listing assumptions and detailed descriptions of what was and was not scripted. One thing that would be valuable would be for there to be an expectation that all model source code be made public (maybe on the Web so that space is not an issue) so that such things could be examined. On the other hand, so few people would bother to examine such code it is not clear how much this would help.

Lebiere and Archer (IMPRINT/ACT-R)

The answer is not easy because the problem is a fundamental one. Human performance is fundamentally not reducible in the same way as the physical world. Human performance models are a complex interaction of architecture, model, and task that cannot be dissolved down to a neat, orthogonal set of equations or assumptions. However, HPM practitioners often complicate matters by adding a lot of details and tweaks to improve the fit of their model to the data. Therefore, the problem and its solution are not primarily about exposition but instead methodology. Architectural assumptions are usually, if not simple, at least fairly well defined and formalized. Models should be constructed in an incremental fashion, with each assumption clearly stated as it is introduced in the model, not post facto when it is time to describe the model. This would encourage modelers to be as parsimonious as possible with the complexity of their model and would play a counterbalancing role to the drive for better and better fits by any means necessary. Other countervailing forces include openness in the form of making all models publicly available, cross-validation where part of the data is withheld for testing model validation and preventing overfitting, and model reuse to encourage generality over specialization.

Corker (Air MIDAS)

The assumptions about human performance need to be made explicit; however, in keeping with the "Who Is the Model User?" section, simply stating the assumptions presumes that the reader or user is sufficiently expert in human performance to understand the implications of the assumptions so stated. We feel this is not a reasonable assumption. In addition to the assumptions (including degree of "scripting") that are associated with a model, I think the modeling community has a responsibility for interpretation for those limitations and assumptions. For instance, to say that certain tasks are constrained in the model to be "serial" in performance (perhaps largely for modeling constraints as opposed to performance constraints) needs to be elaborated to say that this limitation will have impact in underestimating the possibility of dual task interference and overestimating the required time for performance. Further, I feel that the implication of these over and under estimates needs to be provided in terms of the operational impact of their implementation. For example, in this case, there may be an overoptimistic estimate of the impact of information overload and possible confusion leading to error; at the same time, there may be an artificially high constraint on the efficiency of the implemented technology.

Deutsch and Pew (D-OMAR)

We do not believe that models will be made meaningfully transparent at any point in the near future. For a group that is open to using human performance modeling to address a class of problems, we would suggest that the best way to go forward is through collaboration with a model-building person or group. A successful collaboration will require effort by all parties: The modelers will have to make their theory, implementation, and model applicability understood and the domain experts will need to achieve a necessary comfort level with the theory, implementation, and level of flexibility. Basic functional capabilities of the model will need to be evaluated and perhaps revised as domain-specific capabilities are implemented. There will be much to learn for each side in a new collaboration and, for that, there is no substitute for time and effort. An after-action review on such a collaborative effort might well shed more light on just what is needed to help others understand and evaluate a model's capabilities and shortcomings.

Wickens and McCarley (A-SA)

This issue relates closely to question about who the model user is. We see two plausible functions of a model:

- It can be used by professional modelers to understand the cognitive processes that underlie human performance and to answer design questions. In this context, we assume that the modeler will fully state any assumptions that are made in model operations and document the extent to which any arbitrary "scripts" are built into the model exercise.
- It can be used by the nonmodeler in the sort of context described in our answer to who the model user is. This will be feasible to the extent that user-friendly primitives are sufficient to operate the model, and that the model's computational architecture is simple, intuitive, and visible. Detailed documentation of the model's assumptions beyond those that are strictly required to understand and apply the model's predictions may be counterproductive, imposing an undue burden on the user.

MODEL VALIDATION

In the human performance modeling community, there is little consensus as to what validation entails for complex human performance models. What does validation mean to you? What do you consider the minimum requirements for validation? In complex, dynamic scenarios and domains where it is difficult to isolate single factors, is the goal of validation even a reasonable expectation?

Byrne and Kirlik (ACT-R)

Validation is certainly a reasonable expectation as long as the validation criteria are considered in light of the uses to which models will be put. Here we think it may be important to contrast how cognitive models are used in psychology and cognitive science and how they could or should be used in the design and evaluation of human-machine systems. While bad science typically makes bad engineering, this does not mean that the criteria for validating an engineering model are the same as the criteria used in science. These criteria may overlap, but science and engineering each has its own special requirements and some requirements that the other does not have. For example, the need for a scientific model to shed light on theory is one often discussed criterion that may not be required of an engineering model.

At the same time, engineering has special needs too, and these needs will often not be automatically met by applying solely scientific criteria to evaluating models. Perhaps most importantly in the current context is the need for the model or modeler to be able to say something about interesting and important design. This means that the model must have resources capable of systematically demonstrating behavioral variance due to environmental design. Note that this is not necessarily a feature of many cognitive models used for psychological research. We raise this issue to suggest the possibly counterintuitive point that purely scientific criteria for model validity can be both too high *and* too low a bar for evaluating human performance models in engineering contexts.

Lebiere and Archer (IMPRINT/ACT-R)

Validation is a relative term. By definition, no model will ever be a perfect reflection of reality, and the discrepancies between model and reality increase with the complexity of the task. An important aspect of advancing a model is to emphasize which dimension of the model and its predictions are fundamental, and which are incidental and constitute a sometimes substantial approximation for purposes of tractability and because they are largely unrelated to the focus of the model. It is also important to distinguish between qualitative findings, which can be discussed in binary terms, and quantitative results that inherently carry margins of error. Which findings are definite predictions and which carry a necessary amount of imprecision should be made clear at the outset. In that context, validation means ensuring that the human performance falls with a given probability (e.g., 95%) within the bounds of the model's predictions, both in terms of predicted qualitative effects and quantitative performance measures. As such, a model can only be validated against a given data set.

One advantage of cognitive architectures is that their constituent mechanisms have been validated against many data sets and thus bring significant validation weight with them. Their assembly into a given model of a particular task, especially when that task involves dynamic interactions between the constituent mechanisms that might not have been tested before, needs to be validated against a data set of that specific task. However, the constraints of the architecture have reduced degrees of freedom in such a way that the model's validation is likely to be broader than that of an ad hoc model developed entirely for that task.

Corker (Air MIDAS)

Yes, validation (as distinct from verification) can be achieved by establishing a test case against which the simulation can be evaluated. Validation is usually validation of model performance and prediction against some source of information outside the model itself. Validation of performance—against data produced by humans in similar operational contexts, against historical data, or against another model's output—speaks to the output of the model corresponding to the output of some other measured system. The "process" by which this behavioral correspondence is achieved may not be part of the validation effort.

The goal of validation is very important to the utility and acceptability of model results. It is achievable if you look to the output at some level of composition (as opposed to the internal mechanisms of the model). Construct validity is difficult to achieve in computational model paradigms. But performance validity is achievable, assuming that you have a data source against which to compare the model's output.

Deutsch and Pew (D-OMAR)

Simulation-based studies should always be an exercise grounded in pragmatics with respect to goals. Appropriate questions are: What are the goals we are trying to achieve? Can simulation provide a tool to help achieve those goals? That is, is the fidelity of the behaviors of the simulated entities good enough so that the answers provided can reasonably be expected to be representative of real-world situations? More particularly, for human performance models, are the critical aspects of the model a good enough representation of the relevant human behaviors? With respect to validation, the issue of the range of the applicability of a model is not often addressed explicitly. For the present, pragmatics suggests a narrow form of validation with respect to particular model capabilities with respect to particular goals. Rather than validation per se, it might be best to establish the "usefulness" of the model for the proposed purposes of its use as a reasonable criterion.

Given this stance, we would not claim to have produced a validated model; yet, the usefulness criteria were sufficient to lead to valuable contributions. Having previewed the error rates in the HITL taxi scenarios, we were not optimistic about the chances of producing similar rates from the model. As partial compensation for this shortfall, the model and its underlying theory proved to be a valuable tool in understanding the various sources of error that did occur in the HITL trials. On the other hand, the model did prove accurate in comparing pilot performance in the baseline and SVS scenarios; range of applicability is clearly an issue. It further supported the trials with the enhanced SVS. Setting aside the difficulties in establishing a working definition for HPM validation, if HPMs can be judged good enough to help answer the questions at hand, their use can provide an alternative or supplement to HITL studies.

Wickens and McCarley (A-SA)

In our view, validation can be defined on at least two separate levels (Wickens, Vincow, Schopper, & Lincoln, 1997):

- At one level, *correlational validation* generates mean performance predictions across different conditions and then provides product-moment correlation values between predicted and obtained performance (or, in our case, visual attention) scores. Two issues are relevant here. (1) Most such efforts entail only a few conditions (e.g., in our Application 3, there were eight conditions), so statistical significance may be difficult to achieve. The absolute value of r (or r-squared, the variance accounted for) measures, however, is quite important. (2) Models *can* be made to look artificially good by the selection of "unchallenging" validation conditions. For example, almost any workload model can produce a very high correlation if it is validated against a "no load" condition, an easy condition, and an extreme-overload condition.

- At a second level, *time line validation* can be accomplished by predicting the occurrence of events in real time as they unfold. Some form of time series analysis is critical here, and this technique has the advantage of being able to validate the model within a single condition (in contrast to correlational validation). However, there are great challenges for error models here because of the issues noted in our answer to the question regarding modeling error and performance: Actual error occurrences are not reliable estimates of population means.

MODELS: THE PAST AND THE FUTURE

SUCCESSES AND DIFFICULTIES

What did not go well in this research effort? What do you wish you had done differently? Given the hard lessons you have learned, what guidance do you have for others who might consider funding (or participating in) future model comparison efforts? We encourage the authors to air some of their "dirty laundry."

Byrne and Kirlik (ACT-R)

We probably had to spend more time and effort creating our own solutions to coupling the cognitive and environmental components of our model than we would have liked. Although the relatively recent additions of perceptual and motor modules to ACT-R have been a tremendous advance, the current version of ACT-R is tailored pretty much to HCI or experimental psychology applications. In these applications, the range of perceptual and motor units is quite limited, and the bandwidth of communication between the head and the world is reasonably low. As Salvucci's (e.g., Salvucci, Boer, & Liu, 2001) work on driving and our own aviation work demonstrate, there are significant challenges associated with scaling up the world from a desktop computer to an automobile or aircraft.

Lebiere and Archer (IMPRINT/ACT-R)

Probably the most difficult and problematic part of our modeling effort was reproducing the environment, including aircraft, controls, and scenario, in our task network model. This required quite a bit of work with comparatively many degrees of freedom in the representation and manipulation of those constructs. Our experience with a prior model comparison effort—the Agent-based Modeling and Behavior

Representation (AMBR; Gluck & Pew, 2005) Project—established the importance of having every modeling team interact with the same simulation as the one with which the human subjects interacted, for three reasons: to avoid duplication of efforts in reimplementing the task, to remove degrees of freedom in the implementation of the task, and to constrain the cognitive model to interact with the task in the exact same way as human subjects. We have also found it useful to focus the modeling teams' attention on a set of common experimental findings that all teams have to address, avoiding the pick-and-choose approach all too common in modeling efforts.

Corker (Air MIDAS)

We find that "technique" variations are difficult to manage in our model framework. In a goal-based system, the coding of control structures to achieve the same goal in various ways is a difficult thing to do. Why an operator would choose one process of performance over another is not clear and it is not something that subject matter experts are particularly good at explaining. The issue of manifestation of expertise is not something we are particularly well suited to developing. Another area of concern, related to expertise, is the impact, at a systematic level, of emotion, tension, and shortcuts taken in a procedure because of time pressure. We can postulate certain changes of behavior as people move from strategic to tactical to opportunistic behavior, but how that change is controlled is difficult to represent in an appropriately fluid manner. This also leads back to the question of validity: How can the impact of these emotion-based processes be validated?

Deutsch and Pew (D-OMAR)

Our sense was that our research effort went very well. NASA provided two challenging real-world problem spaces and the information needed to get our models operational quickly. Building models that are predictive of error needs more work, yet we felt the use of models as tools to help understand the sources of error was a good contribution. Modeling SVS usage identified an unexpected impact on scan behavior and led to a suggestion for a revised instrument design with subsequent model trials that indicated the revised design would restore the baseline behavior. We were well satisfied with the results of the research effort.

Wickens and McCarley (A-SA)

The biggest frustration was the lack of data across a large number of pilots and subjects. In particular, in our Application 2, where the attentional component of our model could be most directly evaluated via the scanning data, only three pilots' data were available. Furthermore, this challenge was compounded because, in Application 2, there were few opportunities for errors (see our answer to the question regarding modeling error and performance). It was for this reason that we decided to generate the database for Application 3, with eight pilots' data. A second frustration, which was probably unique to our model, was the absence of explicit situation awareness (SA) measures. We assumed scanning to reflect Endsley's (2000) Stage 1 SA, but other measures of SA (Stage 2: understanding; Stage 3: projection) were not available.

FUTURE DIRECTIONS AND CHALLENGES

What are the future directions and challenges for (a) human performance modeling, in general, and (b) human performance modeling in aviation, specifically?

Byrne and Kirlik (ACT-R)

One could conceivably write an entire book on just the answer to the first part of this question. In the interest of not generating that book, just some brief highlights, we think that, in general, cognitive architectures will continue to incorporate a wider diversity of mechanisms, both cognitive and noncognitive (e.g., mechanisms to deal with factors like fatigue) in order to increase empirical coverage. Improvements in software (GUIs, analysis tools, connections to machine vision systems to handle more perceptual issues) are also likely to appear. On the other hand, solutions to the knowledge engineering problem will not be easy, nor will it become possible to master these tools rapidly (perhaps more rapidly than at present, but still not quickly). Human performance modeling appears to be on the upswing; it has had a solid presence at CHI (ACM's computer-human interaction conference) for many years and now seems to be growing in the HFES (Human Factors and Ergonomics Society) community as well. This is good for the field and will hopefully result in a wider variety of available tools.

Whether it can sustain this increased level of interest and investment will almost certainly depend on its ability to demonstrate either some big "wins" or sustained usage in some subdomain. Human performance models of manual control behavior were able to do exactly this, until success at modeling and automatic control became so great that manual control became less prevalent in aviation systems. Our sense is that it is probably beyond the dreams (and nightmares!) of most computational cognitive modelers to think that they will be so successful that they will no longer be needed.

Lebiere and Archer (IMPRINT/ACT-R)

Future directions and challenges for human performance modeling are too numerous even to be listed here. Therefore, we will concentrate instead on entire categories. The first direction is theoretical development. We need to develop cognitive architectures that emphasize model creation through automatic means such as learning (from experience, instructions, etc.) rather than knowledge engineering. Advantages are both theoretical, in terms of removing degrees of freedom that require endless validation, and practical, in terms of making modeling more efficient and available to a broader range of users.

The second direction is the development of models themselves, to provide an ever broadening set of models of human performance. This development will be self-reinforcing, with more available models encouraging reuse, which will in turn increase the power and sophistication of the models—in turn rewarding further model development. This virtuous cycle of infrastructure development has been demonstrated in the software domain in general and most recently and convincingly for the Internet in particular. The third direction of progress is the development of tools and applications that bring human performance modeling out of the laboratory and into the hands of users, providing a new impetus for development.

This phenomenon has been demonstrated on a smaller scale in the development of user communities for cognitive architectures such as ACT-R and Soar (Laird et al., 1986).

Two near-term challenges for human performance modeling in aviation are the development of models of high-level expertise and the integration of human performance modeling tools. Cognitive architectures provide a reasonable representation of basic cognitive, perceptual, and motor capabilities. But, just as an experienced pilot is the culmination of years of rigorous training and flight experience, representing the knowledge and skills embodied in those pilots provides a significant challenge to modeling. Meeting that challenge, however, would yield great dividends because such a model could be applied in a ubiquitous fashion in the aviation domain. As we discussed previously, at a practical level the integration of HPMs into existing aviation tools and simulations can be more work than the model development itself. Providing a general, reusable interface to cut down on the costs of bringing the models to the aviation practitioner is a significant practical, organizational, and theoretical challenge that would also pay broad dividends.

Corker (Air MIDAS)

Monitoring of the aviation system and fast-time simulations technologies are interdependent in establishing verification and validation of risk and performance assessment. With fast-time simulations, risk and behavioral assessment techniques at one (detailed) level of granularity can be highly leveraged in refining the risk assessment process. There is a potential for expansion of fast-time simulations to system-wide assessment, and this has been demonstrated by initial integration efforts. Understanding the dynamic interaction between human performance and system performance is a critical requirement in performance assessment and safety risk assessment because of the issue of increasing cost for HITL simulation at the scales needed to assess system-wide and multiagent operational concepts. Also, fast-time simulations are needed for sensitivity studies and for prediction of hazards in new operations. We propose a multitool approach that supports different query types and analysis at varied levels of system maturity and specificity.

Deutsch and Pew (D-OMAR)

Given the current state of the art in human performance modeling, reasonable models can be developed, particularly for most procedurally driven tasks. At a finer grain, models can be developed that emulate many detailed individual human capabilities. Unfortunately, these models operate within very narrow bounds: Expected cues with unexpected values or novel cues are not usually well accommodated. As in artificial intelligence, where brittleness continues to be a problem, brittleness in models places significant constraints on the usefulness of HPMs. In essence, there is much that we do not know about how, we, as people, routinely produce *robust* performance in doing the work we do. The better understanding of robustness in human performance and of how to build that robustness into model performance is the central challenge facing modelers today.

Commercial aviation has two long-standing challenges: increasing airspace throughput and improving safety. Each involves the design of new equipment and the revision of procedures to utilize the new capabilities. For each, simulation provides a tool to assist the design process and assure that anticipated gains are actually realized. To the extent that human performance models can accurately represent the roles of aircrews and controllers, they will extend the range of design and performance issues that can accurately be addressed using simulation. A near-term challenge is to build more comprehensive models of aircrew and controller capabilities for each flight phase under nominal or near-nominal conditions. The models should be readily extensible in order to address then the evaluation of new equipment and procedure designs in commonly encountered situations, such as severe weather.

Probing aircrew models to better understand the sources of aircrew error is beginning to yield new insights into the mechanisms leading to those errors. As such, modeling provides a new approach to gaining insight into how to provide equipment designs and revised procedures to further reduce the incidence of error and mitigate the effects of error.

As progress is made along these lines, the issue of model transparency should be revisited with the goal of transitioning model use to the aviation user community.

Wickens and McCarley (A-SA)

Validation, validation, validation! More particularly, validation with simulation data from multiple operators. This is true in response to both parts (a) and (b) of the question. In validating models of aviation performance specifically, efforts should be made to use pilots performing realistic tasks in realistic scenarios, and to avoid simplifications imposed in order to create "clean data." Models must be able to handle "dirty data," since it is their goal to predict real-world (extralaboratory) errors and performance.

REFERENCES

Alexander, A. L., & Wickens, C. D. (2005). Flightpath tracking, change detection, and visual scanning in an integrated hazard display. *Proceedings of the Human Factors and Ergonomics Society 48th Annual Meeting* (pp. 68–72). Santa Monica, CA: HFES.

Anderson, J. R., Bothell, D., Byrne, M. D., Douglass, S., Lebiere, C., & Qin, Y. (2004). An integrated theory of the mind. *Psychological Review, 111,* 1036–1060.

Baron, S., Kruser, D. S., & Huey, B. M. (Eds). (1990). *Quantitative modeling of human performance in complex, dynamic systems.* Washington, DC: National Academy Press.

Dekker, S. W. A. (2001). The disembodiment of data in the analysis of human factors accidents. *Human Factors and Aerospace Safety, 1,* 39–57.

Endsley, M. R. (2000). Theoretical underpinnings of situation awareness: A critical review. In M. R. Endsley & D. J. Garland (Eds.), *Situation awareness analysis and measurement* (pp. 3–32). Mahwah, NJ: Lawrence Erlbaum Associates.

Gluck, K. A., & Pew, R. W. (2005). *Modeling human behavior with integrated cognitive architectures: comparison, evaluation, and validation.* Mahwah, NJ: Lawrence Erlbaum Associates.

Hollnagel, E. (1993). Human reliability analysis: Context and control. In B. R. Gaines & A. Monk (Eds.), *Computers and people series.* New York: Academic Press.

Horrey, W. J., & Wickens, C. D. (2004). Focal and ambient visual contributions and driver visual scanning in lane keeping and hazard detection. *Proceedings of the Human Factors and Ergonomics Society 48th Annual Meeting.* Santa Monica, CA: HFES.

Laird, J., Rosenbloom, P., & Newell, A. (1986) *Universal subgoaling and chunking.* Boston, MA: Kluwer Academic Publishers.

Lebiere, C. (2005). Constrained functionality: Application of the ACT-R cognitive architecture to the AMBR modeling comparison. In K. A. Gluck & R. W. Pew (Eds.), *Modeling human behavior with integrated cognitive architectures.* Mahwah, NJ: Lawrence Erlbaum Associates.

Lebiere, C., Archer, R., Hamilton, A., & Schunk, D. (2005). ACT-R/IMPRINT integration: a.k.a. Human behavior architecture. *Paper presented at the 2005 IMPRINT developer's workshop*, Arlington, VA, December 2005. Paper available at: http://www.maad.com/index.pl/imprint

Meyer, D. E., & Kieras, D. E. (1997). A computational theory of executive control processes and human multiple-task performance: Part 1. Basic mechanisms. *Psychological Review, 104,* 3–65.

Mumaw, R., Sarter, N., Wickens, C., Kimball, S., Nikolic, M., Marsh, R., Xu, W., & Xu, X. (2000). *Analysis of pilot monitoring and performance on highly automated flight decks* (NASA Final Project Rep.: NAS2-99074). Moffett Field, CA: NASA Ames Research Center.

Palmer, M. T., Abbott, T. S., & Williams, D. H. (1997). Development of workstation-based flight management simulation capabilities within NASA Langley's Flight Dynamics and Control Division. In R. S. Jensen & L. A. Rakovan (Eds.), *Proceedings of the Eighth International Symposium on Aviation Psychology*, pp. 1363–1368. Columbus: The Ohio State University.

Salvucci, D. D., Boer, E. R., & Liu, A. (2001). Toward an integrated model of driver behavior in a cognitive architecture. *Transportation Research Record, 1779,* 9–16.

Simon, H. A. (1969). *The sciences of the artificial.* Cambridge, MA: MIT Press.

Wickens, C. D., Goh, J., Helleberg, J., Horrey, W., & Talleur, D. A. (2003). Attentional models of multitask pilot performance using advanced display technology. *Human Factors, 45*(3), 360–380.

Wickens, C. D., Vincow, M. A., Schopper, A. W., & Lincoln, J. E. (1997). *Computational models of human performance in the design and layout of controls and displays* (CSERIAC SOAR Rep. 97-22). Wright Patterson AFB, OH.

12 Advancing the State of the Art of Human Performance Models to Improve Aviation Safety

Becky L. Hooey and David C. Foyle

CONTENTS

INTRODUCTION

The state of the art of human performance models (HPMs) has advanced to a level of maturity that they are now considered important tools in the arsenal of design, analysis, and evaluation tools used by aviation safety researchers and system designers. HPMs, used in conjunction with human-in-the-loop (HITL) simulations, allow researchers and designers to evaluate new aviation displays, automation, operations, and procedures and identify candidate concepts that are likely to increase safety and efficiency, as well as to weed out candidate concepts that have the potential for problems, before significant development costs have been incurred.

The National Aeronautics and Space Administration Human Performance Modeling (NASA HPM) project was a large, government-funded research effort that applied five different modeling approaches to two very different aviation domain problems (taxi navigation errors observed in current-day operations and the design and evaluation of displays, procedures, and operations for a synthetic vision system [SVS] still in the conceptual development phase). These two aviation problems were chosen to explore the use of HPMs throughout the entire design and evaluation life cycle by testing both early and mature concepts.

Each of the five modeling tools within this NASA HPM project demonstrated the ability to model the complex environment of aviation, including multiple operators, vehicle dynamics, environmental cues and constraints, and both nominal and off-nominal operations. In doing so, the cognitive component of each of the frameworks (i.e., the model of the pilot) was modified and expanded to include procedural and declarative knowledge regarding the taxi and approach and landing tasks, along with quantification of various parameters of related flight activities. In addition to modeling the pilot at some level of abstraction, representations of the flight deck (displays and controls), the aerodynamics of the aircraft, the physical environment (weather, terrain, and airport), and other interacting agents (air traffic controllers [ATCs] and first officer [FO]) were also instantiated.

In summary of the NASA HPM project, this final chapter will first outline the ways that HPMs and HITL simulations were used synergistically in each phase of the system design and evaluation process, including concept definition, system definition, system evaluation, robustness testing, and system integration. Next, significant augmentations to the state of the art of the HPM tools that were required in order to address the complex aviation problems will be reviewed, including modeling the human–environment interaction in a closed-loop fashion, as well as several approaches that were adopted to model visual attention in supervisory control systems, situation awareness, and human error. Finally, this chapter will conclude with a number of considerations regarding the use of models in complex domains, including issues relating to selecting a model architecture, developing a model simulation, interpreting a model's output, and validating a model.

AN INTEGRATED HITL/HPM APPROACH
TO AVIATION SYSTEM DESIGN

In chapter 2, we introduced a human-centered design and evaluation process (shown again in Figure 12.1) that has been employed successfully in the design and evaluation

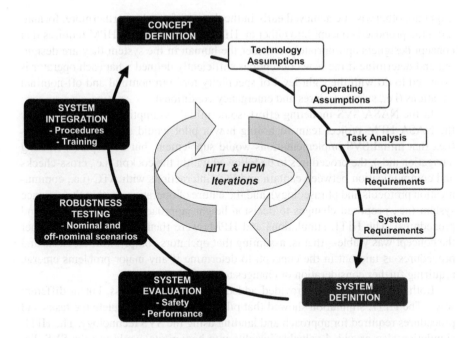

FIGURE 12.1 Integrated HITL/HPM approach to system design and evaluation.

of advanced avionics systems (Foyle et al., 1996; Hooey, Foyle, & Andre, 2002). One major feature of the process is the iterative evaluation/validation loop that integrates both HITL simulations and HPM evaluations toward the development of error-tolerant, human-centered system designs. This NASA HPM project provided an opportunity to explore the synergistic benefits that can be gained from using HPMs and HITL evaluations together. The following section describes how both approaches can contribute in unique ways to each phase of the design and evaluation life cycle and will be illustrated by examples from the NASA HPM project.

CONCEPT DEFINITION

One of the first steps in the design of a new complex system is the development of the concept of operations (CONOPS), which is a specification of the high-level system requirements, including the equipment (e.g., automation, display interfaces, etc.), operational requirements, and procedures. The CONOPS typically takes the form of a scenario (or series of scenarios) that describes what the operator is expected to do within the system, the sequence of tasks to be followed, and any interactions with other operators or equipment. Both HITL simulations and HPMs can be used in the development and evaluation of the CONOPS—ideally before significant time, money, and effort have been expended in developing hardware and software algorithms.

This process itself requires the concept developer to think through the proposed concept in a very methodical manner, which in and of itself is a valuable contribution to the conceptual design process since it requires a level of design specificity that might not otherwise be achieved early in the design life cycle. Furthermore, formalizing the proposed concept into either an HITL simulation or an HPM requires that concept designers appropriately consider the human in the system they are designing and determine if the new concept has sufficiently defined what each operator is expected to do with the right level of specificity for both nominal and off-nominal situations (i.e., system failures and emergency conditions).

In the NASA SVS modeling effort, some initial assumptions were adopted by the NASA HPM project team regarding how a pilot would use the SVS displays (i.e., that initial SVS implementations would supplement, but not replace, existing navigation aids), the procedures to be followed within the cockpit (i.e., cross-checks and communication between captain and FO), interactions with ATC (i.e., communication protocols and phraseology), and the nature of operations within the airspace system (i.e., proposed changes to decision height, approach plates, and go-around procedures). The HITL simulations and HPMs were then used to assess whether the concept was viable—that is, assuming that operators comply with the tasks and procedures as laid out in the concept, to determine if any major problems emerge requiring further consideration or changes to the CONOPS.

Both HITL and HPMs provided added value to this process, but in different ways. The HITL simulation showed that pilots were able to complete the tasks and procedures required for approach and landing using the SVS technology. The HITL simulation also provided valuable insights into how pilots would use the SVS displays, as well as the user acceptance of the technology—both issues that are largely unknowable from HPM efforts. There were no red flags raised from the HITL simulation that would suggest potential problems ahead when fielding the technology; however, the HITL simulation did suggest that there was a great deal of variability among pilots with regard to how they would use the technology, pointing to the need for formalized procedures and training programs before system implementation. This issue of pilot variability is very important and cannot be characterized and assessed without HITL simulations.

The HPM approach, on the other hand, brought to light different issues regarding the viability of the SVS CONOPS. The five modeling teams each demonstrated that the virtual pilots were able to carry out approach and landing procedures with the SVS. One model (Air MIDAS, chapter 7) also demonstrated that lower decision altitudes would be viable with the addition of SVS technologies on board. However, several of the models identified potential areas of concern associated with the assumption within the initial CONOPS that stated that the SVS should be a secondary display and not a primary display. For example, the Air MIDAS effort showed that SVS, when used as a secondary display redundant with the primary flight display (PFD) and navigation display (ND), resulted in small delays of flight control action initiation because of the need to cross-check the information visually on the SVS display with that on the PFD/ND. This finding might never have been uncovered in an HITL study unless the researchers had reason to suspect this was a potential problem and create scenarios and dependent measures targeted to probe this issue. However, now based on this HPM finding, an HITL simulation can be designed with a specific goal to better understand how the flight control action initiation delay manifests in actual crew procedures, particularly in emergency or off-nominal situations, and how it might trade off against other factors such as the assumed benefit associated with redundancy and cross-checking.

System Definition

The next phase of the system development life cycle typically involves increased definition of the system; this includes specifying information requirements, system requirements, and detailed design of the system components. At this stage, the ability to rapidly mock up and test various displays, automation, operations, or procedural concepts is very useful. This is ideally done with inexpensive rapid prototyping, before the actual full-scale system is designed and developed. It is possible to do this with either HITL simulations or HPMs, though the HITL approach requires actual physical mock-ups and often repeated iterations, which can be time consuming and costly, with much of the effort and cost absorbed by technology integration issues (i.e., inserting or emulating a realistic algorithm for automation logic or a particular display component into a full-mission simulation). As such, much of the design and development work has to be advanced to a sufficient stage of maturity before HITL testing can be performed.

On the other hand, HPMs allow for much more rapid prototyping with substantial savings in terms of costs, time, and effort as was demonstrated in the modeling of the SVS display concepts. Several of the modeling teams demonstrated how "what if" scenarios can be used in the design and evaluation of displays, operations, and procedures. The ACT-R team (chapter 5) and the D-OMAR team (chapter 8) used "what if" analyses to determine if different SVS symbology would alter visual scan patterns and provide for more optimal visual attention across information elements. Recall that in the NASA HITL study (chapter 3), the SVS display resulted in less attention paid to the ND than in baseline (no SVS) trials. It was proposed (chapters 5 and 8) that the SVS symbology overlaid on the terrain caused the change in scan pattern.

The ACT-R team (chapter 5) subsequently used the model to explore "what if" the SVS consisted of different symbology sets. For example, they removed the altitude indicator and evaluated the change in visual scan patterns. In doing so, they demonstrated that symbology sets (in terms of information content) can be easily and quickly explored in the model. Similarly, the D-OMAR team (chapter 8) "virtually redesigned" an integrated PFD/SVS that would restore visual scans back to the efficient time-tested scan pattern that all pilots are trained to perform (i.e., baseline, no SVS, scan pattern) while preserving the advantages of the SVS. Indeed, they found that, by creating an integrated display that combined the PFD and the SVS, the scan pattern was restored to baseline levels.

For the HPMs in this NASA HPM project, the largest cost was in producing the first version of the model within a given domain (taxi or SVS); subsequent modifications were relatively inexpensive in both time and effort. The examples from within the NASA project provide evidence that HPMs can indeed be used to shorten the design life cycle, and reserve more costly hardware prototyping and HITL simulations for design prototypes that have reached some level of maturity through HPM testing. This approach may not only save time, but may also be less costly to effect, and it may provide a more thorough and thus better final system design outcome.

SYSTEM EVALUATION

As design specificity increases, evaluations can be conducted to identify the potential for safety vulnerabilities that may put human operators at risk or negatively affect system performance. It is through these evaluations that potential latent error precursors can be identified and addressed through redesign of technology or procedures to mitigate operational errors (see chapter 2). Again, there is a role for both HITL and HPM simulations here, and the two approaches are quite synergistic.

The importance of the interplay between HITL simulations and HPMs in the NASA HPM project was perhaps best exemplified by the taxi error analyses. Many of the modelers involved with the NASA project stated that the modeling efforts would have been futile without the HITL data to inform the model development process. The HITL taxi study provided a wealth of data about the nature of errors, including factors such as the taxiway geometry where errors occurred, the crew tasks that preceded the error, and the relative frequency of each type of error. These data were then used by the modelers to develop a deeper understanding of the error and identify causal factors, including crew procedures and operations that exceeded human capabilities for memory and workload (chapters 6 and 7), taxi clearances that contradicted commonly used decision-making heuristics and expectations (chapters 5 and 8), and complex airport layouts and signage that hindered the development of situation awareness (SA; chapter 9).

Using HITL simulations and HPMs together to understand error offered several advantages over either approach alone. Specifically, the HITL simulations provided necessary data with which to populate and prime the models and allowed the modelers to understand the important issues to pursue within the models (i.e., SA and airport complexity). HPMs, on the other hand, allowed for the systematic manipulation of variables such as memory capacity or decay rate, which cannot be manipulated within human subjects, especially in complex real-world tasks (in contrast with more simple-task psychology memory experiments). With a better understanding of these underlying causal factors that might lead to errors, one can more effectively propose and prioritize mitigating solutions. The results of the taxi research effort suggest that providing memory aids such as data-linked taxi clearances, taxi clearances that conform to pilots' expectations, and improved navigation tools to compensate for complex airport layouts should be considered.

The SVS study also demonstrated the synergy between HPMs and HITL simulations, but in this case, the HPMs were used to inform future HITL studies. The initial HITL study did not reveal serious safety vulnerabilities, but the HPMs did. For example, Air MIDAS (chapter 7) predicted spikes in operator workload that may be indicative of latent error conditions or error-precursor conditions (see chapter 2) by leaving the pilot vulnerable to time-pressure errors, dropped tasks, or other degradations of performance. D-OMAR (chapter 8) was particularly well suited to identify safety vulnerabilities caused by procedural interruptions, or inadequate crew communications and cross-checks. The A-SA model (chapter 9) identified a potential safety threat of attentional tunneling, or attending to elements on a flight display at the expense of elements in the world. These safety vulnerabilities identified by the HPMs were not observed in the HITL simulations. This is a common problem in HITL simulations because the highly skilled pilots who serve as test participants tend to compensate for weaknesses inherent in a system's design or demanding procedural requirements.

System Robustness Testing

The steps in the design and evaluation life cycle discussed thus far have assumed nominal performance; that is, they assumed that the system functions as expected and the human operators complete the tasks and procedures as required. However, for complex systems to be implemented in actual operations, system robustness testing must also be conducted that evaluates the system from the standpoint of potential deviations from "nominal" procedures to determine the impact on the performance of the human and the system. Off-nominal testing, in which unexpected, atypical conditions are evaluated, allows for increased understanding of the human–machine system, may uncover latent errors in system designs that should be addressed before a system is fielded or implemented, and ensures that systems scale up from the laboratory to actual operations (Foyle & Hooey, 2003). Note that although these steps are presented serially here, system robustness testing can, and should, be conducted along with nominal performance testing early in the system development process.

This NASA HPM project showed the value of extending the nominal/off-nominal experimental paradigm (Foyle & Hooey, 2003) to include HPM evaluations and identified many synergistic benefits of the integrated HPM and HITL approach that capitalizes on the benefits of each technique. Because of its fast-time nature, human performance modeling is a powerful technique to assess the impact of latent errors because system design flaws typically induce pilot error only under some low-probability confluence of precursors, conditions, and events. However, the state of the art of human performance modeling is not at a point that enables one to model human responses to truly surprising off-nominal events. This sentiment was voiced repeatedly by the modelers in chapter 11, who cited that many of the models lack the ability to model adequately important factors such as the effect of fatigue, operator training, and pilots' ability to compensate for system errors.

At the same time, evaluating off-nominal scenarios solely with HITL techniques is challenging. HITL studies are often limited to collecting only a single data point per test participant (usually at the end of the study) since expectation and cueing effects, or lowered system trust, can taint subsequent trials—at least for truly surprising events such as near incursions or complete system failures (see Foyle & Hooey, 2003).

An integrated HITL/HPM approach can be used to overcome the individual limitations of each technique. HITL data collected during off-nominal scenarios can be used to populate or prime the models, which are then repeatedly run, systematically varying the timing or value of relevant parameters to assess the consequences of the human response on other operators in the system or on other aspects of system performance. The models, populated by actual HITL data, can then be used to investigate the internal cognitive processes of the operator to determine and assess the impact of behaviors on system safety and, in some instances, extrapolate to proposed display or procedural conditions not tested in the HITL simulation. Coupled together in this manner, HITL testing and human performance modeling can be considered a powerful tool set to identify potential latent error conditions or latent design flaws prior to implementation of a new piece of technology or operational procedure.

SYSTEM INTEGRATION

Another important step in the system design process is to consider how the system will be integrated into actual operations including, for example, developing standard operating procedures, checklists, and pilot training programs. There is a long history within the aviation community of these types of procedural integration efforts; however, using HPMs for this purpose is less common.

One potential use of HPMs for these integration activities is the development of a "gold standard"—a metric describing optimal performance to which pilot performance, technology, or procedures can be compared. Wickens et al. (chapter 9) demonstrated that their model can be used as a gold standard for developing pilot training programs. Specifically, they developed a scanning optimality score for each pilot and provided evidence that optimal scanning led to faster traffic detection, and better flight-path tracking. Wickens et al. conclude that their additive A-SA model might serve as a gold standard by which the pilot visual scan can be trained and evaluated. If further research upholds this assertion, such a metric (conformance to the additive A-SA model) could prove to be a useful standardized measurement characterizing pilot behavior in system evaluations, as well as for assessing pilot training techniques.

AUGMENTATIONS AND APPROACHES FOR MODELING AVIATION TASKS

To implement HPMs in each stage of the system design and evaluation process within the NASA HPM project, significant advances in the state of the art of the five HPM tools were realized. These advances are described next and summarized in Table 12.1.

MODELING HUMAN–ENVIRONMENT INTERACTIONS

The complex demands of evaluating new aviation concepts and systems, such as the SVS in the NASA HPM project, pushed the modeling teams to consider new approaches to modeling environmental factors including aircraft dynamics and the external environment. Two advances that were particularly noteworthy include: (1) modeling the human–environment interaction in a dynamic closed-loop fashion by combining the cognitive model architectures with external desktop simulators, and (2) integrating a task network model of the environment with the cognitive architecture of the human.

Two of the modeling teams, ACT-R and Air MIDAS, spent considerable effort integrating their HPM architectures with desktop aircraft simulators to represent the environment, attesting to the importance of modeling the human interaction with the environment in a high-resolution manner. Byrne, Kirlik, and Fleetwood of the ACT-R team (chapter 5) developed a model that is dynamically sensitive

---

TABLE 12.1

Advances in the State of the Art of Human Performance Modeling

Model Capability	Specific Advance	Modeling Team (Chapter)
Human–environment interactions	Enabled closed-loop behavior by integrating HPM with desktop simulator	ACT-R (5); Air MIDAS (7)
	Integrated a task network model with a cognitive model	IMPRINT/ACT-R (6)
Visual attention	Replicated information-seeking behavior of pilots	ACT-R (5); Air MIDAS (7); D-OMAR (8); A-5A (9)
	Implemented model learning of visual scan patterns	IMPRINT/ACT-R (6)
Situation awareness (SA)	Demonstrated how SA changes as a function of time and distraction	A-SA (9)
Human error	Identified error vulnerabilities due to pilots' heuristics, biases, and strategies	ACT-R (5); D-OMAR (8)
	Identified error vulnerabilities due to memory deficits	IMPRINT/ACT-R (6); Air MIDAS (7)
	Identified error vulnerabilities due to SA deficits	A-SA (9)

to top-down (i.e., task structure and information-seeking goals) and bottom-up (i.e., parameters of visual attention and structure of the information environment) constraints because they believed that one cannot predict performance by considering just the system interface or just the task structure alone. They adopted an integrated, closed-loop computational cognitive modeling approach by connecting their model (ACT-R) to X-Plane, a PC-based flight simulator, using a UDP (user datagram protocol) interface such that the modeled pilot was controlling the simulated aircraft. Similarly, the Air MIDAS team (chapter 7) connected their HPM to a NASA-developed flight simulator (PC Plane) using DLL (dynamic link library) functions to generate aircraft inputs offering improved time synchronization over the UDP approach.

In essence, both of these approaches to closed-loop modeling allowed the modeled pilot to "fly'" the aircraft—that is, to make control inputs or alter visual scan patterns dynamically as a function of the aircraft state. Without this dynamic coupling, the modeled pilot was essentially just monitoring the aircraft as it followed a set of procedures. As noted by Leiden and Best (chapter 10), this closed-loop model capability, which provides closely coupled feedback mechanisms between the pilot–action/aircraft–response, also allows for improved reuse of the model for future tasks and generalization to other research problems.

A second noteworthy approach for modeling the environment was that adopted by the IMPRINT/ACT-R team (chapter 6). They integrated the cognitive architecture (ACT-R) with a task network model (IMPRINT). This approach took advantage of the strengths of each tool in order to increase the fidelity of the HPM, while controlling the complexity of the model. They reported that this provided for a clean, practical, and conceptual separation between models of the environment and of the pilots and prevented unwanted and erroneous dependencies and interactions among the cognitive, physical, and environmental representations.

MODELING VISUAL ATTENTION

The ability to replicate information-seeking behavior of actual human operators is critical for any concept involving supervisory control issues such as how operators monitor, manage, or otherwise interact with displays and automation for any information-rich domain, including military, medical, surface transportation, and nuclear applications. In order to model the SVS concept, the HPMs were augmented to replicate and predict the allocation of visual attention to information elements in the environment. Of note in this HPM project were three different, yet equally valid, approaches used to model supervisory control tasks, and in particular, visual attention.

The first approach, adopted by the Air MIDAS team (chapter 7, Simulation 2), was to rely on an extensive data set from an external HITL study in the literature to develop the background pilot scan pattern while allowing it to be overridden by events in the environment and scenario. Although the HITL data set they used did not fully replicate the NASA HPM SVS scenario conditions (e.g., the experimental scenarios involved approach and landing in low visibility, but without SVS technology), they modeled visual attention to the SVS based on the CONOPS assumption that the pilots would use the SVS instead of the out-the-window (OTW) information; that is, they replaced scans OTW with scans to the SVS. The second approach, adopted by the D-OMAR, ACT-R, and A-SA teams (chapters 8, 5, and 9, respectively), relied on a thorough analysis of the information needs of the pilots using subject matter experts (SMEs) to build their scan patterns based on task analysis. Third, the IMPRINT/ACT-R team (chapter 6) invoked a model learning approach to teach the modeled pilot where to allocate attention. Each of the three approaches relied progressively less on access to HITL data—an important consideration when testing new concepts, operations, or technologies for which HITL data may not be available.

MODELING SITUATION AWARENESS

Given the importance of the construct of SA in aviation research, there is a need for a computational, predictive model of SA. Wickens and colleagues' work (chapter 9) is an important step forward in this pursuit (also see previous work by Shively, Brickner, & Silbiger, 1997; Zacharias, Miao, Illgen, Yara, & Siouris, 1995). Their A-SA model identified safety threats associated with degraded pilot SA as a function of time and task interruption and related this to the likelihood of error when SA degradations occurred during critical decision choice points. The A-SA model assumed that loss of SA occurs because SA degrades over time—slowly when nothing else is occurring and rapidly in the presence of irrelevant distracting events.

Wickens et al. made large advances in the state of the art of computational models to predict SA, but also acknowledge that there is much more to be done. For example, there is a need for empirical validation of the model's attention component coupled with the SA belief-updating module. Also, particularly within the context of aviation modeling, SA models must be expanded to include the concept of shared awareness (i.e., multiple operators having similar SA representations) among crew members and ATC. As systems become increasingly automated, the effect of various stages and levels of automation (Parasuraman, Sheridan, & Wickens, 2000) on operator SA, and how degraded SA in such automated systems can be mitigated through display support (Horrey, Wickens, Strauss, Kirlik, & Stewart, 2006; Strauss & Kirlik, 2006) will become increasingly important. The A-SA model, with some modifications, is poised to address these issues.

Modeling Human Error

The low error rate that is typical of the aviation domain due to the high skill level of the operators and the proceduralized nature of the tasks has proven to be a challenge both for studying the causes of errors within the field of aviation human factors and for developing modeling techniques to address human error. As discussed by the modelers in chapter 11, one of the biggest stumbling blocks in modeling human error in complex environments has been the lack of adequate data sets with which to populate and validate the models. This project presented a unique opportunity to explore approaches to model human error, given the relatively rich HITL simulation data of airport taxi navigation errors (see chapter 3).

The modeling teams not only demonstrated the ability to replicate error patterns inherent in actual HITL data, but also demonstrated the ability to develop and test specific research hypotheses about the role of expectations and heuristics, memory, and SA in the errors observed in the HITL data. The D-OMAR and ACT-R teams developed theoretical explanations for the error behavior, focusing on decision heuristics and strategies that contributed to taxi error. In doing so, they were able to offer a deeper understanding of why these errors occurred and what might be done to reduce them. The D-OMAR team (chapter 8) showed that error could be driven by expectation based on partial knowledge and by habits. The ACT-R team (chapter 5) identified taxi decision heuristics and showed that the heuristic called by the pilot differed as a function of time constraints. They found that pilots followed environmental heuristics, such as "turn toward the gate" or "taxi the shortest distance," when under time pressure.

The Air MIDAS and IMPRINT/ACT-R teams developed hypotheses regarding the role of memory in the taxi errors. Air MIDAS (chapter 7) identified the effect of memory capacity and decay rate, while IMPRINT/ACT-R (chapter 6) identified errors of omission that occurred when a segment of the taxi clearance was not recalled because of time-based decay or activation noise, and two errors of commission (turn on wrong taxiway or wrong direction) caused by interference, similarity-based partial matching, priming, or activation noise. These advancements of the state of the art in error modeling are important for understanding error causation and are a fundamental requirement toward designing and testing error-mitigating solutions.

The A-SA model (chapter 9) showed promise as a predictive model of error. It was based on the assumption that if pilots had low SA at a given choice point, the probability of a navigation error would be high. If pilots had high SA at the choice point, the pilot would be more likely to recognize the correct response option and behave appropriately. Their analysis also showed that cockpit displays such as electronic moving maps would enable pilots to maintain a higher level of SA across the entire taxi route, thus increasing the probability of correct taxi decisions along the way. Although further research is required to validate the predictive nature of the A-SA model, it certainly shows promise as a model that can be used to predict errors that occur due to degradations in SA or loss of SA.

Despite these advances, we still lack the ability to predict when an error will occur, or under which set of conditions (e.g., operator state, task demands, environment characteristics) an error is more or less likely. One difficulty associated with error prediction in the aviation domain (as discussed in chapter 11) is that pilots are highly flexible and adaptive and readily develop coping strategies that allow them to compensate for poor design or high workload. Although the HPM community is not quite ready to claim success in the arena of error *prediction*, significant advances were made in modeling error.

CONSIDERATIONS FOR MODELING COMPLEX AVIATION TASKS

Because of the relatively unique opportunity to apply multiple HPMs to two different aviation-domain problems at different phases of the design life cycle, the project revealed several important considerations regarding the utilization of the models for aviation system design and evaluation. Specifically, important considerations related to model selection, development, interpretation, and validation were observed:

- With regard to selecting a model, the philosophies, approaches, and underlying assumptions of the models differ widely and these factors must be considered in the selection of a model.
- With regard to model development, it was observed that models of complex environments require intensive knowledge engineering and would be aided greatly by the availability of task analysis techniques and approaches aimed at populating models with relevant input including not only task sequences, but also operator strategies.
- There was a clear need for visualization and documentation tools to enable easier interpretation of the underlying model assumptions and model results to ensure the model output is understood and useful for the end-user.
- It was evident that the validation of complex aviation HPMs, especially for novel systems in the concept development phase, presents a number of challenges. Several validation techniques focused on different end goals and employed in different phases of the model development efforts are presented.

Each of these four considerations is discussed in turn.

SELECTING A MODEL ARCHITECTURE

It is clear, after completing this multiyear, multimodel research effort, that selecting the "best" model is not straightforward. One cannot choose a model based solely on the statistical goodness of fit to determine which model produced the best level of validation with actual HITL data. Although we tend to gravitate toward quantitative data to make these types of decisions, it is clear that this is not sufficient for many reasons. Just as in choosing the appropriate level of fidelity for an HITL simulation, the choice of HPM depends on the research questions being asked. Before choosing a model, one must understand the domain and the specific research problems: What must the model predict or explain? Is a closed-loop modeling environment necessary? Are interactions among multiple operators important? Is the time scale at the level of milliseconds, hours, or days? What are the cognitive processes at play—memory, decision-making, vision, etc.?

It is important to consider the answers to questions such as these with an understanding of the capabilities inherent in each of the modeling tools. Without this analysis, one risks, as the old saying goes, "with a hammer in hand, turning every problem into a nail." Each model is tuned to answer certain research questions. That is, a model that emphasizes decision-making processes may focus on pilot decision making, while ignoring other factors that may be equally, or more, important. Of the models discussed in this volume, Air MIDAS is arguably best suited to answer questions related to operator workload and A-SA would be the model of choice to investigate SA issues. D-OMAR would be a good choice to investigate crew procedures and communications issues. The two ACT-R models are ideally suited to answer questions that required deep understanding of the human cognitive processes such as memory.

Each of these model architectures was developed with these constructs as focal points. However, this is not to say that, for example, an SA mechanism could not be incorporated into D-OMAR or that A-SA cannot be augmented to manage shared SA issues across multiple operators. However, the emphasis has not been placed on these aspects, and as such these aspects of the model architecture are not as well developed at this point in time. Furthermore, it is not sufficient to know that a cognitive architecture has, for example, a memory model, but an understanding of the level of fidelity of the memory representation within each model is also necessary, as this can vary widely. As an example, the reader is referred to Leiden and Best's description in chapter 10 of the memory models across the various cognitive architectures discussed in this volume.

Just as models all specialize in different aspects of human behavior, they also adopt very different approaches and philosophies; understanding these differences is a critical step in choosing a model for any given application. For example, ACT-R is not only a modeling tool, but also a theory of human cognition (Anderson & Lebiere, 1998). The construction of the ACT-R programming language reflects assumptions about human cognition as derived from fundamental psychology experiments (for example, ACT-R contains models of memory for text and for lists of word and language comprehension). By selecting ACT-R, one is implicitly also accepting this underlying theory; the two cannot be separated. Air MIDAS also has built-in micromodels (for example, memory, vision, and workload); however, the Air MIDAS approach has been to select the best micromodels for integration, rather than to develop and implement a single unified theory of cognition. These two approaches have the advantage of embodying deep knowledge to a level beyond any one person's expertise, as pointed out by Byrne, Kirlik, and Fleetwood (chapter 5).

At the other end of the continuum is the approach adopted in D-OMAR. This model, instead of asserting an underlying cognitive theory, is a simulation language that allows modelers to define and evaluate their own theory. Deutsch and Pew take the stance that "there is still much to learn about cognitive architectures and there is much to be gained by leaving room to explore alternative approaches to their construction" (chapter 8).

Researchers must understand how the model addresses a particular issue and then decide whether they agree with the approach taken. There are several resources available by which to do this. For the five models reviewed here, the interested reader is referred to chapter 10 for a detailed comparison of individual model components, including error prediction, external environment, crew interactions, scheduling and multitasking, memory, visual attention, workload, situation awareness, learning, and resultant or emergent behavior. Other resources include Ritter and colleagues' (2003) review of selecting models, Pew and Mavor's (1998) review of models used in military and organizational behavior, and Gluck and Pew's (2005) agent-based modeling and behavior representation (AMBR) model comparison project.

DEVELOPING MODELS FOR AVIATION TASKS

This NASA HPM project successfully demonstrated that HPMs can be useful tools to answer aviation safety research questions and address system design, development, and integration issues. It is believed that this success is owed, in part, to the availability of rich data sources, including HITL experimental studies, task analyses informed by SMEs, and the close collaboration between the NASA research team and the five modeling teams. Harkening back to the old adage, "garbage in, garbage out," this observation highlights the importance of populating models with appropriate and valid domain knowledge in order for a model to be of true value for the aviation community.

Even as we strive toward general-purpose cognitive architectures that are capable of modeling any domain environment, the population of the architectures for a given research or development problem will always require input by domain experts and end-users (in this case, pilots and air traffic controllers). One of the first steps in most HPM projects, and this one was no exception, is to conduct a detailed task analysis that identifies the tasks conducted, by whom, for how long, and other model-specific parameters, which may include estimated workload and task interruptibility. Towards this end, the NASA HPM project office commissioned detailed task analyses for the various flight tasks to be modeled (see chapter 3) and each was developed with the assistance of SMEs. While this was necessary for model development, it was apparent that each modeling effort also required further domain knowledge specific to its own model development needs. For example, Byrne, Kirlik, & Fleetwood (chapter 5) worked with commercial pilots to better understand the heuristics that pilots use when navigating on the airport surface because pilot strategy was not fully represented in the task analysis to the degree required for building their ACT-R model. In fact, across many of the modeling teams, discussions of pilot strategies and how to model them were raised. As readily observed in many aviation research domains, actual task performance may differ from published standard operating procedures due to a range of factors, including pilot experience, motivation, fatigue, environmental conditions such as weather and traffic, and even organizational pressures, economic factors, and on-time ratings.

Card, Moran, and Newell (1983) adopted a common engineering technique for addressing this kind of uncertainty by modeling a nominal value as well as low and high values. They defined three versions of a model: *slowman*, in which all the parameters were set to give the worst performance; *fastman*, in which all parameters were set to give the best performance; and *middleman*, which provided nominal performance. Kieras and Meyer (1998) also found success with this approach to accommodate the fact that task performance is very heavily influenced by what they called "optional aspects of task strategy" and that human operators tend to adopt different strategies, making it difficult to create a model that captures human behavior. They adopted a bracketing heuristic approach, similar to that of Card et al., (1983), in which they defined a "nominal" set of tasks for any given scenario, and also a slowest reasonable version and a fastest possible version. They cited examples from military combat domains, in which soldiers can normally be expected to work rapidly; yet, in some circumstances, soldiers might opt to slow performance to be able to sustain their level of effort for long durations. Kieras and Meyer note that these strategic differences are often difficult to determine as such strategy issues involve aspects of human metacognition that are not well understood.

It was proposed (Card et al., 1983; Kieras & Meyer, 1998) that this "fastman/slowman/middleman" method may be useful for system design in many ways. First, it can be used to obtain performance predictions for a proposed system, despite uncertainties about how an operator will perform a task, and produce a range of outcomes that may represent differences in levels of training, experience, and motivation. Second, when absolute requirements for overall speed of performance are needed, the bracketing technique can be used to ensure that all reasonably expected behaviors fall within the absolute requirements to ensure the acceptability of design. In our opinion, this approach warrants important consideration, particularly when applying models of complex environments such as aviation. From our experiences in the NASA HPM project, we agree with Kieras and Meyer that a crucial future objective for both the scientific and practical application of cognitive architectures is to develop a priori strategy-identification methods that are powerful, accurate, and reliable enough for practical application.

This requires task analysis tools and techniques that not only guide the systematic recording of tasks that are performed, but also capture which aspects of the task are mandatory and which are optional, the different ways that the task could be performed, and how the tasks are prioritized by the operator. Additionally, much more research is required to create meaningful bracketed-model solutions in the aviation realm (or any other complex domains) for each strategy or performance-degrading factor of interest. For example, to create a nominal model bracketed by a general aviation pilot model and a commercial airline captain model would require knowledge of which parameters are likely to be degraded or enhanced in each case and by how much. It is likely that not all parameters will be affected, and certainly not affected equally, as was the case in the models of Card et al. For example, there is no a priori reason to expect memory capacity to differ between the two models;

however, workload and situation awareness likely would differ. This area is ripe for research and model augmentations.

INTERPRETING MODEL OUTPUT

Since the goal of the NASA HPM project was to model actual aviation domain problems to improve aviation safety, considerable time and effort were spent to understand and interpret the model results and determine their implications for aviation safety. As "outsiders" to the individual model development teams, the NASA HPM project team experienced the need for model transparency first hand and noted the importance of guarding against the recipient or end-user of the model blindly accepting the output without fully understanding the underlying assumptions and constraints—or worse, (mis)applying the model findings to make important decisions beyond the scope of the model's applicability.

The modernization of the adage "garbage in, garbage out" to "'garbage in, gospel out" refers to the human tendency to accept the results from computer systems with unquestioning faith (Oettinger, 1964). This is a particularly relevant concern for the field of human performance modeling as, all too often, models are seen as black boxes and the output is accepted as fact. There are two reasons for this "black box" view: (1) lack of transparency of modeled tasks and procedures (e.g., what procedures and tasks did the modeled operator complete, and what data were used to populate the model?), and (2) lack of transparency of the underlying theory or cognitive structures (e.g., does this model architecture have a memory model? What is the basis for the workload computation?).

Transparency of Modeled Tasks and Procedures

It is important to understand how a model was populated and the underlying data and assumptions on which it was built. There are several approaches one could take to populate a model. As examples, one could derive the means and standard deviations for task times from the existing literature or HITL studies; use established, validated tables for lower level behaviors (i.e., time to push a button, scan the environment, etc.) and assume they apply to the relevant domain; or make rough estimations or educated guesses (perhaps using SMEs). In practice, for many models there is no easy way to ascertain the value of any given parameters such as mean, standard deviation, or data distribution of each variable, or which parameters were populated by objective data and which were assumptions made by the modeler. Since each of these decisions can influence the outcome of the model and the conclusions drawn, it is important that end-users (e.g., the user requesting or funding the modeling effort and/or those with specific domain knowledge) understand the assumptions and impact of parameter values used for model input before being asked to accept the model output.

Run-time visualizations that illustrate exactly what the modeled operator is doing, when, and how—such as those provided by D-OMAR and ACT-R in the current NASA HPM effort; those developed and reported in Gluck and Pew (2005, p. 411); those by NASA's MIDAS, which incorporates an anthropometric figure carrying out tasks and run-time plots of workload and SA (Gore & Smith, 2006); and those by the

SA panel developed by Councill, Haynes, and Ritter (2003) for a complex cognitive model (TacAir-Soar)—represent important tools toward mitigating both the input and output portions of the "garbage in, gospel out" effect. Tools such as these allow an observer to inspect the model and identify (potentially erroneous) tasks or task sequences that may have contributed to the generation of false or suspect conclusions. For example, consider the (fictional) case in which a model of pilot taxi error predicts a very high error rate and thus the case is made to augment cockpits with display technology to mitigate these navigation errors. However, on inspection of the model run-time visualization, one might notice that the modeler failed to include a full read-back of the taxi clearance in the modeled tasks of the FO, thus removing a source of error cross-check and artificially inflating the error rate.

Without some visualization or trace of the modeled operator's task, catching the omitted task (clearance read-back) in the model would be very unlikely, even by the modeler, given the complexity of computer code required to implement a complex model. An external observer or recipient of the model outcome, such as aircraft or avionics manufacturers; government policy makers such as NASA and the FAA; airline management; or those who make decisions regarding aircraft equipage, operational procedures, and training programs, would never have the ability to discover such an error if they do not have the skill, knowledge, or access required to read the model code or to interrogate the model in a hands-on fashion.

Transparency of Underlying Theory or Cognitive Structures

With the advancements of HPMs, we have moved from simple mathematical models (e.g., Markov processes) to more complex, cognitive architectures. In their paper, "The Art of Model Development and Testing," Shiffrin and Nobel (1997) point out that because of the complexities and interacting components in today's models, even conceptual analysis—that is, analyzing models without actually exercising them in specific diagnostic ways—may not be possible. Such complex instantiations of theories impede the ability to understand the underlying theory or model because of logistical roadblocks, especially so when pitting two models against one another to infer necessary cognitive constructs within the models. Historically, this is how theories and their instantiations in models are evaluated, assessing how well the two models or their variants predict the data—not just in quantitative goodness of fit, but also in regard to elegance, generalizability, and sensibility of assumptions (Shiffrin & Nobel, 1997). The formal pitting of model variants that Wickens et al. did with A-SA in chapter 9 attests to the value of such an approach. However, with the newest generation of complex computational HPMs, this is difficult for logistical reasons.

In order to understand all that is embodied within an HPM, researchers now need detailed computer programming knowledge to interpret the code and cognitive science knowledge to interpret the underlying theory of the model (or access and funds for someone who has that knowledge). It should also be noted that this adds a potentially fallible extra communication layer: The cognitive theorist must correctly communicate to the programmer the theory to instantiate, and the programmer must communicate back to the theorist that the instantiation sufficiently embodied the theory. This proves to be difficult when the researcher is the "owner" of the theory; imagine doing so for someone else's cognitive theory.

Similarly, Shiffrin and Nobel (1997) note that the modeling process typically progresses with multiple failed modeling attempts early in the development cycle. This might be because of an unnecessary theoretical construct, the details of how it was implemented, or the manner in which it is assumed to interact with other constructs. They note that it is typical for these early model variants to go unreported in the literature. The researcher may not even keep track of these variants formally, possibly because the determination as to "what is working" is made on an intuitive basis, rather than on a formal basis. Each of these unreported changes in assumptions and construct interactions is yet another piece of the puzzle that is missing preventing anyone other than the theorist or modeler from gaining a deep understanding of the underlying theory (see also Baron, Kruser, & Huey, 1990, regarding the "opaque nature" of model parameters).

Relatedly, one could engage or disengage particular micromodels within a modeling tool. For example, if behaviors become too constrained by some rule, the model developer can turn off this constraint, thereby allowing the model to proceed to completion within a given architecture. Alternatively, the model developer can remove the effect of certain elements of human performance (presumably deemed to be nonrelevant) by turning elements of a model on or off. For example, to remove the effect of memory on human performance, one might disengage the memory model, thereby ensuring perfect memory. (See chapter 10 for a discussion of two distinct approaches taken by the two ACT-R teams in this NASA HPM project; one engaged the base-level learning module and one disengaged it.) It is often difficult to ascertain which micromodels were operating during run-time and which were not, and what the resultant effects on predicted behavior would be. This also impedes the deep understanding of the underlying cognitive theory.

Increasing Model Transparency

It is imperative that the modeling community strive to make the underlying assumptions inherent in each model—and the consequences of those assumptions—accessible to others with a range of knowledge and expertise in computer science and cognitive science. This suggestion is by no means a novel one; Gluck and Pew (2005, p. 410) highlighted the need to make internal performance more transparent and, in fact, this was a contractual requirement within their government-funded modeling research program. However, in targeting the use of these cognitive models in more applied settings, such as for aviation system, operations, and procedural design and development, we extend the challenge to make models more understandable to a wider range of users such as pilots, air traffic controllers, airline managers, aviation policy makers, and avionics manufacturers, who would not necessarily be expected to possess the expertise in domains such as human performance modeling, cognitive science, and computer science. That is, even if developing and exercising HPMs requires very high levels of expertise in cognitive science and computer programming, their output must still be understandable to those without these skills in order for their value to be realized fully by the aviation community.

This highlights the need for model simplicity and visualization techniques that enable users to understand and "see" the underpinnings of the model. Councill et al. (2003) also recognized the need for providing "explanation facilities to make the rationale for model behavior (both external, with respect to the context of a simulation domain, and internal, with respect to the structure and cognitive processes of models) more accessible" (p. 91). They acknowledge that difficulty in understanding model behavior is partially due to model complexity and that this complexity is exacerbated by model development tools that do not provide a structured approach to model creation and fail to provide features for exposing and supporting exploration of a model's state by nonprogrammers.

There is little disagreement among researchers on the need for model transparency or what has also been called model interpretability (Diller, Gluck, Tenney, & Godfrey, 2005) or model transpicuity (Love, 2005). However, there is less agreement as to what constitutes appropriate model transparency and how to capture the relevant information, including the assumptions inherent in the model and their implications. It is particularly important to document these parameters and assumptions if model reuse is intended as many parameters may depend on the context of the modeled task and environment.

Research in design rationale capture (e.g., Moran & Carroll, 1996) has attempted to develop documentation tools to archive the design trace of systems and products. The issues they face are also relevant for documenting the assumptions, constraints, and limitations of HPMs: how to capture the information naturally and spontaneously during the model development process so as not to burden the modeler and increase project costs; how to best represent trade-offs and compromises (i.e., the decision to model certain elements at lower or higher levels of fidelity) made because of model architecture constraints or other factors such as time and resources; how to track the state of the model and elements that need to be refined and articulated as system design progresses; and how to track assumptions, particularly implicit assumptions, and their consequences on model output. These are important questions that, to date, have not been adequately addressed by the modeling community.

MODEL VERIFICATION, VALIDATION, AND CREDIBILITY

The NASA HPM project faced model validation challenges similar to those that have been observed and discussed in other complex-task/complex-model validation efforts (e.g., Glenn, Stokes, Neville, & Bracken, 2005; Gluck & Pew, 2005; Sargent, 1998). The U.S. Defense Modeling and Simulation Office (DMSO) defined validation as "the degree to which a model and its associated data are an accurate representation of the real world from the perspective of the intended uses of that model" (U.S. Department of Defense, 2001). While model validation may be straightforward when the models are simple—correlating inputs and outputs for both model and data (Shiffrin & Nobel, 1997)—as models become more complex, so does the complexity of the validation approach as well as the specific validation efforts. Several reasons for the increased complexity of the validation approach are discussed next.

One challenge associated with validation efforts of complex systems that was observed repeatedly throughout this NASA HPM project (and also cited by Gluck & Pew, 2005) was the issue of individual differences in performance among the pilots using the actual system (HITL data), as was observed in the NASA SVS study. The three pilots adopted very different usage strategies that resulted in widely varying performance data in terms of percent fixation time on the SVS and OTW during some phases of flight. These individual differences posed a challenge to each of the modeling teams to determine whether to validate against a single pilot (and if so, which one) or against the average of all the pilots.

Indeed, this is a decades-old question that has plagued modelers across many domains. Recalling his seminal work from the 1960s, Estes (2002) offers a review of the dilemma of averaging data, stating that, on the one hand, averaged data can help bring out trends in the data by reducing unbiased statistical error, but that, on the other hand, averaging can also be a source of distortion. That is, averaging data may give the impression of a process or function that is only seen in the group data, while the individual data imply different processes. In Estes's classic example, each individual subject's data may be in the form of a step function, but with each individual having a different value where that step occurs. When averaging across subjects, an ogive, or S-shaped psychometric function, appears in the group data; however, no individual subject's data reflected such a function.

Another challenge associated with validating models of complex-domain tasks is the potential for interactions among the models' subcomponents (e.g., workload, memory, attention, and perception), even if each subcomponent has been previously independently validated. Baron et al. (1990) point out that the integration of the many validated subcomponent models, each with its own set of assumptions, may negatively interact. One's ability to perform a particular task may depend on the nature of the other tasks for which he or she is responsible; some combinations of tasks may degrade performance and others enhance performance. Baron et al. provide the example that sequentially moving among regulation, recognition, and problem-solving tasks could lead to degraded performance relative to conducting each single task alone. On the other hand, they suggest that there may be a natural relationship between tasks, in terms of information requirements, that can lead to positive transfer from one to the other and result in enhanced performance. These sorts of interactions are ignored when validating models only at the subcomponent level.

The requirement that a model must be validated from the *perspective of the intended uses* (Campbell & Bolton, 2005, refer to this as "application validation") further adds to the challenges of model validation for complex-domain tasks. Campbell and Bolton noted that the intended application of the model has the potential to impact significantly the processes, metrics, and requirements associated with assessing the validity of the model. Validating HPMs intended for complex-domain tasks is often difficult because a multitude of domain- and task-specific assumptions are required. Details and nuances regarding the operator's task are made, slight variations in model structures may be implemented, and some subcomponents of a model may be engaged or disengaged (as discussed earlier). All of this makes the model and its validation a moving target.

Results validation (as discussed in chapter 10, also see Sargent, 1998), or tests that establish that model output closely resembles the output data from the actual system, are frequently considered to be the most definitive test of a model's validity (Law & McComas, 2001). However, relying on a single outcome measure (or a small number of outcome measures) to assess model validity should not be considered a strong test of the validity of a complex HPM, particularly when one considers that each HPM may have hundreds of parameters, including the tuned "hidden parameters," modified assumptions, and the more visible input parameters. "Hidden parameters" are those model characterizations or assumptions that may have originally been "tweaked" or tuned in the early stages of development but that, during later development, were set constant, or "hard coded" for the application (Shiffrin & Nobel, 1997). An example of a common hidden parameter is the size of the working memory buffer, which is potentially variable but often set early in the modeling stage.

Given this "combinatorial explosion" of de facto parameters in complex HPMs, results validation of a single or small number of outcome measures, by itself, may not be a sufficient test of a model's validity. Campbell and Bolton (2005) point out that a close fit between a model's output and HITL data can be derived from the fact that the model is valid or that the model is powerful enough to fit (or overfit) any set of data. For example, if an HPM performs well in the prediction of an outcome result (for example, predicting a pilot's decision to land or go around, or predicting the percent of visual fixation on the SVS display in the SVS study), there could be literally hundreds of combinations of the parameters that could produce the outcome; however, they may not accurately represent actual human behavior.

The challenge, then, for developers of HPMs is to demonstrate an appropriate match between the model and actual HITL data on dimensions that represent the relevant human processes for the specific task. A model should be required not only to predict an *outcome* similar to human behavior, but also to be able to demonstrate that the *process* by which that outcome is achieved is also similar to human behavior. However, this too poses problems for complex HPMs, as it is not possible to predict all possible levels of internal constructs matched to those observed in the operator across a range of complex operational domains. Given that a simulation model of a complex system is only intended to be an approximation of the actual system, absolute model validity is neither possible nor desired (Law & McComas, 2001). As Law and McComas point out, the more time and money spent on a model, the more valid the model should be, in general. However, the challenge to the modeler involved in actual system design and evaluation projects is to strike a balance between cost effectiveness and model validity. That is, tests and evaluations must be conducted until sufficient confidence is obtained that a model can be considered valid for its intended application (Sargent, 1998). It is for this reason that it is commonly argued that validation should not be based on a single comparison of model predictions to HITL data made at the end of a model development project. Rather, it should be an iterative process throughout the model development and iteration cycle (see Law & McComas, 2001; Sargent).

Verification and Validation Techniques Used in the NASA HPM Project

With the preceding considerations in mind and recognizing that to validate a model one must consider the specific intent and purpose of the model, no specific techniques or criteria were asserted by the NASA HPM project. Rather, the modeling teams were challenged to identify or generate the techniques that best demonstrated the validity of their complex HPMs, given the specific aviation safety and system development issues that they addressed in their individual efforts. As seen in chapters

TABLE 12.2

Validation Techniques Employed by the NASA HPM Project Teams

Sensitivity analyses

Compare high-level outcomes of model to NASA HITL outcomes

Compare model task sequences and pilot procedures (model traces) to NASA HITL task sequences

Calibrate using higher level processes and validate against lower level processes within same NASA
  HITL data set

Calibrate using external HITL data and validate against NASA HITL data, making assumptions where
  conditions and tasks differ

Validate against independent HITL data with identical conditions and tasks

---

5 through 9 of this volume, numerous methods and techniques (see Table 12.2) were adopted by the different modeling teams, each of which depended on the research questions asked, the system development issues under investigation, and the stage of model development. It is important to consider that each modeling team used a combination of techniques throughout the iterative model development process. As the model development efforts progressed, the modelers chose increasingly more rigorous and quantitative techniques. A subset of the techniques aimed at highlighting how different techniques were deemed appropriate for different stages of the model development process and for different model goals follows.

Sensitivity analysis is one approach to model verification and validation that was used early in the model development process by two of the modeling teams (IMPRINT/ACT-R and A-SA). Sensitivity analyses are frequently used to determine which model factors have a significant impact on the desired measures of performance (Law & McComas, 2001). This technique consists of changing the values of the input and internal parameters of a model to determine the effect upon the model's behavior and its output, and determine that the relationship observed in the model occurs in the real system (Sargent, 1998). This allows the modeler to focus efforts on the parameters most likely to cause change in the model's output. The IMPRINT/ACT-R team (chapter 6) conducted sensitivity analyses on five main parameters (visual speed, manual speed, communications, decision consistency, and procedural speed) by varying the values of these parameters over a range of possible quantities; the team evaluated their influence on performance (percentage of correct landings) across 100 Monte Carlo runs for each parameter combination.

The A-SA team (chapter 9) also conducted sensitivity analyses by exercising their model through various iterations to show how SA and choice accuracy would wax and wane over time, and to demonstrate how these changes would be influenced by changes in model parameters such as varying the impact of salience or SA decay rate. They used this approach early in their model development process, before they had access to HITL SA estimates for comparison. They showed the loss of SA associated with the passage of time and with distraction, the recovery of SA following relevant events, the sensitivity of SA to different assumed decay rates, and the benefit of SA improvement with a display aid.

In the early phases of model development, efforts tend to focus on *verifying* that the model is behaving properly and *validating* that the modeled pilot behaves the same as a human pilot would, albeit often at a very high, macro level. The IMPRINT/ ACT-R team (chapter 6) adopted this approach by comparing high-level outcomes between the model and the HITL data (e.g., did the model perform a go-around maneuver under the conditions in which the human pilots did?). Although, to some, this technique may fall squarely under the category of verification, Lebiere et al. (chapter 6) explained that it is a matter of validation, given that correct performance is predicated upon the modeled pilot arriving at the decision altitude at the appropriate distance from the airfield through controlling the simulation. They argued that the model must be validated first at this level to ensure that the model performance falls within the bounds determined by human performance, before continuing on to explore other, more detailed validation efforts.

As mentioned previously, it is not a sufficiently strong test of a model's validity to reduce the performance of a complex model to a single outcome measure. Therefore, another validation technique adopted by two of the NASA HPM teams (ACT-R/IMPRINT and D-OMAR) was to compare model traces of task sequences and pilot procedures of the modeled pilot to those of the HITL pilots. Of note here are the tools developed by the D-OMAR team (chapter 8) for the comparison of model traces, including graphical model traces of events and Gantt charts of goals and procedures. This is an important aspect of validation as it provides some level of confidence that, in addition to producing an outcome that matches the HITL data, the modeled pilot also behaved in a manner similar to that of an actual pilot. This task trace analysis is particularly useful early in the model development phase, as well as continually throughout the repeated model iterations. Kemper and Tepper (2005) argue that this type of validation allows the modeler to gain more insight into the behavior of the system under study and to better understand the outcome.

Later in the model development process, a more quantitative and objective validation technique was employed by the ACT-R team (chapter 5). Citing concerns associated with the small HITL data set and the sequential, dependent nature of the data making a traditional split-half reliability technique tenuous at best, they chose to calibrate the visual attention aspects of their model at a high level of granularity (using percent dwell time to each display) and then validate at a lower level of granularity (using the eye dwell transition matrix). They stated that this is a fairly stringent test of the model's ability to match the human pilots at a fairly fine grain of analysis—an advantage of ACT-R, since it is a process-oriented cognitive model. This approach to validation provided good evidence that the model faithfully represented the eye-scan pattern of the actual pilots from the HITL data. This was critical for their model application—specifically addressing SVS system definition and evaluation questions such as determining the impact of alternative display formats and information content on pilots' visual scan patterns.

Finally, traditional, quantitative approaches were employed by the Air MIDAS and A-SA teams, both of which utilized independent data sets outside those provided by NASA. It is frequently recognized that verification and validation of models of complex systems, such as in the aviation domain, are difficult because of the lack of available data with which to populate and validate models (Law & McComas, 2001).

This is a common problem that many practitioners face when working with models for real-world applications and for systems in the concept development phase, for which no prototypes or HITL studies exist.

The Air MIDAS team (chapter 7) was able to make use of data available in the literature that provided percent dwell time for approach and landing in order to populate and calibrate their model's parameters, and then validate the predictions against the NASA SVS data. Using this approach, the Air MIDAS team had more data with which to populate the model, but was required to make assumptions since these data were from an HITL simulation with a different cockpit display configuration (most notably, without SVS) and different scenarios. It was interesting to note the difference between the model pilot and the HITL pilots in their use of the SVS display. In the model, an assumption was made that the pilot would use the SVS according to the original CONOPS adopted by the NASA HPM project that specified the SVS be used as a secondary display only. In the actual HITL simulations, the three pilots used the SVS display in a manner that was more consistent with a PFD. This highlights the difficulty associated with quantitative validation using small amounts of HITL data—particularly where individual differences and strategic differences are likely.

The A-SA team conducted an independent HITL research study in order to ensure control over the variables and scenarios required for their validation effort. They validated their model's ability to predict operationally important performance levels in a multitask simulation including (1) awareness and detection of traffic, and (2) aircraft control (lateral and vertical flight path) of the aircraft. Clearly, this is the most conventional approach to validation, but, of course, it comes with additional resource costs. This type of validation was appropriate given the potential uses of the A-SA model proposed in chapter 9, including using the model as a gold standard for training pilots' visual scan, as an objective measure of cognitive tunneling, and to predict operator SA. It is worth mentioning that the A-SA model is the least complex of the models discussed in this volume, making the prospect of this quantitative validation much more feasible than for the more complex HPMs that may contain hundreds of parameters (see previous discussion about "hidden parameters") that interact in complex ways.

It is clear that when complex models in the aviation domain are validated, many issues arise that challenge traditional validation techniques, arguing for multiple approaches throughout the entire model development and iteration cycle. There are many ways of validating complex models—some more appropriate than others for a given application and for a given model development phase. It is emphasized again that the intended purpose of the model must be considered during validation. That is, for example, if one's goal is to understand and predict specific operator task sequences, then it may very well be reasonable to require that such a model *for that purpose* be validated by matching predicted model task traces with actual operator task traces. We hope that other researchers in the same position may glean some insights from the approaches adopted in this project; however, there is still a need to continue to identify validation techniques and understand the circumstances under which each produces a reliable and meaningful result.

Model Credibility

As the state of the art of HPMs continues to mature, alternative validation techniques will continue to be proposed and eventually be adopted by the modeling community. In the meantime, however, perhaps a more realistic short-term goal to strive for is *model credibility*. Model credibility is concerned with developing in (potential) users the confidence they require in order to use a model and in the information derived from that model (Sargent, 2004). Baron et al. stated:

> A model may have demonstrated an adequate level of practical utility by repeatedly producing satisfactory answers to real-world engineering questions. Ultimately it is the user, not the model developer, who decides if the model has sufficient utility. To determine whether or not this is true, the user needs access to comparisons of model predictions and experimental results relevant to the applications of interest. (1990, p. 80)

Model credibility includes issues of model verification and validation, but also considers factors such as the model recipients' (e.g., researchers, managers, decision makers, and policy makers who receive and use the model output) understanding of and agreement with the model's assumptions, the model development and use history, the quality of documentation, and the degree to which limitations of the model and its findings are known and stated explicitly (Law & Kelton, 2000). Model *credibility*, as used here, differs in formality from model *accreditation*, which is "an official certification that a model or simulation, or federation of models and simulations and its associated data are acceptable for use for a specific purpose" (U.S. Department of Defense, 2001).

One consideration in model credibility is the extent to which a model has demonstrated adequate levels of practical utility by repeatedly producing satisfactory answers to actual engineering questions. Baron et al. (1990) also discuss this as a requirement for model validation. That is, credibility is built over time as modeling exercises are completed in a variety of domains and for a variety of purposes and are published for inspection. This highlights the importance of providing access to comparisons of model predictions and experimental results. In keeping with this sentiment, we offer this volume as just one source of information with which to make decisions regarding the credibility and utility of these HPMs (readers are also referred to Gluck and Pew, 2005, for a comprehensive comparison of models as applied to an ATC task).

Unfortunately, one barrier toward developing model credibility is that these models are continually undergoing modification and augmentation. Even the more established models discussed in this volume required augmentations to address the specific NASA aviation domain problems. Other barriers, already discussed to some extent in this chapter, include inadequate documentation of model parameters and assumptions that led to a general lack of awareness of the range of validity and applicability (Baron et al., 1990) and the lack of transparency of both the modeled tasks and procedures and the underlying cognitive architectures. Not only do these hinder the development of model credibility, but they may also lead to the potential for misuse and misunderstanding of HPMs.

FINAL WORDS

The NASA HPM project set out to develop and advance the state of the art of HPMs for use on aviation problems, to investigate and inform specific solutions to actual aviation safety problems, and to explore methods for integrating HPMs into the system design and evaluation process. Through a series of closely integrated HITL and HPM efforts, it was demonstrated that, when coupled with HITL simulations, HPMs can be useful tools throughout the entire system design and evaluation process and can positively impact concept definition, system definition, system evaluation, system robustness testing, and system integration. The NASA HPM project demonstrated the value of coupling HITL simulation with human performance modeling to provide deep understanding of the aviation problem addressed; assess the efficacy of training, procedures, and operational concepts; determine the potential for underlying error-precursor conditions; and conduct "what if" system redesigns in an efficient manner. Both HPM and HITL techniques bring unique and important contributions to the process, producing a synergistic combination that is more powerful and effective than either approach alone.

Significant augmentations to the HPM tools were realized throughout this project to enable modeling of closed-loop human–environment interactions, visual attention and supervisory control tasks, situation awareness, and human error. These modeling capabilities are critical for modeling most aviation-domain tasks, but should also prove to be valuable tools in other complex-system domains such as process control, medical applications, surface transportation, and military command and control domains, which share similar human–system interaction issues. Along the way, several challenges for the human performance modeling community were identified, including the need for methods and techniques to aid modelers in capturing the complexity of the operators' tasks in aviation environments; improving model transparency and assumptions documentation; addressing individual differences and operator strategies in models; and methods and techniques for model verification, validation, and credibility.

These challenges highlight the importance of close collaboration among aviation domain experts, aviation safety researchers, and human performance modelers and the need for new tools, techniques, and approaches that transcend domains. These are required to enable both the development of relevant, valid, and useful models and the appropriate use and interpretation of model output. It is only through this close collaboration that HPMs can truly reduce the design life cycle and contribute to the development of new human-centered technologies and operations that will improve the safety and efficiency of the future aviation system.

The field of aviation has only recently celebrated its 100th anniversary, as marked from the Wright Brothers' first powered flight of a heavier-than-air aircraft in 1903. As we progress during this second century of modern aviation, we find ourselves in the early stages of another revolution involving the application of advanced technology into all phases of aviation. Just as our everyday lives have undergone a radical shift because of technological wizardry (e.g., e-mail, cell phones, automobile navigation systems), aircraft flight decks, ATC towers, and the very airspace (i.e., the National Airspace System in the United States) are also undergoing a technological revolution of the same magnitude. In the very near future, likely during the next 20 years, traffic flow of both airborne and taxiing aircraft will be directed by intelligent systems optimizing flow, minimizing fuel consumption, and enabling safer operations.

Although it is just a bit beyond our "visibility horizon" to predict where modern aviation will be as it marks the Wright Brothers' 200th anniversary in 2103, at least for the near term, such systems will still require human operators. Pilots will still operate the aircraft, but will place an even greater emphasis on monitoring flight deck automation and interpreting the advanced visionics displays for terrain and traffic awareness; ATC personnel will monitor advanced displays and automated systems and intervene when necessary; and all systems and operators will likely communicate electronically via data link (i.e., using asynchronous electronic communication).

The magnitude of such changes and the stringent requirements of system reliability and safety in aviation require new methods and technologies to effect the development and implementation of such a change. The NASA HPM project has positioned us favorably to address these changes. The integrated, iterative HITL/ HPM method, the specific HPM models developed, and the actual aviation safety impacts described in this volume will aid aviation system designers in understanding the pilot's performance, as well as that of the ATC operator, in these future complex aviation systems. Human performance models, like those within the NASA HPM project, will enable the development of the evolving next-generation aviation technologies and procedures, allowing aviation systems to be easier to use and less susceptible to error and helping to produce a safer, more reliable, more efficient aviation system.

ACKNOWLEDGMENTS

The authors wish to acknowledge Alex Kirlik, Dick Pew, and Brian Gore for technical reviews of an earlier version of this chapter. We are also grateful to the members of the five modeling teams; many of the issues raised in this chapter were identified through their modeling expertise and experiences, their frank assessment of what worked and what did not, and their collegial discussions and debates over the 6-year program.

REFERENCES

Anderson, J. R., & Lebiere, C. (1998). *The atomic components of thought*. Mahwah, NJ: Lawrence Erlbaum Associates.

Baron, S., Kruser, D. S., Huey, B. (Eds.), (1990). *Quantitative Modeling of Human Performance in Complex Dynamic Systems*. Washington, DC: National Academy Press.

Campbell, G. E., & Bolton, A. E. (2005). HBR validation: Integrating lessons learned from multiple academic disciplines, applied communities, and the AMBR project. In K. A. Gluck & R. W. Pew (Eds.), *Modeling human behavior with integrated cognitive architectures* (pp. 365–396). Mahwah, NJ: Lawrence Erlbaum Associates.

Card, S. K., Moran, T. P., & Newell, A. (1983). *The psychology of human–computer interaction*. Hillsdale, NJ: Lawrence Erlbaum Associates.

Councill, I. G., Haynes, S. R., & Ritter, F. E. (2003). Explaining Soar: Analysis of existing tools and user information requirements. In F. Detje, D. Doerner, & H. Schaub (Eds.), *Proceedings of the Fifth International Conference on Cognitive Modeling* (pp. 63–68). Bamberg, Germany: Universitats-Verlag Bamberg.

Diller, D. E., Gluck, K. A., Tenney, Y. J., & Godfrey, K. (2005). Comparison, convergence, and divergence in models of multitasking and category learning, and in the architectures used to create them. In K. A. Gluck & R. W. Pew (Eds.), *Modeling human behavior with integrated cognitive architectures* (pp. 307–350). Mahwah, NJ: Lawrence Erlbaum Associates.

Estes, W. K. (2002). Traps in the route to models of memory and decision. *Psychonomic Bulletin & Review, 9*(1), 3–25.

Foyle, D. C., Andre, A. D., McCann, R. S., Wenzel, E. M., Begault, D., & Battiste, V. (1996). Taxiway navigation and situation awareness (T-NASA) system: Problem, design philosophy and description of an integrated display suite for low-visibility airport surface operations. *SAE Transactions: Journal of Aerospace, 105,* 1411–1418.

Glenn, F., Stokes, J., Neville, K., & Bracken, K. (2005). *An investigation of human performance model validation* (Tech. Rep. AFRL-HE-WP-TR-2006-0002). Fort Washington, PA: CHI Systems, Inc.

Gluck, K. A., & Pew, R. W. (Eds.). (2005). *Modeling human behavior with integrated cognitive architectures.* Mahwah, NJ: Lawrence Erlbaum Associates.

Gore, B. F., & Smith, J. D. (2006). Risk assessment and human performance modeling: The need for an integrated approach. *International Journal of Human Factors of Modeling and Simulation, 1*(1), 119–139.

Hooey, B. L., Foyle, D. C., & Andre, A. D. (2002). A human-centered methodology for the design, evaluation, and integration of cockpit displays. In *Proceedings of the NATO RTO SCI and SET Symposium on Enhanced and Synthetic Vision Systems,* September 10–12, 2002. Ottawa, Canada.

Horrey, W. J., Wickens, C. D., Strauss, R., Kirlik, A., & Stewart, T. R. (2006). Supporting situation assessment through attention guidance and diagnostic aiding: The benefits and costs of display enhancement on judgment skill. In A. Kirlik (Ed.), *Adaptive perspectives on human–technology interaction* (pp. 55–71). New York: Oxford University Press.

Kemper, P., & Tepper, C. (2005). Traced based analysis of process interaction models. In M. E. Kuhl, N. M. Steiger, F. B. Armstrong, & J. A. Joines (Eds.), *Proceedings of the 37th Winter Simulation Conference* (pp. 427–436). Orlando, FL: Winter Simulation Conference.

Kieras, D., & Meyer, D. E. (1998). *The role of cognitive task analysis in the application of predictive models of human performance* (EPIC Rep. No. 11, TR-98/ONR-EPIC-11).

Law, A. M., & Kelton, W. D. (2000). *Simulation modeling and analysis* (3rd ed.). New York: McGraw–Hill.

Law, A. M., & McComas, M. G. (2001). How to build valid and credible simulation models. In B. A. Peters, J. S. Smith, D. J. Medeiros, & M. W. Rohrer (Eds.), *Proceedings of the 2001 Winter Simulation Conference* (pp. 22–29). Piscataway, NJ: Institute of Electrical and Electronic Engineers.

Love, B. C. (2005). *In vivo* or *in vitro*: Cognitive architectures and task-specific models. In K. A. Gluck & R. W. Pew (Eds.), *Modeling human behavior with integrated cognitive architectures* (pp. 351–364). Mahwah, NJ: Lawrence Erlbaum Associates.

Moran, T. P., & Carroll, J. M. (Eds.). (1996). *Design rationale: Concepts, techniques, and use.* Mahwah, NJ: Lawrence Erlbaum Associates.

Oettinger, A. G. (1964). A bull's eye view of management and engineering information systems. *Proceedings of the 19th ACM National Conference* (pp. 21.1–21.14). New York: ACM Press.

Parasuraman, R., Sheridan, T. B., & Wickens, C. D. (2000). A model for types and levels of human interaction with automation. *IEEE Transactions on Systems, Man, and Cybernetics—Part A: Systems and Humans, 30*(3), 286–297.

Pew, R. W., & Mavor, A. S. (Eds.). (1998). *Modeling human and organizational behavior: Application to military simulations.* Washington, DC: National Academy Press.

Ritter, F. E., Shadbolt, N. R., Elliman, D., Young, R. M., Gobet, F., & Baxter, G. D. (2003). *Techniques for modeling human performance in synthetic environments: A supplementary review.* Wright-Patterson Air Force Base, OH: Human Systems Information Analysis Center.

Sargent, R. G. (1998). Verification and validation of simulation models. In D. J. Medeiros, E. F. Watson, J. S. Carson, & M. S. Manivannan (Eds.), *Proceedings of the 30th Winter Sinulation Conference.* (pp. 121–130). Los Alamitos: IEEE Computer Society Press.

Sargent, R. G. (2004). Verification and validation of simulation models. In M. E. Kuhl, N. M. Steiger, F. B. Armstrong, and J. A. Joines, (Eds.), *Proceedings of the 37th Winter Simulation Conference.*

Shiffrin, R. M., & Nobel, P. A. (1997). The art of model development and testing. *Behavior Research Methods, Instruments, and Computers, 29*(1), 6–14.

Shively, R. J., Brickner, M., & Silbiger, J. (1997). A computational model of situational awareness instantiated in MIDAS. *Proceedings of the Ninth International Symposium on Aviation Psychology* (pp. 1454–1459). Columbus, OH: The University of Ohio.

Strauss, R., & Kirlik, A. (2006). Situation awareness as judgment II: Experimental demonstration. *International Journal of Industrial Ergonomics, 36*, 475–484.

U.S. Department of Defense. (2001). *VV&A recommended practices guide glossary.* Washington, DC: Defense Modeling and Simulation Office. Retrieved November 6, 2006, from http://vva.dmso.mil/Glossary/Glossary-pr.pdf.

Zacharias, G. L., Miao, A. X., Illgen, C., Yara, J. M., & Siouris, G. M. (1995). *SAMPLE: Situation awareness model for pilot in-the-loop evaluation* (Final Rep. R95192). Cambridge, MA: Charles River Analytics.

Index

For Product Safety Concerns and Information please contact our EU representative GPSR@taylorandfrancis.com Taylor & Francis Verlag GmbH, Kaufingerstraße 24, 80331 München, Germany